Engineering Drawing and Design

Fifth Edition

Workbook

Kevin Standiford

DELMAR
CENGAGE Learning™

Australia • Brazil • Canada • Mexico • Singapore • United Kingdom • United States

DELMAR
CENGAGE Learning™

Engineering Drawing and Design, Workbook, Fifth Edition
Kevin Standiford

Vice President, Editorial: Dave Garza

Director of Learning Solutions: Sandy Clark

Senior Acquisitions Editor: James Devoe

Managing Editor: Larry Main

Senior Product Manager: Sharon Chambliss

Editorial Assistant: Cristopher Savino

Vice President, Marketing: Jennifer Baker

Executive Marketing Manager: Deborah S. Yarnell

Marketing Manager: Kathryn Hall

Production Director: Wendy Troeger

Production Manager: Mark Bernard

Senior Content Project Manager: Michael Tubbert

Senior Art Director: Casey Kirchmayer

Technology Project Manager: Christopher Catalina

Production Technology Analyst: Joe Pliss

Cover Illustration: ATV Illustration © Jim Hatch, www.hatchillustration.com, Client: Honda, Ad Agency: Vreeke and Associates.

ISBN-13: 978-1-111-30958-9

ISBN-10: 1-111-30958-2

Delmar
5 Maxwell Drive
Clifton Park, NY 12065-2919
USA

Cengage Learning is a leading provider of customized learning solutions with office locations around the globe, including Singapore, the United Kingdom, Australia, Mexico, Brazil, and Japan. Locate your local office at: www.**cengage.com/global**

To learn more about Delmar, visit **www.cengage.com/delmar**

Purchase any of our products at your local college store or at our preferred online store **www.cengage.com**

Notice to the Reader
Publisher does not warrant or guarantee any of the products described herein or perform any independent analysis in connection with any of the product information contained herein. Publisher does not assume, and expressly disclaims, any obligation to obtain and include information other than that provided to it by the manufacturer. The reader is expressly warned to consider and adopt all safety precautions that might be indicated by the activities described herein and to avoid all potential hazards. By following the instructions contained herein, the reader willingly assumes all risks in connection with such instructions. The publisher makes no representations or warranties of any kind, including but not limited to, the warranties of fitness for particular purpose or merchantability, nor are any such representations implied with respect to the material set forth herein, and the publisher takes no responsibility with respect to such material. The publisher shall not be liable for any special, consequential, or exemplary damages resulting, in whole or part, from the readers' use of, or reliance upon, this material.

Printed at CLDPC, USA, 08-21

CONTENTS

Preface xii

SECTION 1
Introduction to Engineering Drawing and Design 01

Chapter 1 Introduction to Engineering Drawing and Design 03

Problem 1-1 Definitions 03
Problem 1-2 Engineering and Drafting Fields 05
Problem 1-3 Copyright 07
Problem 1-4 Copyright and Trademarks 09
Problem 1-5 ADDA Certification 11
Problem 1-6 Drafting Employment Sites 13
Problem 1-7 Job-Seeking Strategies 15

Chapter 2 Drafting Equipment, Media, and Reproduction Methods 17

Problem 2-1 Problem Solving 17
Problem 2-2 Technical Research 19
Problem 2-3 Drafting Machine Parts 21
Problem 2-4 Drafting Instruments 23
Problem 2-5 Drafting Instruments 25
Problem 2-6 Protractors and Angle Reading 27
Problem 2-7 Drafting Machines 29
Problem 2-8 Drafting Machines 31
Problem 2-9 Drafting Scales 33
Problem 2-10 Drafting Scales 35
Problem 2-11 Dimension Lines 37
Problem 2-12 Scales 39
Problem 2-13 Scales 41

Chapter 3 Computer-Aided Design and Drafting (CADD) 43

Problem 3-1 AutoCAD Screen Identification 43
Problem 3-2 Rectangular Coordinates 45
Problem 3-3 Rectangular Coordinates 47
Problem 3-4 Point Identification 49
Problem 3-5 Point Identification 51
Problem 3-6 Point Identification 53
Problem 3-7 AutoCAD Drawing Project 55
Problem 3-8 AutoCAD Drawing Project 57
Problem 3-9 AutoCAD Drawing Project 59
Problem 3-10 Graphing Using AutoCAD 61
Problem 3-11 AutoCAD Commands 63
Problem 3-12 AutoCAD Commands 65
Problem 3-13 AutoCAD Commands 67
Problem 3-14 AutoCAD Commands 71

Chapter 4 Manufacturing Materials and Processes 75

Problem 4-1 Drill and Boring Tools 75
Problem 4-2 Manufacturing Processes 77
Problem 4-3 Alloys 81
Problem 4-4 Steels 83
Problem 4-5 Casting 85
Problem 4-6 Forging 87

SECTION 2
Fundamental Applications 89

Chapter 5 Sketching Applications 91

Problem 5-1 Sketching 91
Problem 5-2 Sketching 93
Problem 5-3 Isometric Sketching 95
Problem 5-4 Isometric Sketching 97
Problem 5-5 Isometric Sketching 99
Problem 5-6 Isometric Sketching 101

Chapter 6 Lines and Lettering 103

Problem 6-1 Line Exercise 103
Problem 6-2 Section Lines 105
Problem 6-3 Section Lines 107
Problem 6-4 Spur Gears 109

Problem 6-5 Line Types 111
Problem 6-6 Lettering 113
Problem 6-7 Lettering 115
Problem 6-8 Lettering 117

Chapter 7 Drafting Geometry 119

Problem 7-1 Guided Practice Bisecting an Angle 119
Problem 7-2 Guided Practice Constructing a Triangle 125
Problem 7-3 Guided Practice Constructing an Equilateral Triangle 129
Problem 7-4 Guided Practice Constructing an Isometric Equilateral Triangle 141
Problem 7-5 Guided Practice Constructing an Isometric Square 151
Problem 7-6 Guided Practice Constructing a Hexagon Using AutoCAD 157
Problem 7-7 Geometric Construction 163
Problem 7-8 Perpendicular 165
Problem 7-9 Transfer Shapes 167
Problem 7-10 Geometric Construction 169
Problem 7-11 Geometric Construction 171
Problem 7-12 Geometric Construction 173
Problem 7-13 Geometric Construction 175

SECTION 3
Drawing Views and Annotations 177

Chapter 8 Multiviews 179

Problem 8-1 Orthographic Questions 179
Problem 8-2 Isometric Matching 181
Problem 8-3 Add the Missing Lines 183
Problem 8-4 Add the Missing Lines 185
Problem 8-5 Complete the View 187
Problem 8-6 Complete the View 189
Problem 8-7 Complete the View 191
Problem 8-8 Orthographic 193
Problem 8-9 Sketching 195
Problem 8-10 Orthographic Sketching 197
Problem 8-11 Orthographic Sketching 199
Problem 8-12 Complete the Missing Views 201
Problem 8-13 Hand Sketching 203
Problem 8-14 Hand Sketching 205
Problem 8-15 Orthographic Projection 207
Problem 8-16 Orthographic Projection 209
Problem 8-17 Orthographic Projection 211
Problem 8-18 Orthographic Projection 213
Problem 8-19 Orthographic Projection 215
Problem 8-20 Orthographic Projection 217

Chapter 9 Auxiliary Views 219

Problem 9-1 Auxiliary Views 219
Problem 9-2 Auxiliary Views 221
Problem 9-3 Auxiliary Views 223
Problem 9-4 Auxiliary Views 225
Problem 9-5 Auxiliary Views 227
Problem 9-6 Auxiliary Views 229
Problem 9-7 Auxiliary Views 231
Problem 9-8 Auxiliary Views 233
Problem 9-9 Auxiliary Views 235
Problem 9-10 Auxiliary Views 237
Problem 9-11 Auxiliary Views 239
Problem 9-12 Auxiliary Views 241
Problem 9-13 Auxiliary Views 243
Problem 9-14 Auxiliary Views 245
Problem 9-15 Auxiliary Views 247
Problem 9-16 Auxiliary Views 249
Problem 9-17 Auxiliary Views 251
Problem 9-18 Auxiliary Views 253
Problem 9-19 Auxiliary Views 255
Problem 9-20 Auxiliary Views 257
Problem 9-21 Auxiliary Views 259

Chapter 10 Dimensioning and Tolerancing 261

Problem 10-1 Dimensioning 261
Problem 10-2 Dimensioning 263
Problem 10-3 Dimensioning 265
Problem 10-4 Dimensioning 267
Problem 10-5 Dimensioning 269
Problem 10-6 Dimensioning 271
Problem 10-7 Dimensioning 273
Problem 10-8 Dimensioning 275
Problem 10-9 Dimensioning 277

Problem 10-10 Dimensioning 279
Problem 10-11 Dimensioning 281
Problem 10-12 Tolerancing 283
Problem 10-13 Tolerancing 285
Problem 10-14 Tolerancing 287
Problem 10-15 Tolerancing 289

Chapter 11 Fasteners and Springs 291

Problem 11-1 Spring and Fasteners 291
Problem 11-2 Fasteners 293
Problem 11-3 Complete the Views 295
Problem 11-4 Machine Screws 297
Problem 11-5 Machine Screws 299
Problem 11-6 Thread Notes 301
Problem 11-7 Springs 303
Problem 11-8 Springs 305
Problem 11-9 Springs 307
Problem 11-10 Springs 309

Chapter 12 Sections, Revolutions, and Conventional Breaks 313

Problem 12-1 Section Identification 313
Problem 12-2 Section Identification 315
Problem 12-3 Full Sections 317
Problem 12-4 Half Sections 319
Problem 12-5 Sections 321
Problem 12-6 Sections 323
Problem 12-7 Freehand Sketch Section 325
Problem 12-8 Section 327
Problem 12-9 Section 329

Chapter 13 Geometric Dimensioning and Tolerancing 331

Problem 13-1 Geometric Tolerancing 331
Problem 13-2 MMC and LMC 333
Problem 13-3 Geometric Characteristic 335
Problem 13-4 Geometric Tolerancing 337
Problem 13-5 Geometric Tolerancing 339
Problem 13-6 Geometric Dimensioning 341
Problem 13-7 Geometric Dimensioning 343
Problem 13-8 Adding Dimensions 345
Problem 13-9 Tolerancing 347
Problem 13-10 Geometric Tolerancing 349
Problem 13-11 Geometric Tolerancing 351
Problem 13-12 Geometric Tolerancing 353
Problem 13-13 Geometric Tolerancing 355

Chapter 14 Pictorial Drawings and Technical Illustrations 357

Problem 14-1 Using Dividers 357
Problem 14-2 Identified Perspective 359

SECTION 4
WORKING DRAWINGS 361

Chapter 15 Working Drawings 363

Problem 15-1 Working Drawing Definitions 363
Problem 15-2 Monodetail Drawing Assemblies 365

Chapter 16 Mechanisms: Linkages, Cams, Gears, and Bearings 367

Problem 16-1 Linkages 367
Problem 16-2 Displacement Diagram 369
Problem 16-3 Cam Follower 371

Chapter 17 Belt and Chain Drives 373

Problem 17-1 Chain Drives 373
Problem 17-2 Belt Drives 375

Chapter 18 Welding Processes and Representations 379

Problem 18-1 Welding Symbols 379
Problem 18-2 Welding Symbols 381
Problem 18-3 Changing the Casting Weldment 383
Problem 18-4 Changing the Casting Weldment 385
Problem 18-5 Welding Symbols 387
Problem 18-6 Welding Symbols 389
Problem 18-7 Fill in the Missing Information 391
Problem 18-8 Picnic Table 393

SECTION 5
Specialty Drafting and Design 395

Chapter 19 Precision Sheet Metal Drafting 397

Problem 19-1 Seams and Hems ... 397
Problem 19-2 Point of Intersection ... 399
Problem 19-3 Bend Allowances ... 401
Problem 19-4 Development ... 403
Problem 19-5 Development ... 405
Problem 19-6 Development ... 407
Problem 19-7 Development ... 409

Chapter 20 Electrical and Electronics Drafting 411

Problem 20-1 Schematic ... 411
Problem 20-2 Start–Stop Control Circuit ... 413
Problem 20-3 Find the Errors ... 415
Problem 20-4 Electrical Layout ... 417
Problem 20-5 Electronic Identification ... 419
Problem 20-6 Electronic Circuits ... 421

Chapter 21 Industrial Process Piping 423

Problem 21-1 Plumbing Fittings ... 423
Problem 21-2 Plumbing Fittings ... 425
Problem 21-3 Hydraulic Circuits ... 427
Problem 21-4 Chemical Pot Feeder ... 429
Problem 21-5 Complete the Missing Information ... 431

Chapter 22 Structural Drafting 433

Problem 22-1 Structural Steel Fabrication Drawing ... 433
Problem 22-2 Site Development ... 435
Problem 22-3 Trusses ... 437
Problem 22-4 Trusses ... 439

Chapter 23 Heating, Ventilating, and Air-Conditioning (HVAC), and Pattern Development 441

Problem 23-1 Purpose of a Furnace ... 441
Problem 23-2 Steps in Designing an Offset Development ... 443
Problem 23-3 Drawing an HVAC System ... 445

Chapter 24 Civil Drafting 447

Problem 24-1 Azimuths, Grade, and Slope ... 447

SECTION 6
Engineering Design 457

Chapter 25 The Engineering Design Process 459

Problem 25-1 Design Analysis ... 459

PREFACE

The *Engineering Drawing and Design* workbook is intended to assist the student in the development of the technical skills necessary to meet the challenges of today's changing global technical community. For an individual to make the transition from student to engineering professional, he or she must have a thorough understanding of the very fabric that binds the engineering community together: the ability to communicate. For without this ability, it would have been difficult, if not impossible, for humankind to have reached the level of civilization we currently enjoy.

In the engineering profession, communication is defined as the conveyance of technical information from one individual to another using either oral or written instructions. Although oral instructions are often quicker and easier for the engineering professional to generate, they often lead to misinterpretation and confusion. It is for these reasons that the engineering community has placed great emphasis on the development of graphical representations of engineering information. This includes the development of standard practices and procedures used to develop and ensure the accuracy of engineering drawings.

The workbook is intended to be used in conjunction with the *Engineering Drawing and Design* textbook to give the student additional practice working with the standard practices used in the engineering profession. This is achieved by providing additional problems in the areas of Sketching and Drawing, Geometric Construction, Multiview Drawings, Auxiliary Views, Descriptive Geometry, Manufacturing Material and Processes, Dimensioning and Tolerancing, Fasteners and Springs, Sections, Revolutions, Geometric Tolerancing, Mechanisms, Belt and Chain Drive, Working Drawings, Pictorial and Technical Illustrations, Solid Modeling, Welding, Processes Piping, Structural Drafting, HVAC, and Electrical. The problems are arranged in accordance to the chapters in which the topic is originally presented in the textbook. All of the problems, including the ones in which drawings need to be made, can be completed within each page, eliminating the need for additional vellum or mylar®. Each page is perforated and can be detached if necessary. Each problem that requires lettering is meant to serve as a lettering exercise and can therefore be graded as such.

NEW TO THIS EDITION

The following chapter topics and subtopics are new to this edition:

Chapter 1 Introduction to Engineering Drawing and Design

Chapter 4

- Alloys
- Steels
- Castings
- Forging

Chapter 23 Heating, Ventilating, and Air-Conditioning (HVAC), and Pattern Development

- Purpose of a Furnace
- Steps in Designing an Offset Development
- Drawing an HVAC System

Chapter 25 The Engineering Design Process

ABOUT THE AUTHORS

Kevin Standiford, author, contributor, and consultant, has been in the technology fields of manufacturing processes, HVACR, process piping, and robotics for more than 20 years. While attending college to obtain his bachelor of science in mechanical engineering technology, he worked for McClelland Consulting Engineers as a mechanical designer, designing HVAC, complex processing piping, and cogeneration systems for commercial and industrial applications. During his college years, he became a student member of the American Society of Heating, Refrigerating, and Air-Conditioning Engineers, where he developed and later wrote a paper on a computer application that enabled the user to simulate, design, and draw heating and cooling systems by using AutoCAD. The paper was entered into a student design competition and became the first-place winner for the region and state.

After graduation, Kevin worked for Pettit and Pettit Consulting Engineers, one of the leading HVAC engineering firms in the state of Arkansas, as a mechanical design engineer. While working for Pettit and Pettit, Kevin designed and selected equipment for large commercial and government projects by using manual design techniques and computer simulations.

In addition to working at Pettit and Pettit, Kevin started teaching part-time evening engineering and design courses for Garland County Community College in Hot Springs, Arkansas. He later stopped working full-time in the engineering field and started teaching technology classes, including heat transfer, duct design, and properties of air. It was also at this time that Kevin started writing textbooks for Cengage Delmar Publishers. The first textbook was a descriptive geometry book, which included a section on sheet metal design. Today Kevin is a full-time consultant working for both the publishing and engineering industries and a part-time instructor. In the publishing industry, Kevin has worked on numerous e-resource products, mapping, and custom publications for Delmar's HVAC, CAD, and plumbing titles.

DEDICATION

This textbook is dedicated to my loving wife, Debrah; my daughter, Cyan; and my son, Kylar. Without their support, and understanding, this project would not have been possible.

Kevin Standiford

SECTION 1

Introduction to Engineering Drawing and Design

CHAPTER 1

Introduction to Engineering Drawing and Design

Problem 1-1 Definitions

Using your textbook, define the following terms:

A. Design application

B. Rapid prototyping (RP)

C. Mechanical drafting

D. Manual drafting

E. Computer-aided drafting

F. Cartesian coordinate system

G. Descriptive geometry

COURSE ________	STUDENT ________________________	DATE ________	PROBLEM 1-1

Problem 1-2 Engineering and Drafting Fields

Using your textbook and the Internet, list ten different fields in which a drafter can become employed and the professional organization in which the drafter can obtain technical support and assistance. For example, the automotive industry employs mechanical drafters who can obtain technical assistance and support from the American Society of Mechanical Engineers.

1.

2.

3.

4.

5.

6.

7.

8.

9.

10.

COURSE________ STUDENT______________________________ DATE________ PROBLEM 1-2

Problem 1-3 Copyright

Using the Internet and the U.S. Copyright Office (www.copyright.gov), list the steps for registering a copyright over the Internet in the following space.

COURSE________ STUDENT________________________________ DATE________ PROBLEM 1-3

Problem 1-4 Copyright and Trademarks

Using the textbook, list the differences among trademarks, patents, and copyrights in the following space.

COURSE ________	STUDENT ______________________________	DATE ________	PROBLEM 1-4

Problem 1-5 ADDA Certification

Using the textbook and the Internet, list the steps in becoming ADDA certified in the following space.

COURSE ________	STUDENT ________________________________	DATE ________	PROBLEM 1-5

Problem 1-6 Drafting Employment Sites

The Internet is a valuable place to seek employment. Hundreds of Web sites are available to help you prepare for and find a job. Explore the Internet and in the following space list five different Web sites that can be used to secure employment in the drafting field.

COURSE________ STUDENT________________________ DATE________ PROBLEM 1-6

Problem 1-7 Job-Seeking Strategies

From the Internet, select one of the sites listed in the previous six problems, and in the following space list the job-seeking strategies the site has listed or resources it makes available to help drafters find employment.

COURSE________ STUDENT______________________________ DATE________ PROBLEM 1-7

CHAPTER 2

Drafting Equipment, Media, and Reproduction Methods

Problem 2-1 Problem Solving

Problem solving is an important aspect in engineering and engineering design. According to the textbook, the problem-solving process is a three-step process. In the following space, list the three steps used to solve technical problems.

1. ______________________________

2. ______________________________

3. ______________________________

Brainstorming is a problem-solving method that allows individuals to voice their thoughts and ideas regarding the specific topic, problem, or project at hand. List four suggestions for working in a brainstorming session.

1. ______________________________
____

2. ______________________________
____

3. ______________________________
____

4. ______________________________
____

COURSE ________	STUDENT ______________________	DATE ________	PROBLEM 2-1

Problem 2-2 Technical Research

In the following space, list the steps in performing technical research.

COURSE ________	STUDENT ______________________	DATE ________	PROBLEM 2-2

Problem 2-3 Drafting Machine Parts

Review the Supplemental Material for Chapter 2 on the CD that accompanies the textbook. Name the drafting machine parts by lettering in the guidelines provided below. Use a single-stroke vertical capital letter.

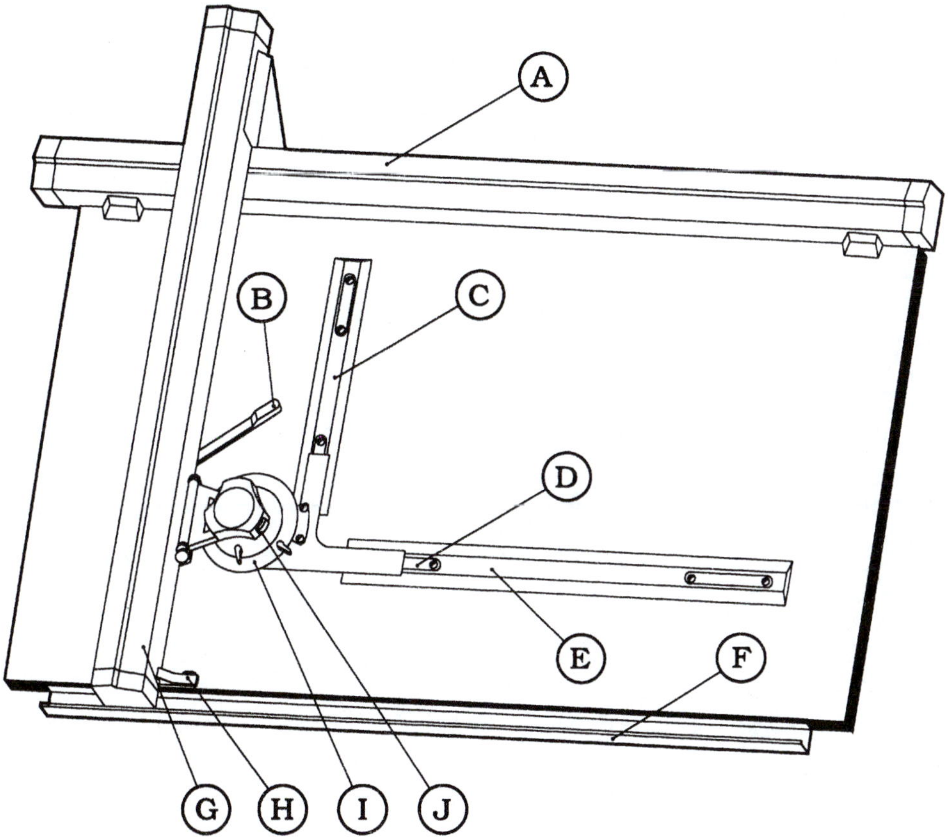

A ______________________

B ______________________

C ______________________

D ______________________

E ______________________

F ______________________

G ______________________

H ______________________

I ______________________

J ______________________

COURSE________ STUDENT______________________________ DATE________ PROBLEM 2-3

Problem 2-4 Drafting Instruments

Name the drafting instruments by lettering in the guidelines provided below. Use a single-stroke vertical capital letter.

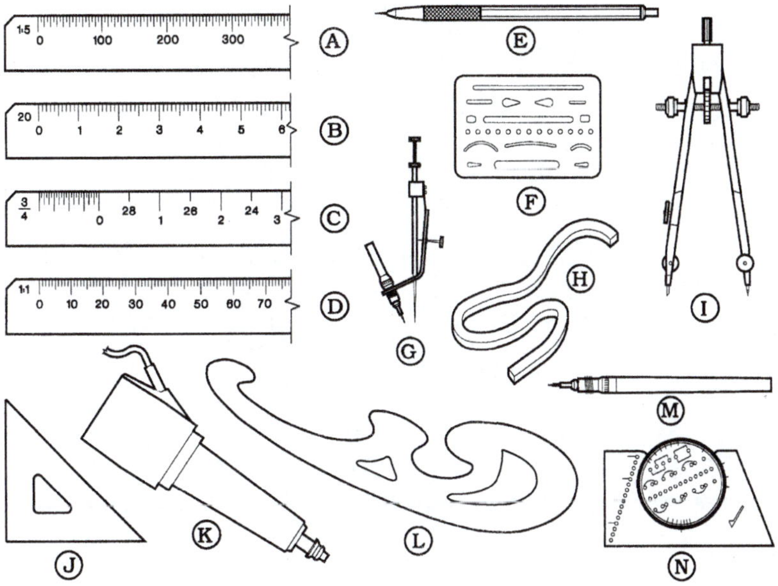

A ____________________

B ____________________

C ____________________

D ____________________

E ____________________

F ____________________

G ____________________

H ____________________

I ____________________

J ____________________

K ____________________

L ____________________

M ____________________

N ____________________

COURSE ________	STUDENT ____________________	DATE ________	PROBLEM 2-4

Problem 2-5 Drafting Instruments

Name the drafting instruments by lettering in the guidelines provided below. Use a single-stroke vertical capital letter.

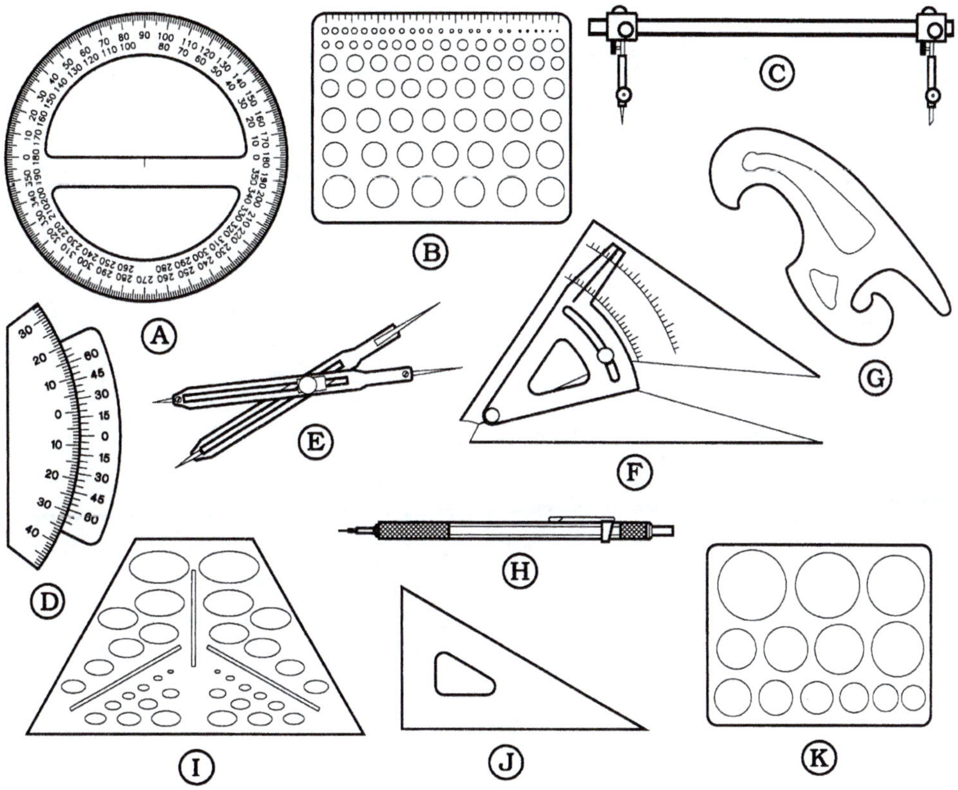

A ______________________

B ______________________

C ______________________

D ______________________

E ______________________

F ______________________

G ______________________

H ______________________

I ______________________

J ______________________

K ______________________

COURSE ________ STUDENT ______________________ DATE ________ PROBLEM 2-5

Problem 2-6 Protractors and Angle Reading

Review the Supplemental Material for Chapter 2 on the CD that accompanies the textbook. Given the following protractors, determine the angular readings. Use a single-stroke vertical capital letter.

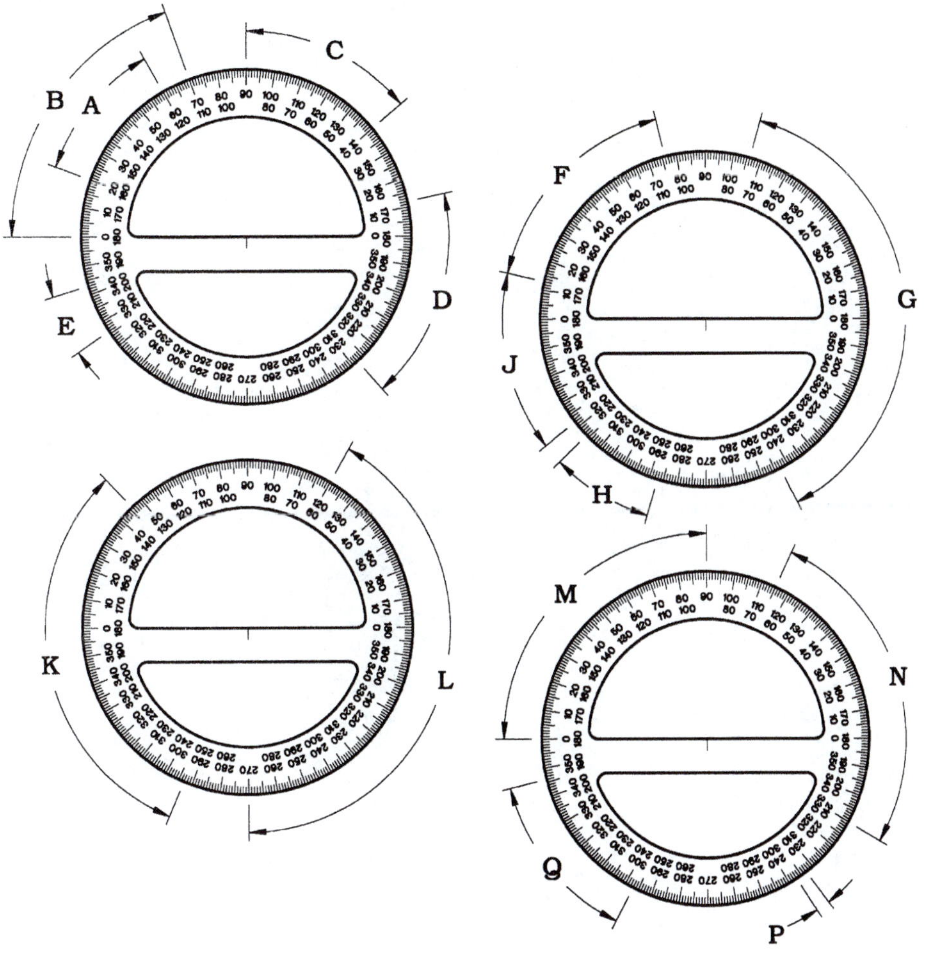

A ________________________

B ________________________

C ________________________

D ________________________

E ________________________

F ________________________

G ________________________

H ________________________

J ________________________

K ________________________

L ________________________

M ________________________

N ________________________

P ________________________

Q ________________________

COURSE________	STUDENT____________________	DATE________	PROBLEM 2-6

Problem 2-7 Drafting Machines

Review the Supplemental Material for Chapter 2 on the CD that accompanies the textbook. Given the following drafting machine protractors, determine the angular readings. Use a single-stroke vertical capital letter.

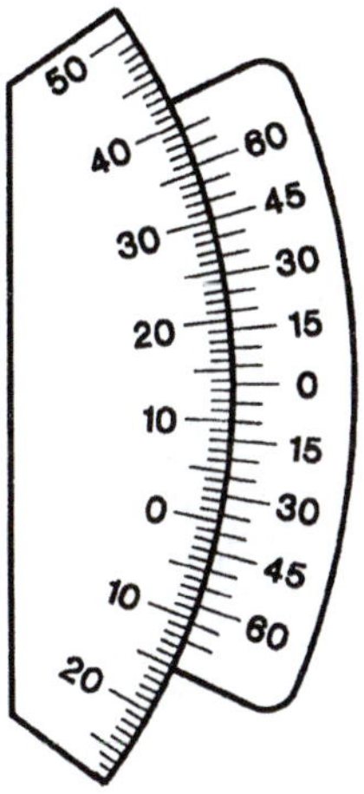

A ____________

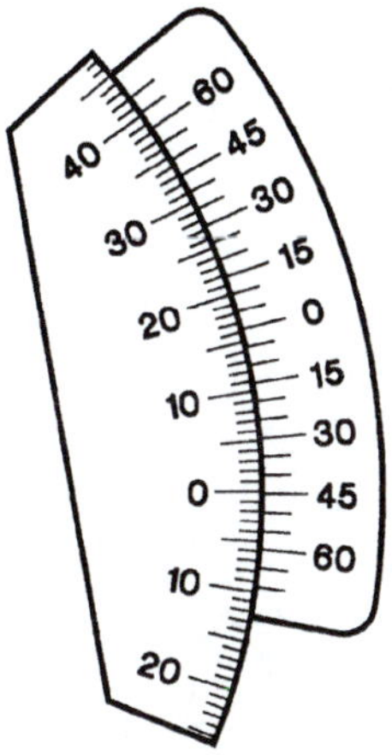

B ____________

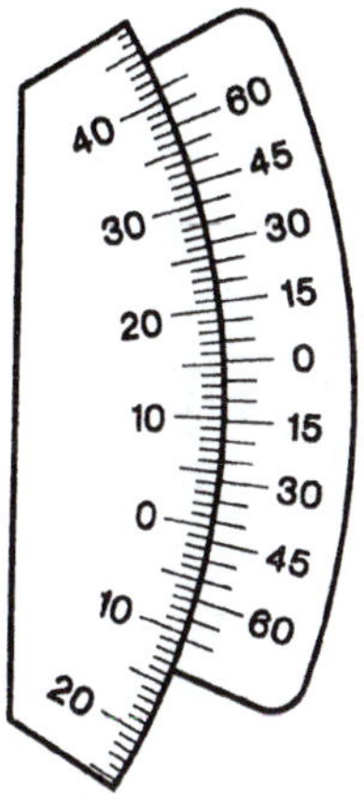

C ____________

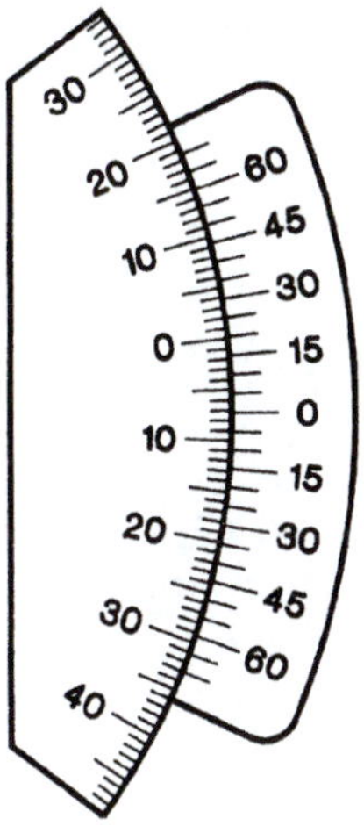

D ____________

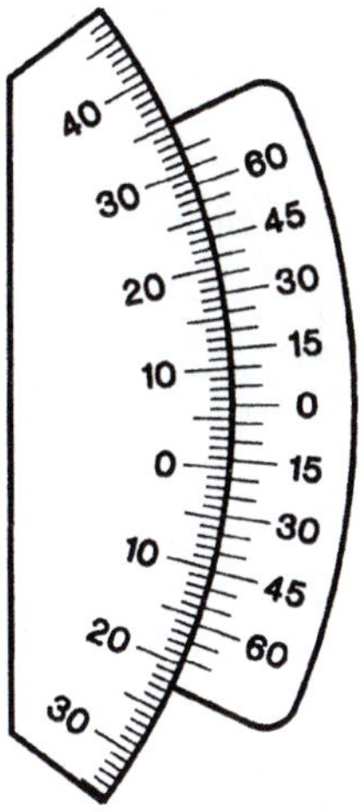

E ____________

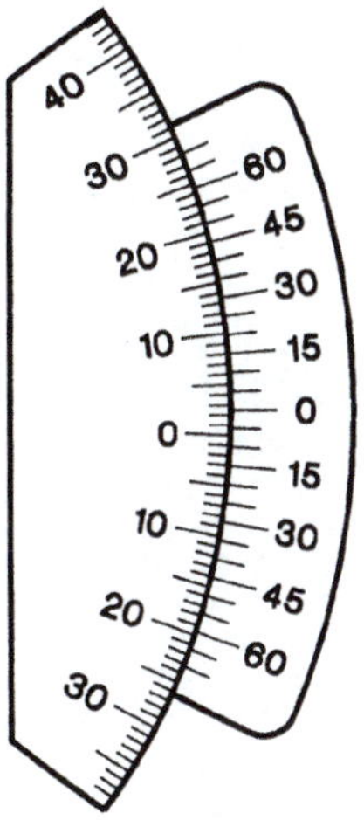

F ____________

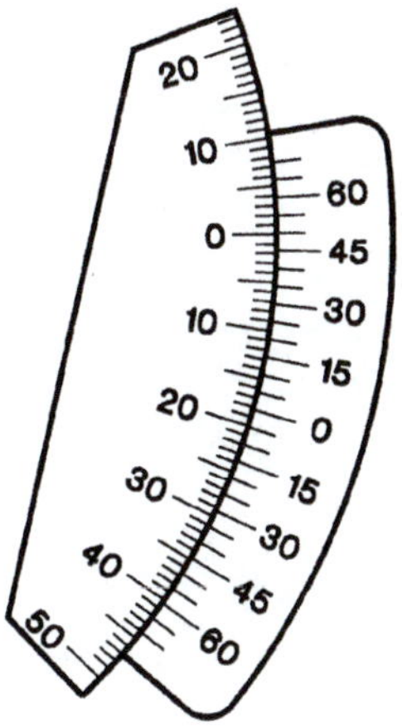

G ____________

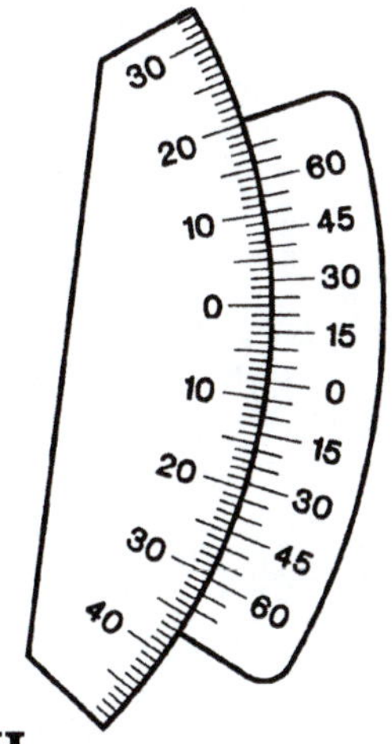

H ____________

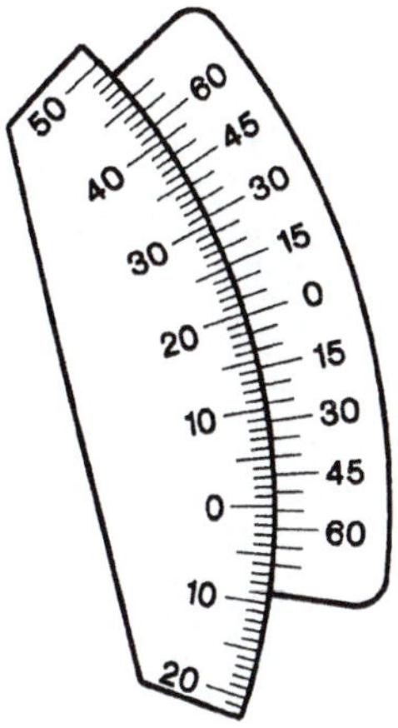

I ____________

COURSE ________	STUDENT ______________________	DATE ________	PROBLEM 2-7

Problem 2-8 Drafting Machines

Given the following drafting machine protractors, determine the angular readings. Use a single-stroke vertical capital letter.

A ________

B ________

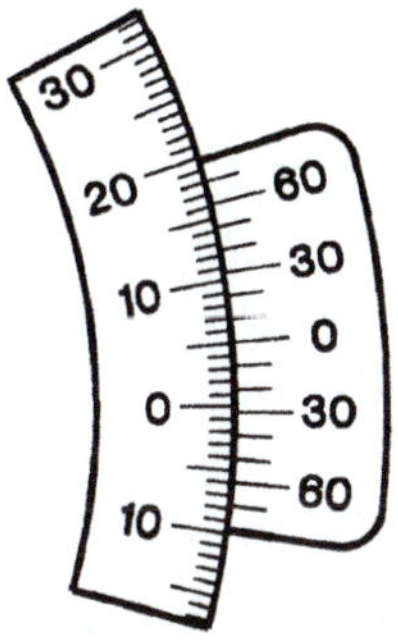

C ________

D ________

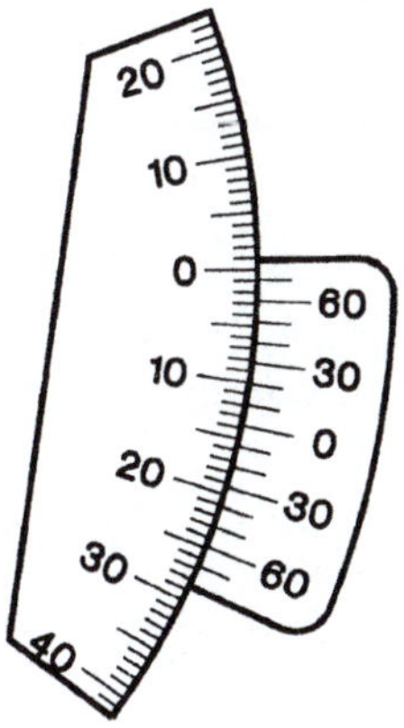

E ________

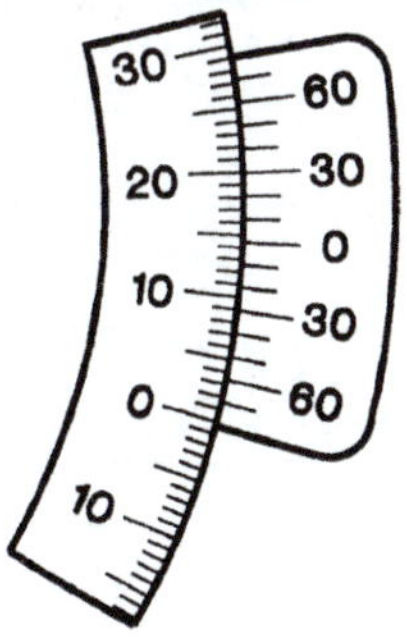

F ________

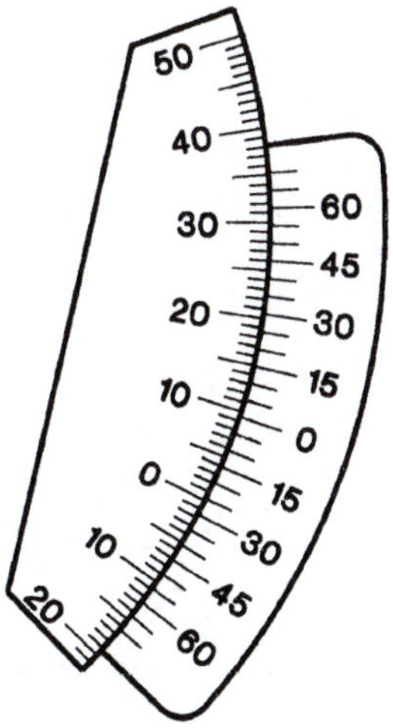

G ________

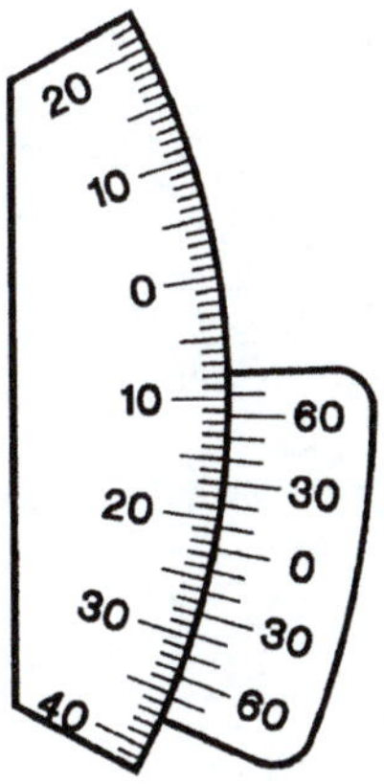

H ________

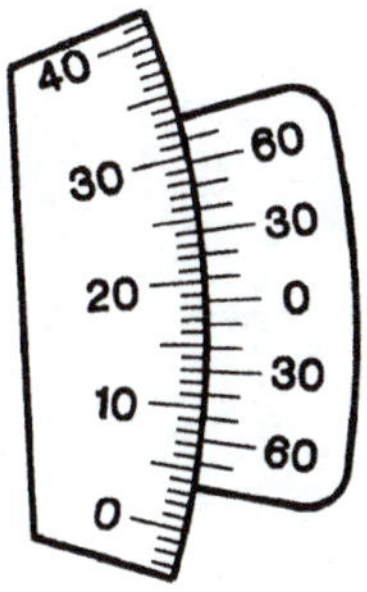

I ________

COURSE ________ STUDENT ________________ DATE ________ PROBLEM 2-8

Problem 2-9 Drafting Scales

Given the following scales, determine the indicated readings. Below each scale, identify the type of scale that is shown.

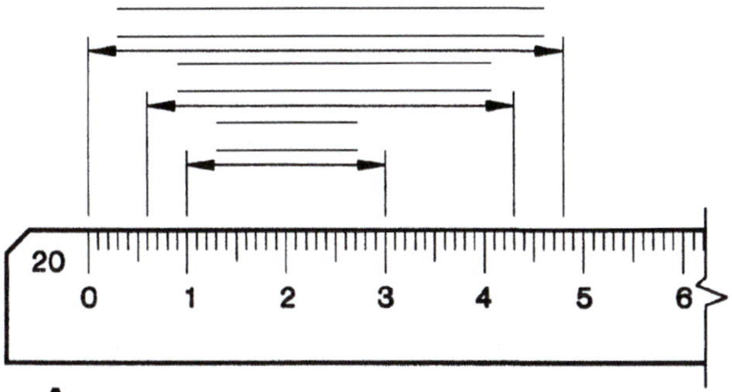

A ______________________

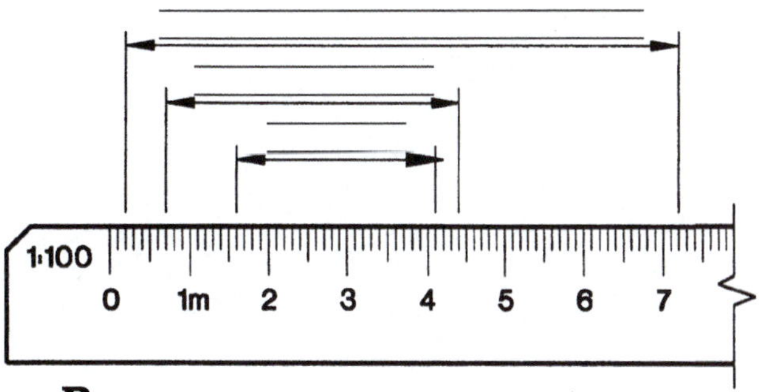

B ______________________

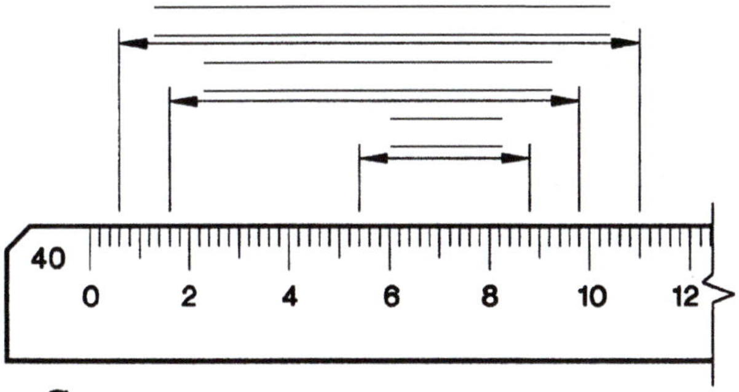

C ______________________

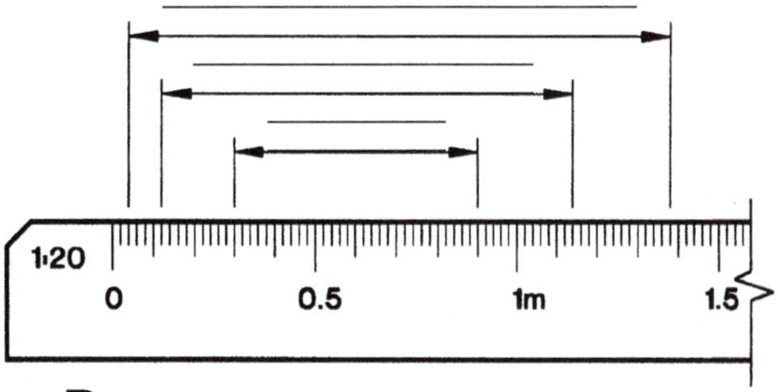

D ______________________

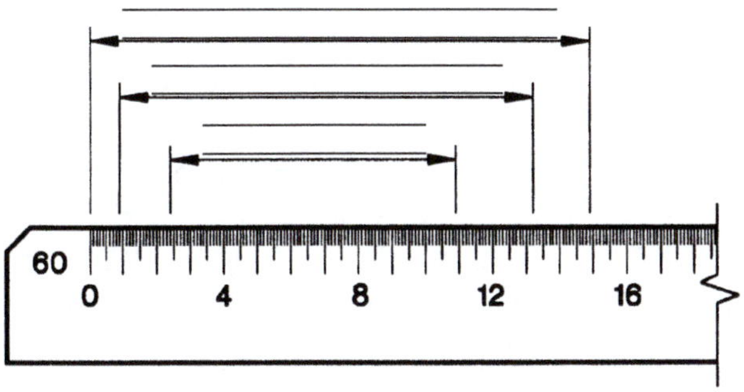

E ______________________

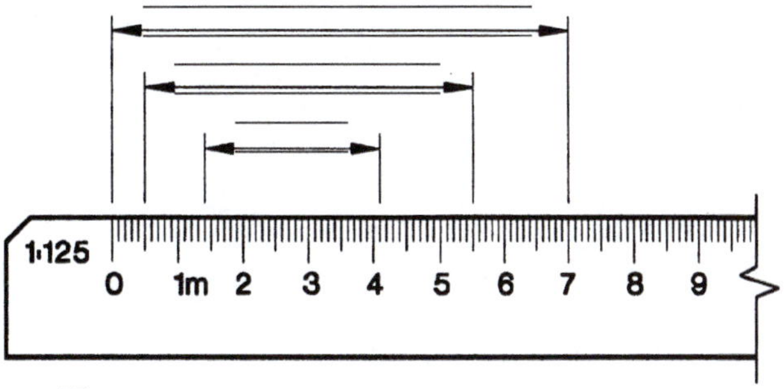

F ______________________

COURSE ________	STUDENT ______________________	DATE ________	PROBLEM 2-9

Problem 2-10 Drafting Scales

Given the following scales, determine the indicated readings. Below each scale, identify the type of scale that is shown.

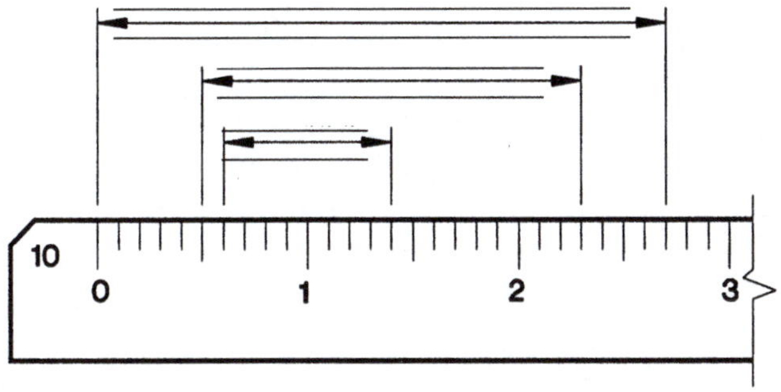

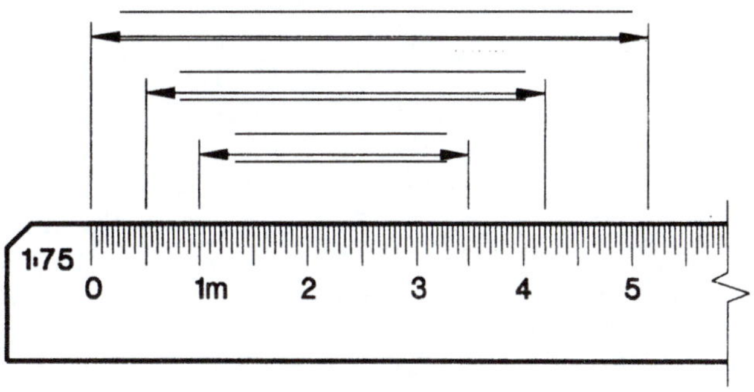

A ______________________________

B ______________________________

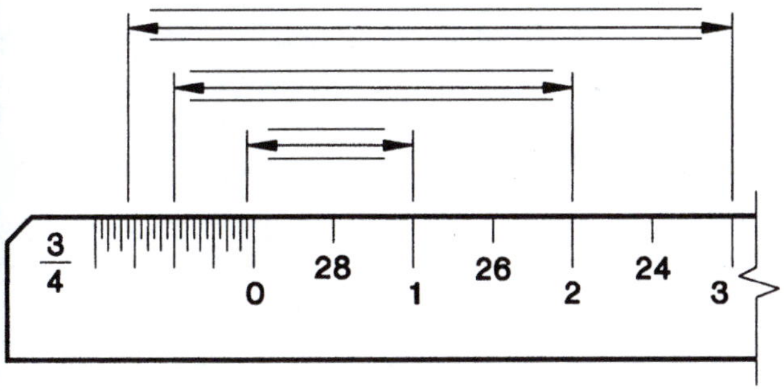

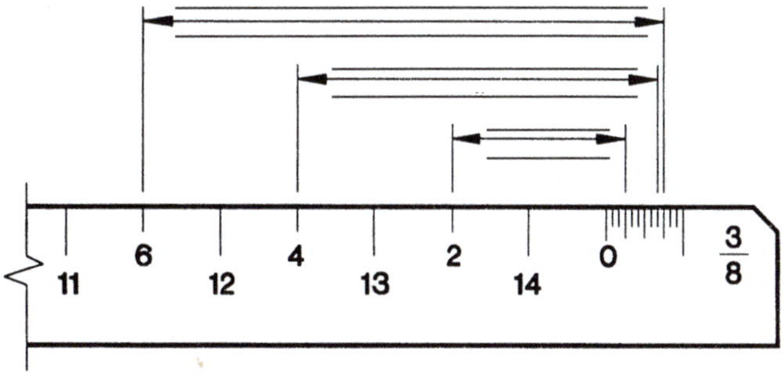

C ______________________________

D ______________________________

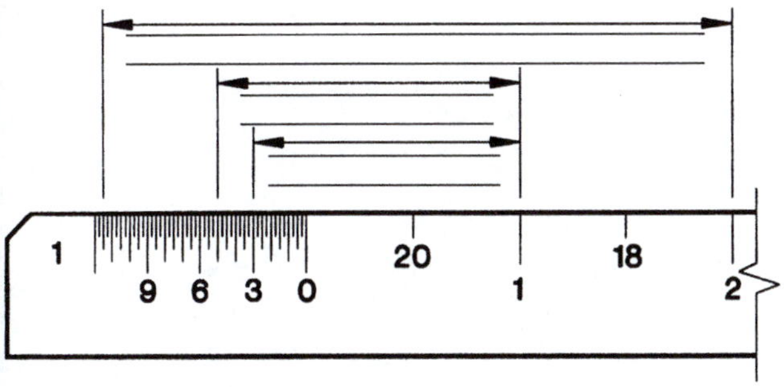

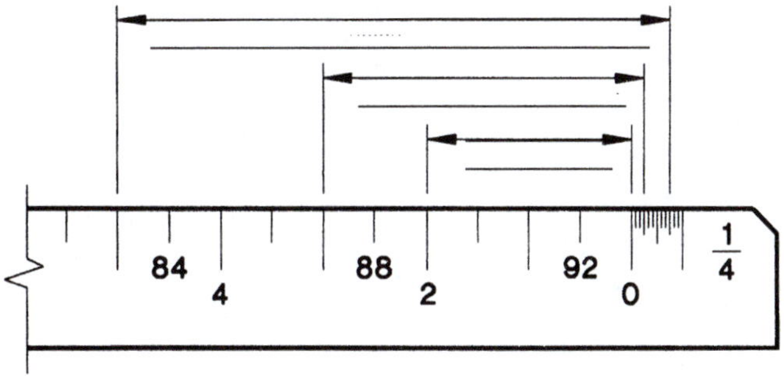

E ______________________________

F ______________________________

COURSE________	STUDENT______________________________	DATE________	PROBLEM 2-10

Problem 2-11 Dimension Lines

1. In the space to the right, record the measurements of the dimension lines.

A 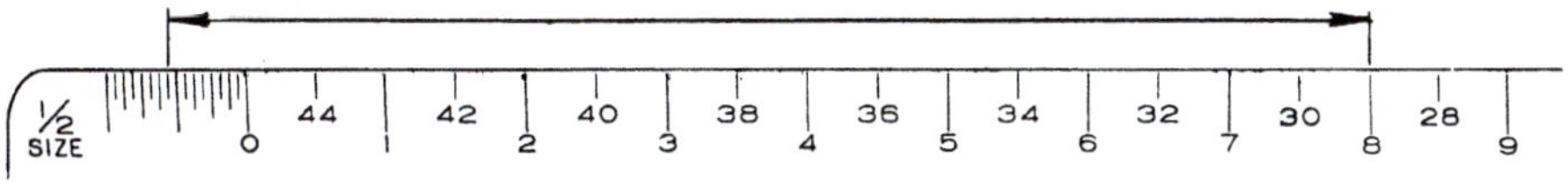

____

B 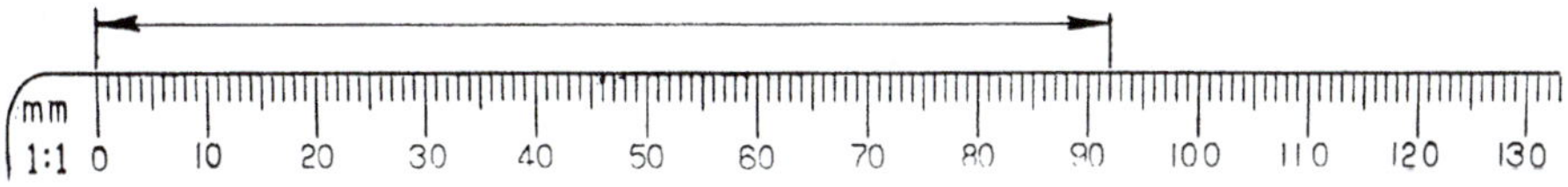

____

C 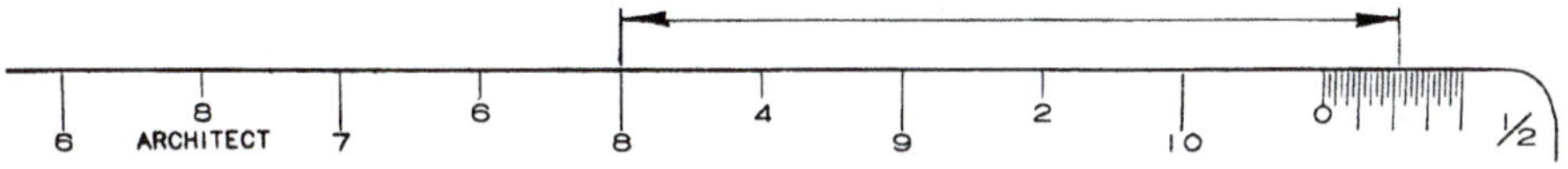

____

D 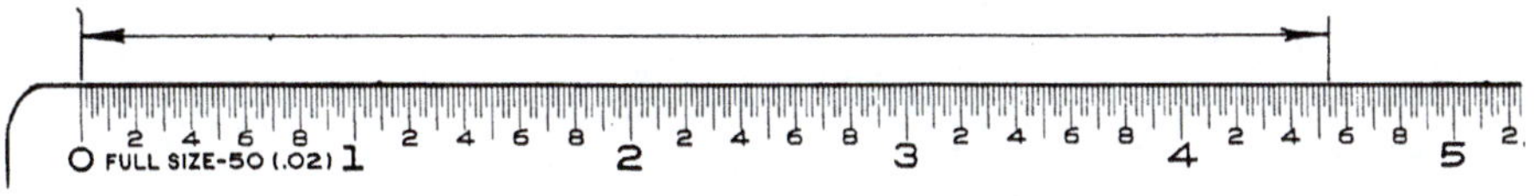

____

2. Add dimension lines to represent the measures that are listed on the right.

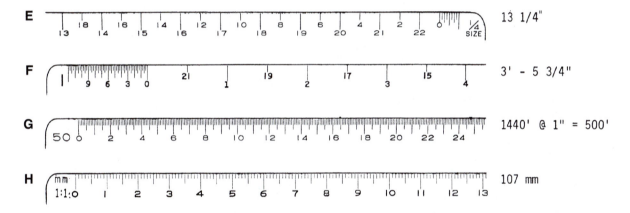

E 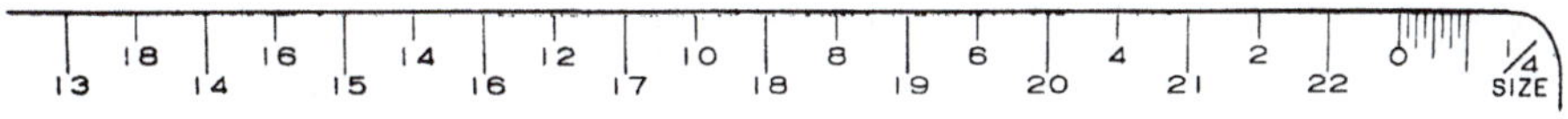

13 1/4"

F 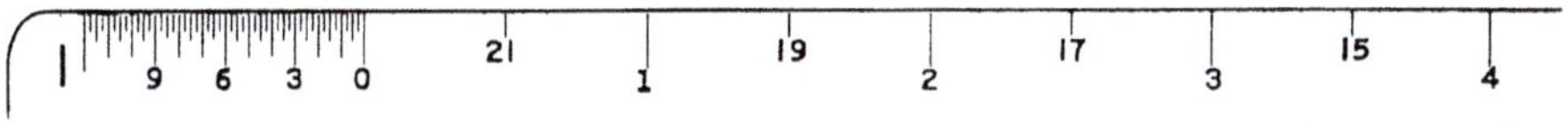
3' - 5 3/4"

G 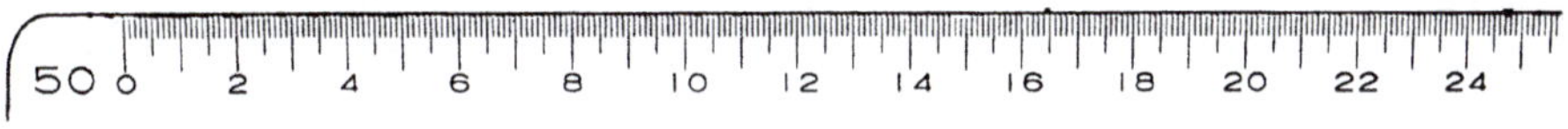

1440' @ 1" = 500'

H 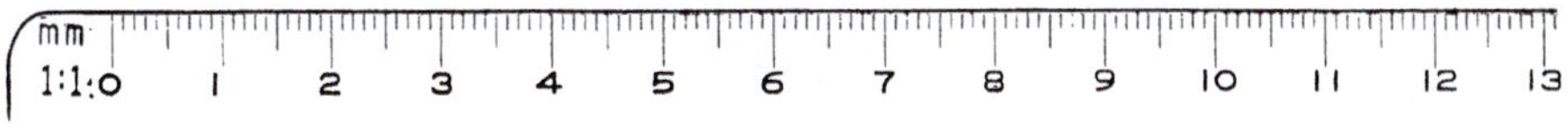

107 mm

3. Measure from the left end of the long line and mark the distance with a sharp, dark, vertical line. Use the distance and scale given at the right of each line.

I	2.77	decimal inch
J	102.5	metric 1:1
K	14 13/16	1/4 size
L	0'-11 1/2"	1/4" = 1'-0"
M	47.0 mm	metric 1:2
N	6 1/2"	half size
O	205.5'	1" = 50'-0"

COURSE________ STUDENT________________ DATE________ PROBLEM 2-11

Problem 2-12 Scales

All the scales on the metric and engineer's scales can be changed by a factor of 10 by simply moving the decimal point. The metric scale is used primarily for engineering drawings. The engineer's scale is used primarily for engineering drawings, but it is also used for diagrams, velocity and acceleration polygons, graphs, and topographic maps.

Using the following example measurements: (1) determine the unknown dimensions and record the values where indicated; (2) lay off the given distances or quantities and record the values over the dimension lines.

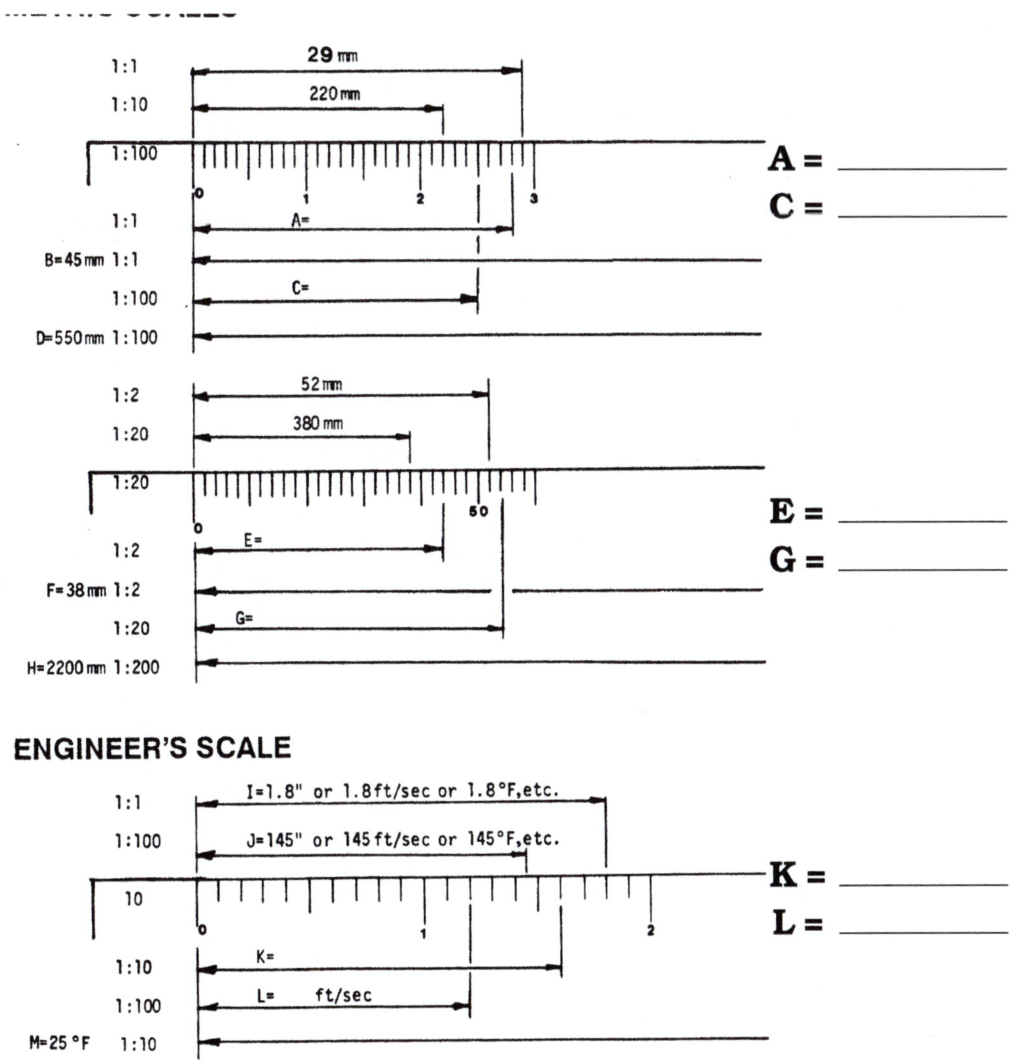

COURSE______	STUDENT______________	DATE______	PROBLEM 2-12

Problem 2-13 Scales

Using the example measurement given below: (1) determine the unknown dimensions and record the values where indicated; (2) lay off the given distances or quantities and record the distances over the dimension lines.

The two scales below are illustrations of the architect's scale. The major divisions on each scale represent 1 foot. One of the 1-foot divisions is subdivided into 12 equal parts to represent the number of inches in 1 foot. Some are further divided into fractions of an inch. The 1 1/2"=1'-0 scale is 1/8 size, and the 1/4'=1'-0 is 1/48 size. Architects' scales are used for structural drawings, commercial buildings, dwellings, topographic maps, and some machinery drawings.

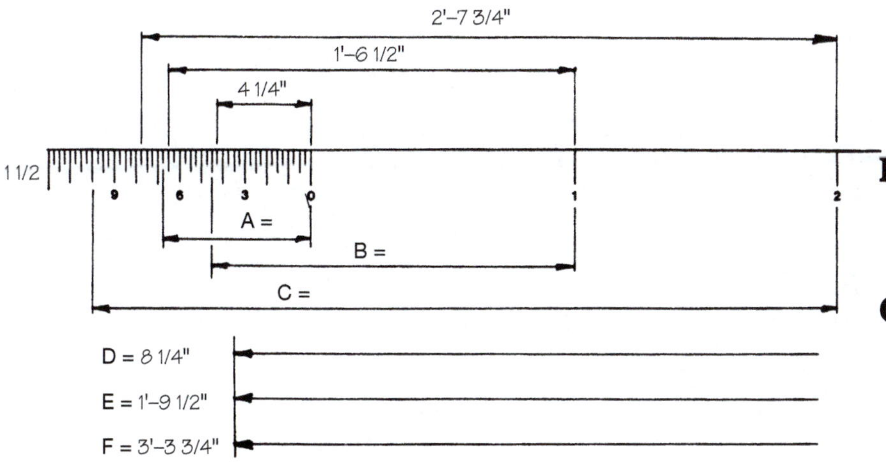

B = ____________

C = ____________

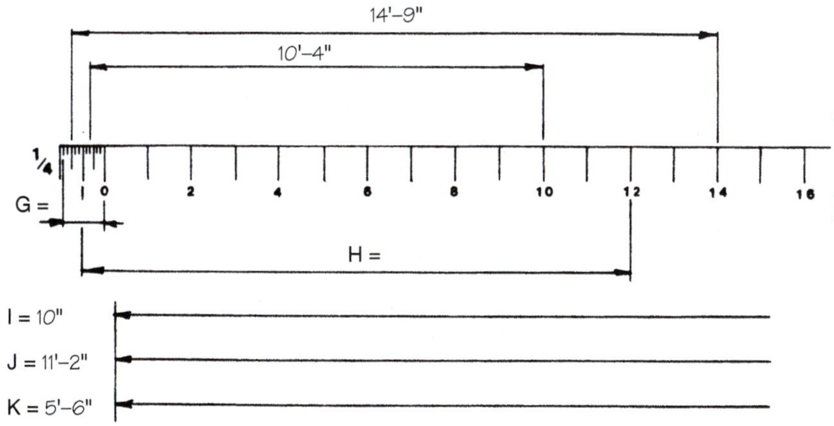

G = ____________

H = ____________

COURSE________ STUDENT______________________________ DATE________ PROBLEM 2-13

CHAPTER 3

Computer-Aided Design and Drafting (CADD)

Problem 3-1 AutoCAD Screen Identification

Identify the following areas of the AutoCAD garphic screen.

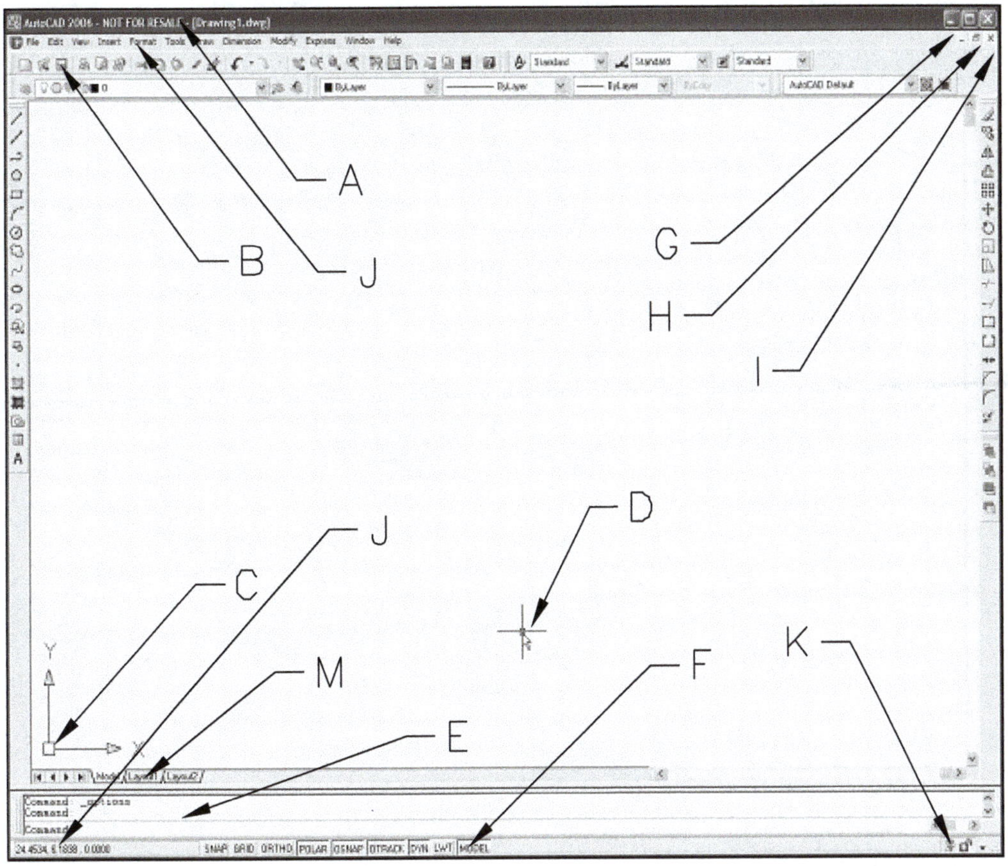

A ____________________

B ____________________

C ____________________

D ____________________

E ____________________

F ____________________

G ____________________

H ____________________

I ____________________

J ____________________

K ____________________

L ____________________

M ____________________

COURSE __________ STUDENT ____________________ DATE __________ PROBLEM 3-1

Problem 3-2 Rectangular Coordinates

Locate each point on the rectangular coordinate system. Connect the point as you progress.

CARTESIAN COORDINATE EXERCISE		
POINT	X	Y
1	2.5	7.5
2	2.5	4.5
3	12.5	4.5
4	12.5	7.5
5	9.5	7.5
6	8.0	6.0
7	7.5	8.0
8	7.0	6.0
9	5.5	7.5
10	2.5	7.5

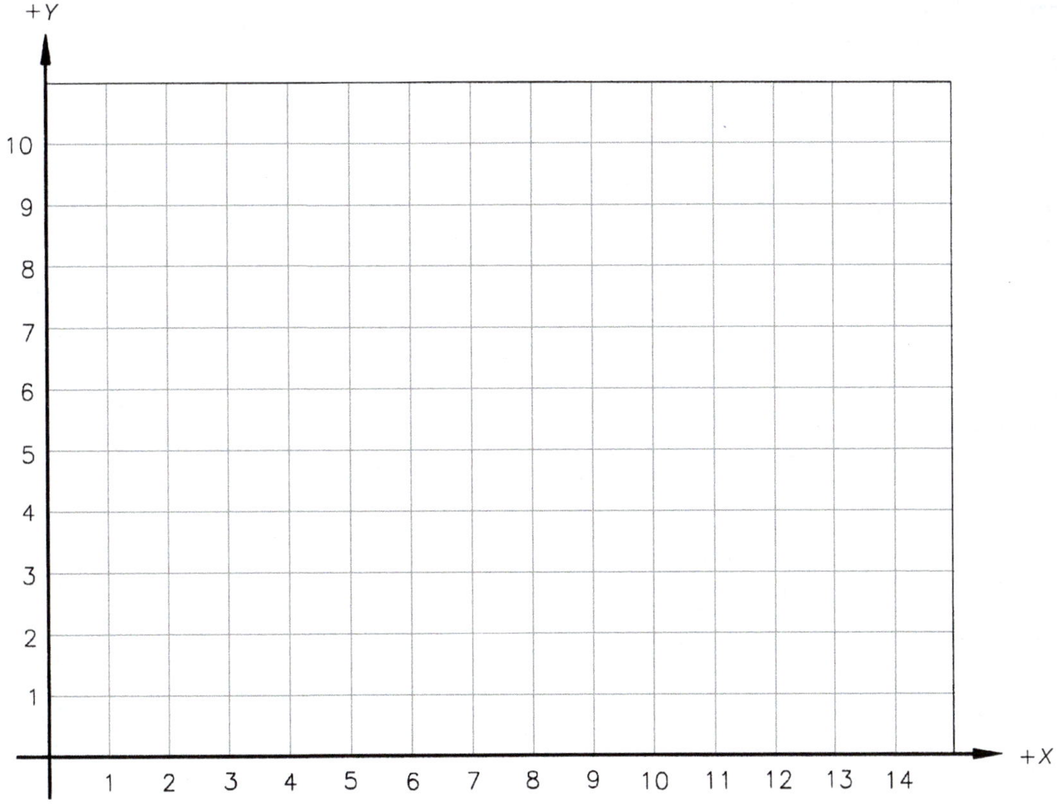

COURSE________ STUDENT____________________ DATE________ PROBLEM 3-2

Problem 3-3 Rectangular Coordinates

Locate each point on the rectangular coordinate system. Connect the points as you progress.

CARTESIAN COORDINATE EXERCISE		
POINT	X	Y
1	4.8	4.1
2	7.9	4.1
3	10.7	5.3
4	10.7	6.6
5	9.5	7.6
6	7.8	7.6
7	6.8	6.7
8	6.3	5.8
9	4.8	5.3
10	4.8	4.1

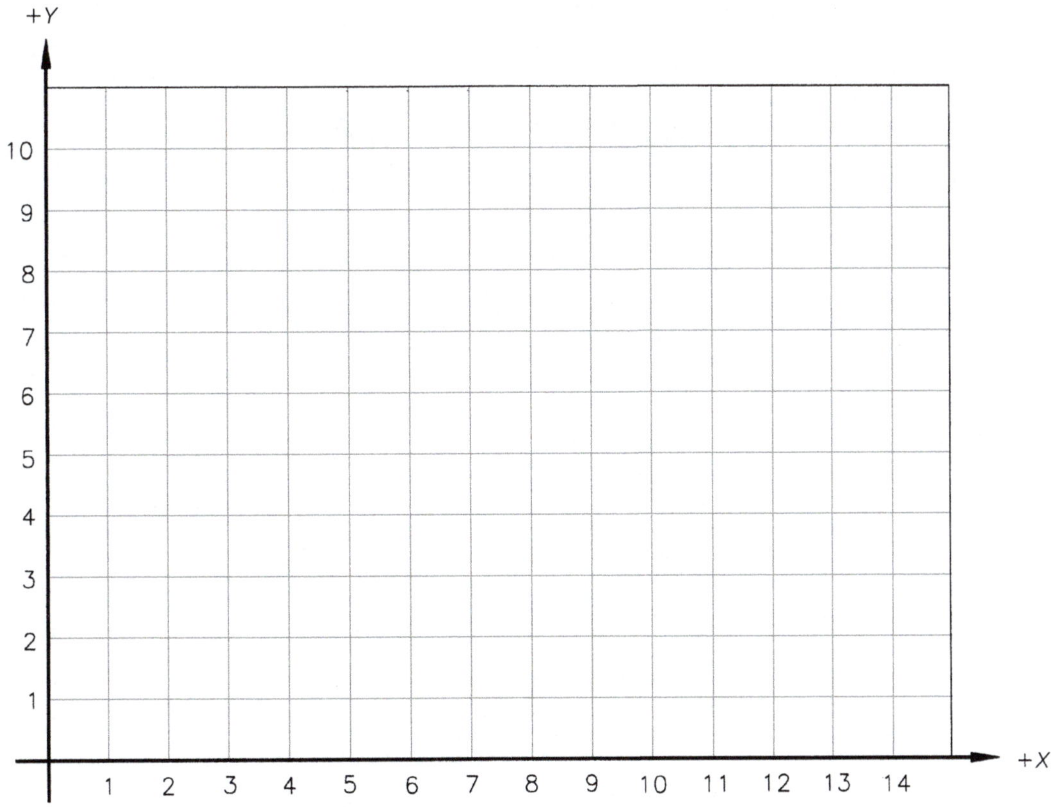

COURSE________ STUDENT______________________________ DATE________ PROBLEM 3-3

Problem 3-4 Point Identification

Using the 1/8" grid and assuming the object's lower left-hand corner is located at 0,0,0, identify each point labeled on the drawing.

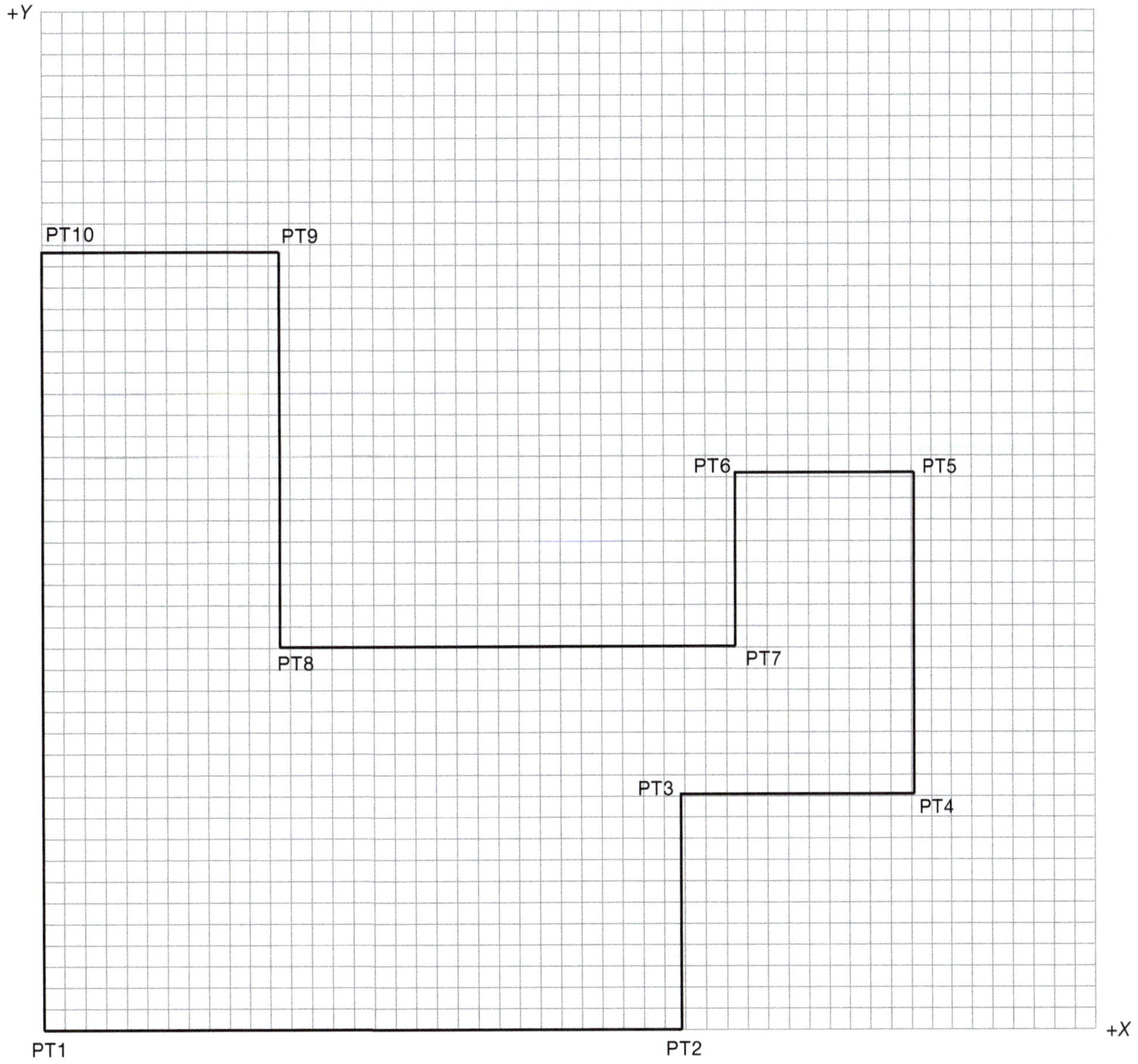

COURSE________ STUDENT________________________________ DATE________ PROBLEM 3-4

Problem 3-5 Point Identification

Using the 1/8" grid and assuming the object's lower left-hand corner is located at 0,0,0, identify each point labeled on the drawing.

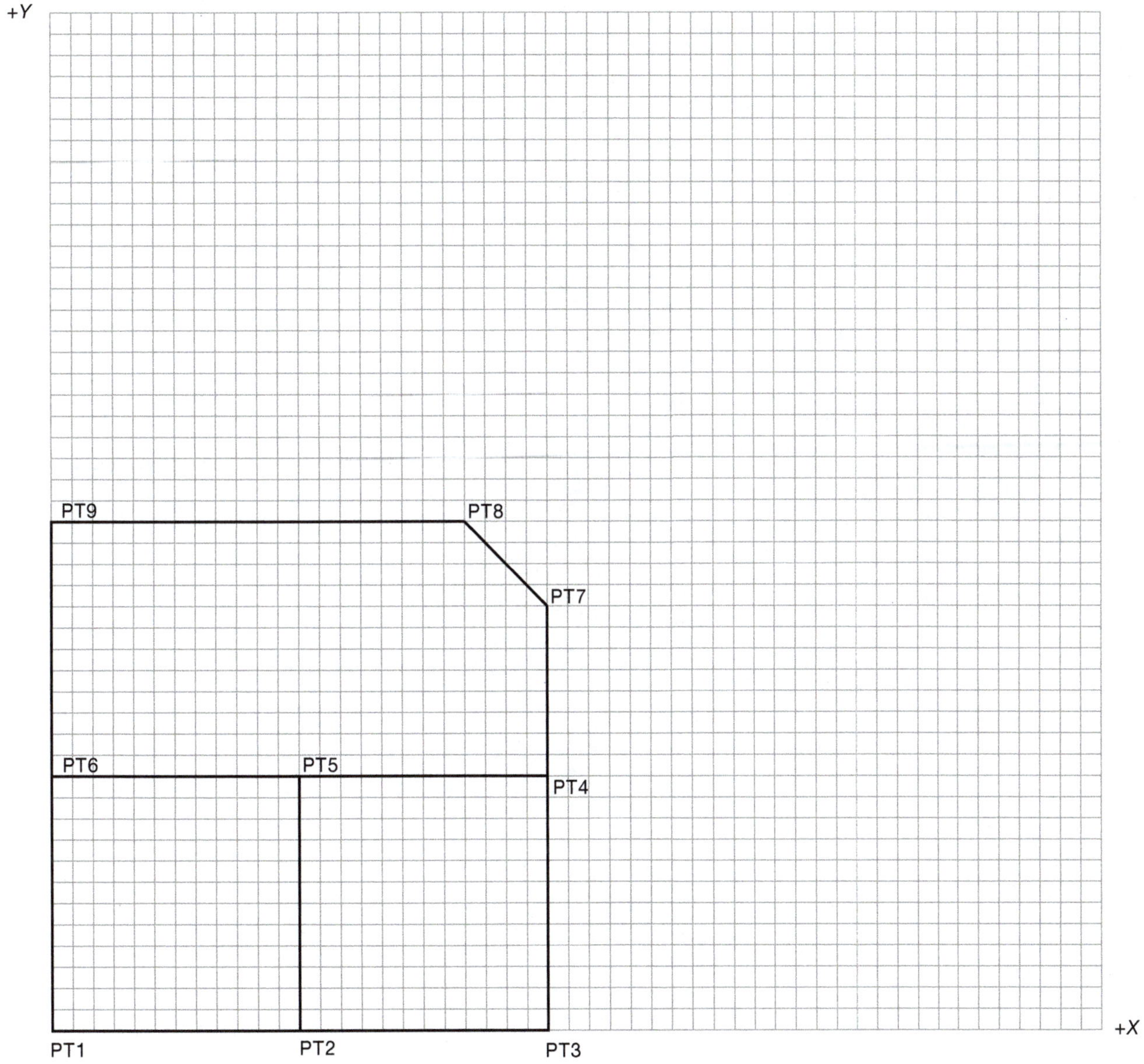

COURSE________ STUDENT________________________________ DATE________ PROBLEM 3-5

Problem 3-6 Point Identification

Using the 1/8" grid and assuming the object's lower left-hand corner is located at 0,0,0, identify each point labeled on the drawing.

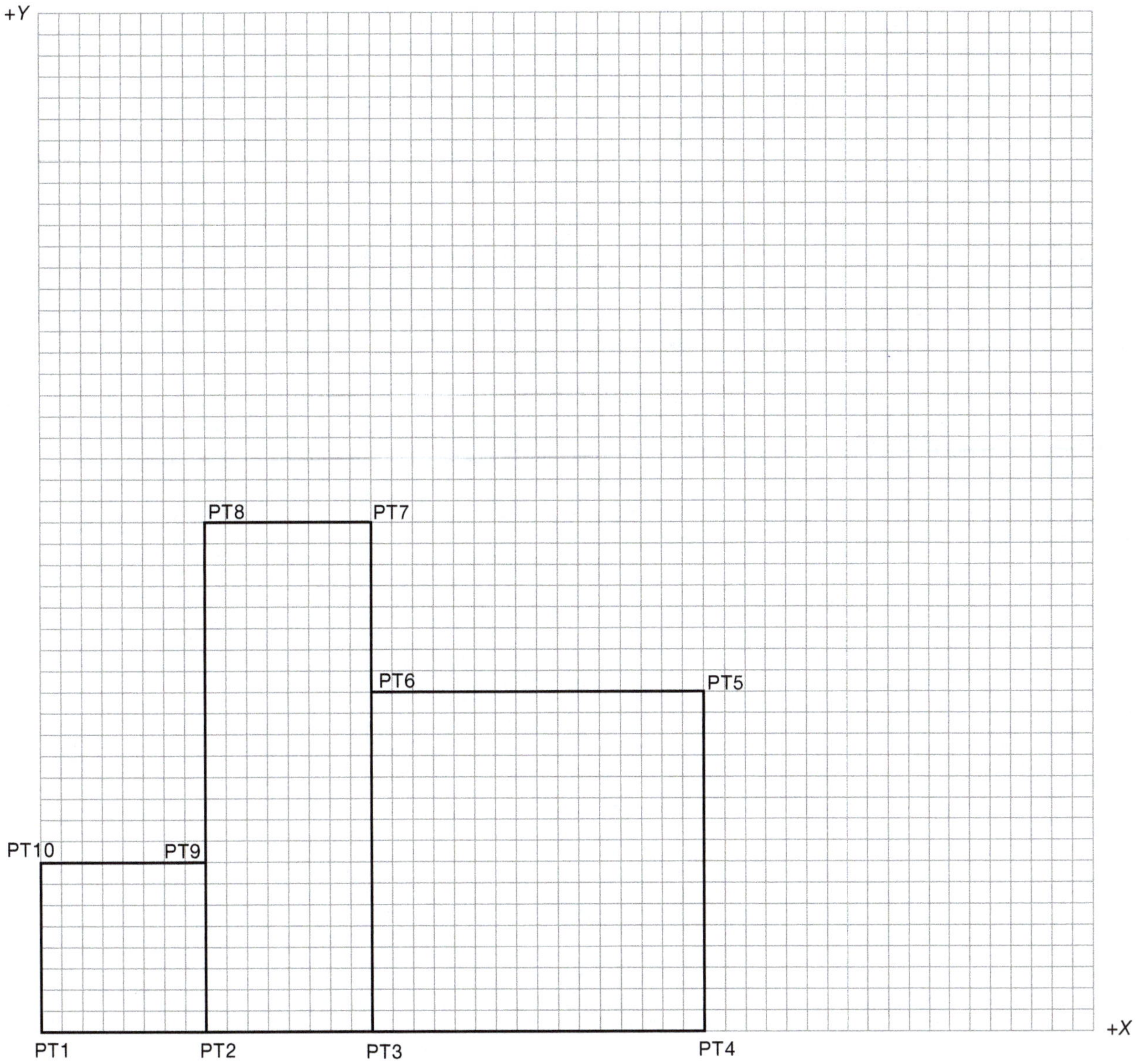

COURSE ________	STUDENT ________________________________	DATE ________	PROBLEM 3-6

Problem 3-7 AutoCAD Drawing Project

Create the following drawing using AutoCAD.

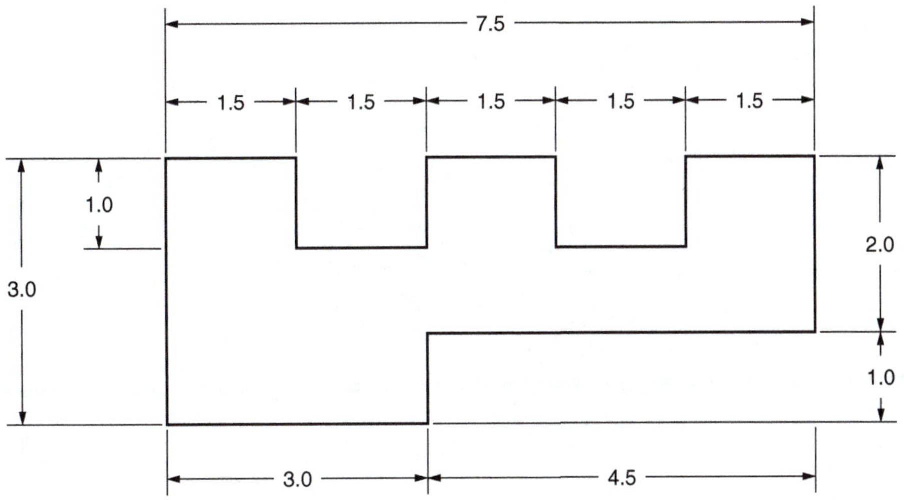

COURSE________ STUDENT______________________________ DATE________ PROBLEM 3-7

Problem 3-8 AutoCAD Drawing Project

Create the following drawing using AutoCAD.

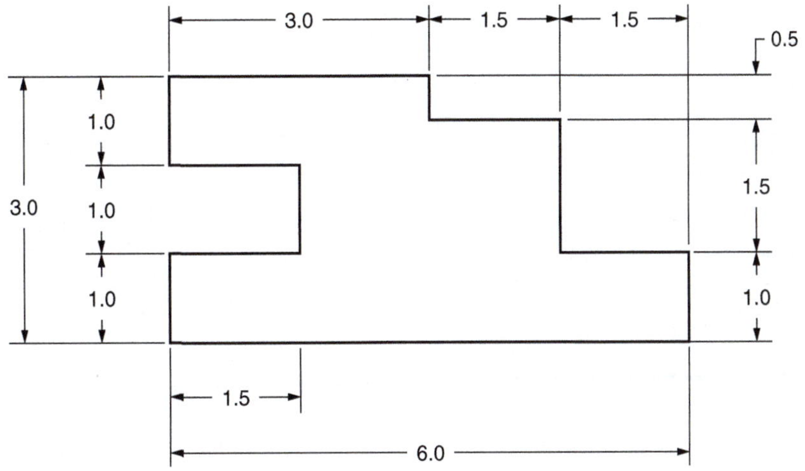

COURSE ________	STUDENT ________________	DATE ________	PROBLEM 3-8

Problem 3-9 AutoCAD Drawing Project

Create the following drawing using AutoCAD.

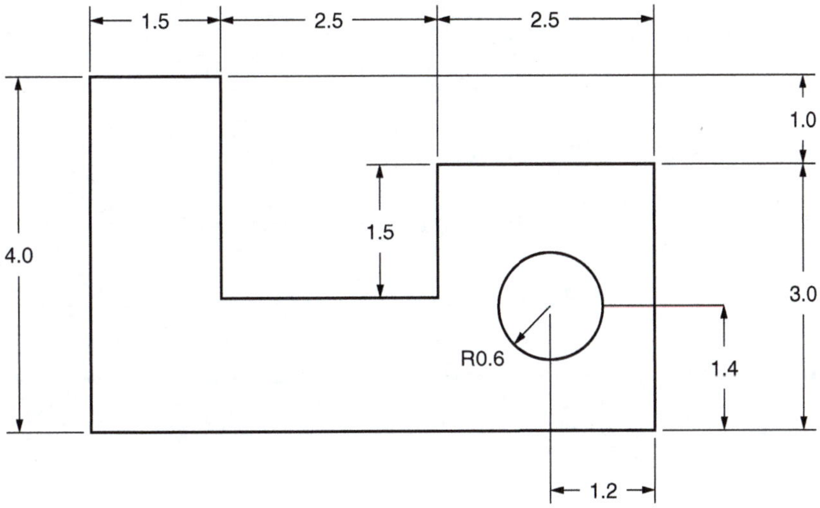

COURSE________ STUDENT________________________________ DATE________ PROBLEM 3-9

Problem 3-10 Graphing Using AutoCAD

Use the LINE command to draw a line graph of the coordinates listed in the following table.

X (Age)	*Y* (Rate $)
25	14
30	15
35	19
40	23
45	37
50	59
55	88
60	130
65	194

COURSE________ STUDENT________________________ DATE________ PROBLEM 3-10

Problem 3-11 AutoCAD Commands

In this exercise, you will be using AutoCAD to create various geometric shapes.

A. Start AutoCAD. (If the Start New Drawing dialog box appears, click Cancel.) With the LINE command, draw several shapes:

1. Rectangle
2. Triangle
3. Irregular polygon (any shape you like)

B. Use the U command. What happens to the last object you drew?

C. Press function key F9 and then look at the status line. Does the button look depressed (pressed in)?

D. Repeat the LINE command and again try drawing these shapes:

1. Rectangle
2. Triangle
3. Irregular polygon

Did you find it easier?

____

____

E. Draw a rectangle with the RECTANGLE command using these parameters:

Corner 2, 3

Other corner 6, 7

F. Draw a circle with the CIRCLE command using these parameters:

Center point 4, 5

Radius 0.75

Is the circle drawn "inside" the box?

____

____

COURSE ________	**STUDENT** ____________________	**DATE** ________	**PROBLEM 3-11**

Problem 3-12 AutoCAD Commands

1. What are AutoCAD drawings constructed with?
2. Name three areas in which commands are entered:
   a.
   b.
   c.
3. Describe the purpose of the "Command:" prompt.
4. How do you exit from print preview mode?
5. What is the purpose of the LINE command?
6. How is the U command helpful?
7. Describe how to cancel commands.
8. What does the mouse control?
9. What is an icon?
10. Name two methods by which you can determine the function of toolbar buttons:
    a.
    b.
11. What are tooltips?
12. What are flyouts?
13. Can toolbars be moved around in the AutoCAD window?
14. Which commands close drawings without exiting AutoCAD?
15. Which commands exit AutoCAD?
16. Which command saves drawings?
17. Which command starts new drawings?
18. Describe how snap and grid are useful:
    Snap
    Grid
19. Is the grid plotted?
20. List three things AutoCAD needs to know, as a minimum, before plotting drawings:
    a.
    b.
    c.
21. Explain the advantage of the PLOT command's Fit to Scale option.
22. Why is the Full Preview option environmentally friendly?
23. How do you exit from print preview mode?

COURSE________ STUDENT________________________________ DATE________ PROBLEM 3-12

Problem 3-13 AutoCAD Commands

In the following exercises, watch how the coordinates and cursor react to different modes.

Step 1. Move the cursor around with your mouse or digitizer.

Question 1. Do the coordinates at the lower-left corner of the screen move?

____

____

Step 2. Press F6. Move the cursor again, and watch the coordinates.

Question 2. What is the difference?

____

____

Step 3. Press F8. Check the status bar: Is ortho mode turned on? Move the cursor.

Question 3. Do you notice a difference?

____

____

Step 4. Press F9 to turn on snap mode. Move the cursor.

Question 4. Now is there a difference in cursor movement?

____

____

In this exercise, you practice using menus and keystrokes that affect commands.

Step 5. Move the crosshairs to the top of the drawing area and into the menu bar.

Question 5. Does the cursor change its shape?

____

____

Step 6. Move the arrow from left to right over the menu bar.

Question 6. What happens as the cursor moves over each word?

____

____

COURSE ________	STUDENT ______________________	DATE ________	PROBLEM 3-13

Step 7. Position the cursor over Draw and then click.

Question 7. Does a menu drop down into the drawing area?

____

____

Step 8. Click on Draw a second time.

Question 8. What happens?

____

____

Step 9. Choose the Draw menu again and then choose Line. Draw some lines. To exit the command, right-click and select Cancel from the shortcut menu. Press the Enter key.

Question 9. Did the LINE command repeat? Clear the command line with ESC. Press the spacebar. What happens?

____

____

Step 10. Press F2.

Question 10. Do you see the Text window? Press F2 again to return to the Drawing window.

____

____

Step 11. Press F1.

Question 11. Do you see the Help window?

____

____

Step 12. Use the LINE command to draw a line graph of insurance premiums for nonsmoking men. Enter the coordinates listed in the following table:

X (Age)	*Y* (Rate)
25	262
35	285
45	365
55	735
65	1681
75	2580

COURSE__________ STUDENT______________________ DATE__________ PROBLEM 3-13

Problem 3-14 AutoCAD Commands

The LIST command shows the following coordinates for a circle: *X*, 12.45; *Y*, 23.65; *Z*, 10.00. Which coordinate is equivalent to the circle's elevation?

____

____

1. Along which axis direction is an extrusion normally projected?

   ____

   ____

2. Briefly explain the following terms:

   Elevation ______________________________

   Thickness ______________________________

3. How could you change the thickness of existing objects?

   ____

   ____

4. After rotating the viewpoint, how do you return to the plan view?

   ____

   ____

5. Describe the change made by thickness to the following objects. (For example, a point becomes a line.)
   a. Line
   b. Circle
   c. Donut
   d. Polygon

6. How do you apply thickness to rectangles drawn by the RECTANGLE command?

   ____

   ____

7. How do you change the elevation when the Elevation option does not appear in the Properties window?

   ____

   ____

COURSE ________	STUDENT ______________________	DATE ________	PROBLEM 3-14

8. Can you use the following commands to change the elevation of objects already in the drawing?
   a. ELEV
   b. CHANGE
   c. CHPROP
   d. PROPERTIES
   e. MATCHPROP
9. Can grips editing change the elevation and thickness of objects?

   ____

   ____

10. Can arcs be drawn with different Z-coordinates at each end?

    ____

    ____

11. Which of the following commands allow editing in 3-D space?
    a. MOVE
    b. OFFSET
    c. COPY
    d. ROTATE
12. Which "object snap" allows you to snap to an intersection that look as if it intersects from a 3-D viewpoint?

    ____

    ____

13. Describe two methods of converting 3-D drawings to 2-D.

    ____

    ____

COURSE________	**STUDENT**____________________	**DATE**________	**PROBLEM 3-14**

CHAPTER 4

Manufacturing Materials and Processes

Problem 4-1   Drill and Boring Tools

Draw a line from the tool to its corresponding hole. Fill in the blanks with the name of each tool.

A ______________________________   D ______________________________

B ______________________________   E ______________________________

C ______________________________   F ______________________________

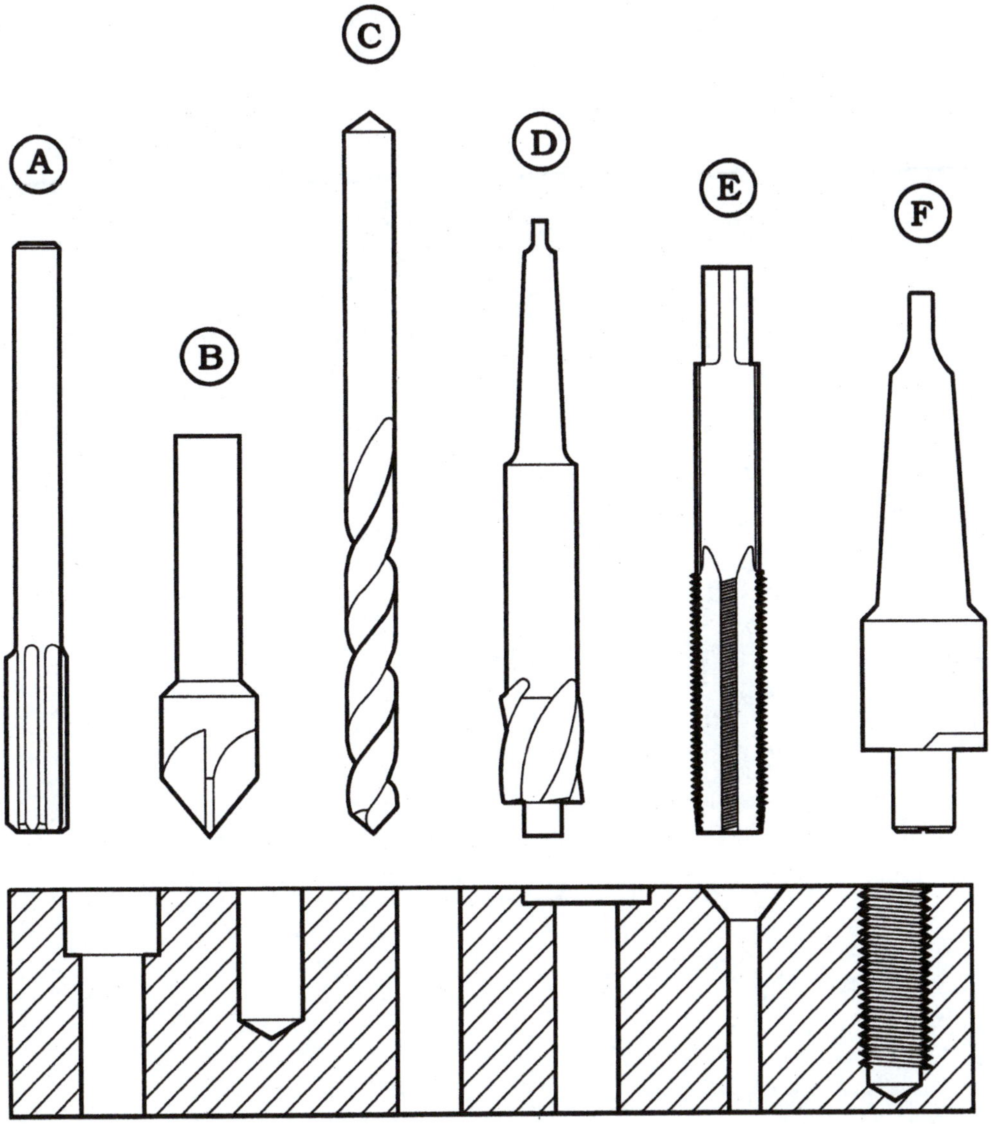

COURSE ________	STUDENT ______________________________	DATE ________	PROBLEM 4-1

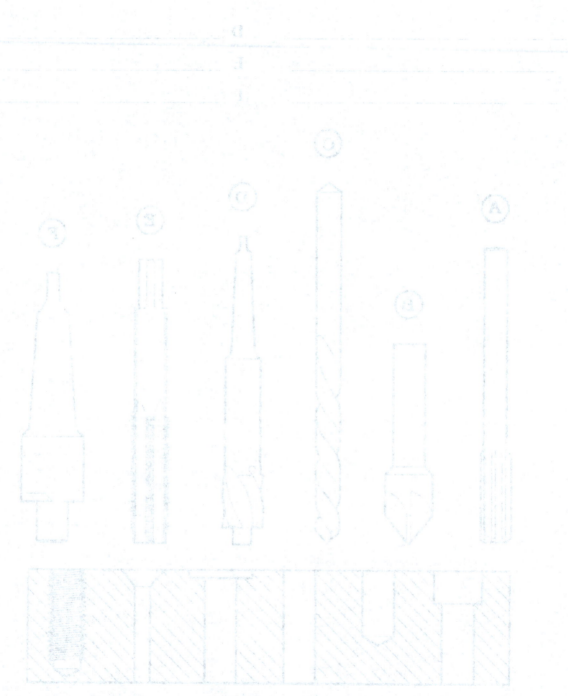

Problem 4-2 Manufacturing Processes

Provide the word or phrase that best fits each of the following statements:

A. The most commonly used method of molding castings.

____

B. An ancient method of forming metals into desired shapes.

____

C. A machine used to drill holes, ream, bore, countersink, counterbore, and tap.

____

D. A cold-forming process used to uniformly roughen a cylindrical or flat surface with a diamond or straight pattern.

____

E. The cutting away of the sharp external or internal corner of an edge.

____

F. A process that uses chemicals to accurately remove material.

____

G. One of the most versatile machine shop tools, this machine uses a rotary cutting tool to remove material from the work.

____

H. An iron that is primarily an alloy of iron and 1.7–4.5% of carbon with varying amounts of silicone, manganese, phosphorous, and sulfur.

____

I. A process in which a direct current is passed through an electrolyte solution between an electrode and the work piece.

____

J. A small-radius outside corner formed between two surfaces.

____

COURSE________	STUDENT______________________	DATE________	PROBLEM 4-2

K. Used as fasteners to hold parts together, to adjust parts in alignment with each other, or to transmit power.

____

L. A note that should be used when machine fragments are to be removed from a part after machining.

____

M. A small radius formed between the inside angle of two surfaces.

____

N. A slot with angled sides that may be machined at any depth and width.

____

COURSE________	STUDENT______________________________	DATE________	PROBLEM 4-2

Problem 4-3 Alloys

Complete the following chart regarding alloys and their characteristics.

Alloy	Characteristics	Common Uses
3103/3003		Vehicle paneling, structures exposed to marine atmosphere, mine cages.
5251/5052 5454		Pressure vessels and road tankers. Transport of ammonium nitrate, petroleum. Chemical plants.
6063		Used for complex profiles. Architectural extrusions, window frames, irrigation pipes.
7075		Airframes.
2014A		Airframes.
5083/5182		A superior alloy for cryogenic use (in annealed condition). Pressure vessels and road-transport applications below 65 °C. Ship building structure in general.
6005A		Thin-walled wide extrusions.
1050/1200		Food and chemical industry.
7020		Armored vehicles, military bridges, motor cycle and bicycle frames.
6061/6082		Stressed structural members, bridges, cranes, roof trusses, beer barrels.

COURSE ________ STUDENT ______________________________ DATE ________ PROBLEM 4-3

Problem 4-4 Steels

In the following space, define the following terms:

Steel

Mild steel (MS)

Medium-carbon steel

High-carbon steel

Hot-rolled steel (HRS)

Cold-rolled steel (CRS)

COURSE ________ STUDENT ______________________________ DATE ________ PROBLEM 4-4

Problem 4-5 Casting

In the following space, discuss the process for casting parts.

COURSE________ STUDENT________________________ DATE________ PROBLEM 4-5

Problem 4-6 Forging

In the following space, discuss the process for forging parts.

COURSE ________ STUDENT ______________________________ DATE ________ PROBLEM 4-6

SECTION 2

Fundamental Applications

CHAPTER 5

Sketching Applications

Problem 5-1 Sketching

Construct all views necessary to fully describe the object.

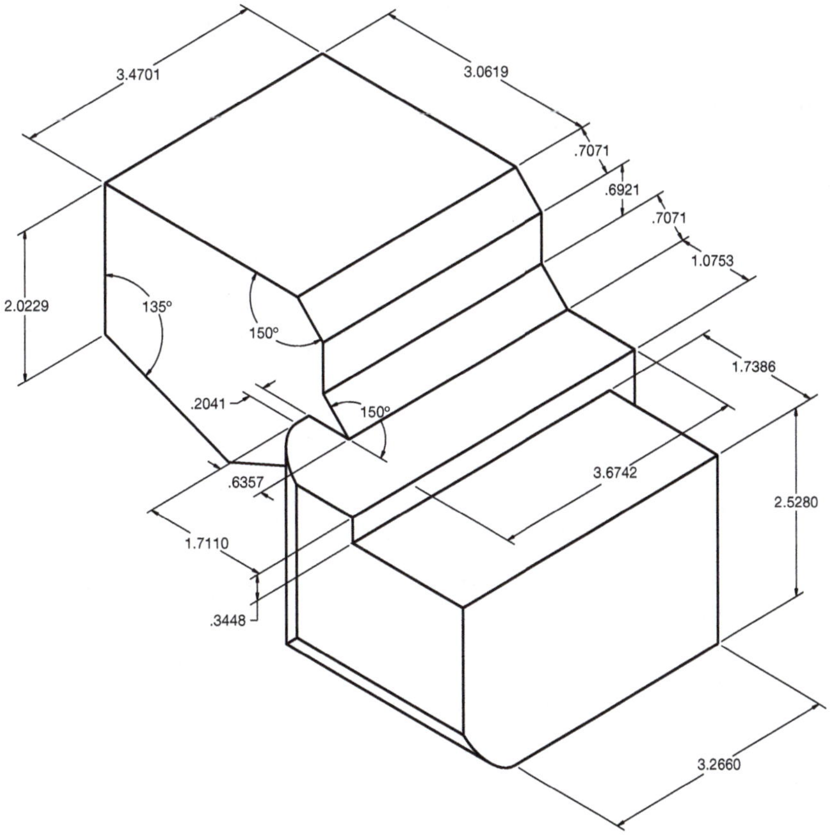

COURSE ________	STUDENT ______________________	DATE ________	PROBLEM 5-1

Problem 5-2 Sketching

Construct all views necessary to fully describe the object.

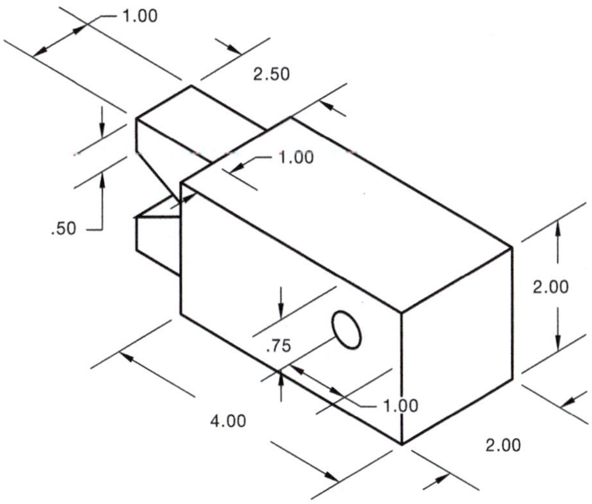

COURSE________ STUDENT________________________________ DATE________ PROBLEM 5-2

Problem 5-3 Isometric Sketching

Sketch the isometric view in the grid, using the following multiviews. Sketch at approximately twice the scale (2:1). Use ANSI standards.

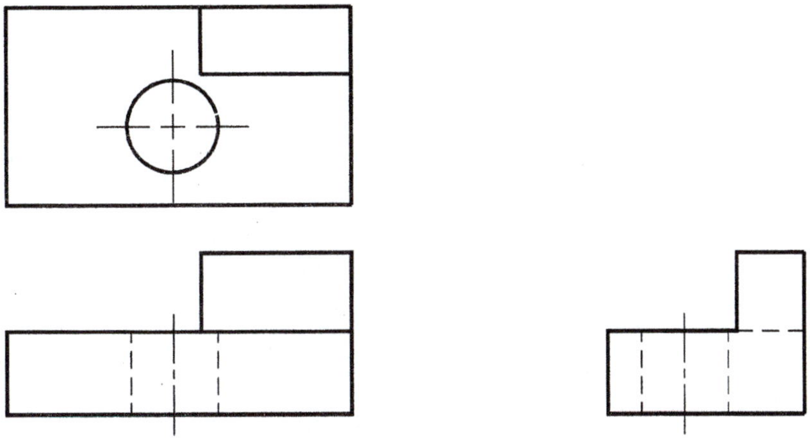

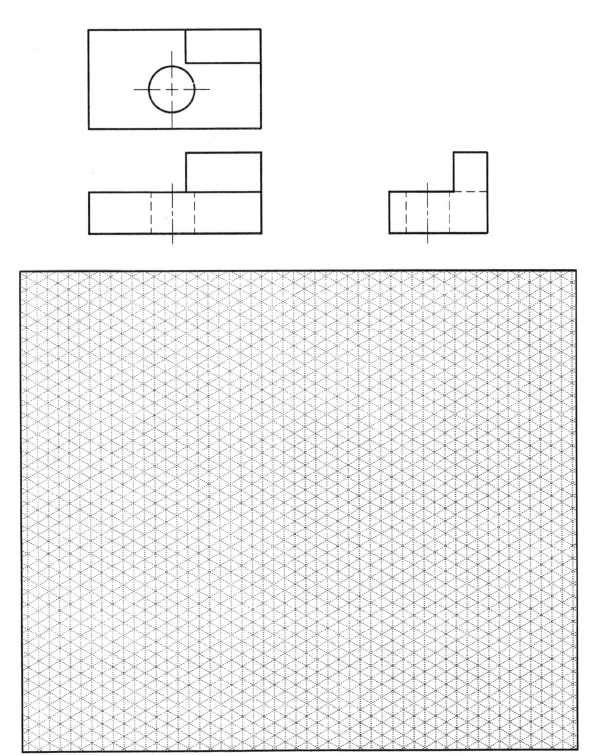

COURSE ________ STUDENT ______________________________ DATE ________ PROBLEM 5-3

Problem 5-4 Isometric Sketching

Using the following isometric grid, sketch the given drawing at twice the scale (2:1). Re-create the line thicknesses shown.

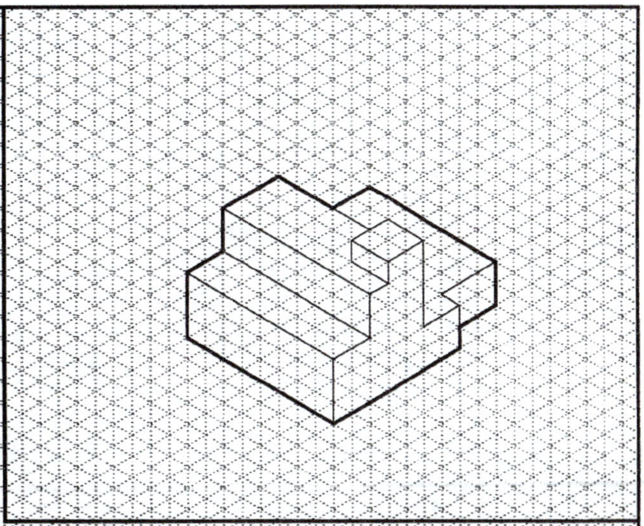

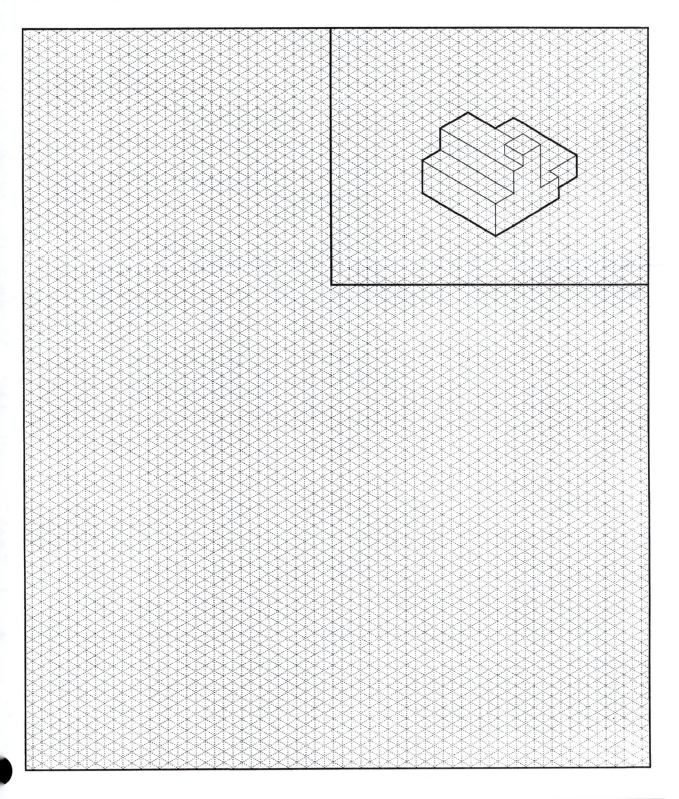

COURSE ________ STUDENT ______________________________ DATE ________ PROBLEM 5-4

Problem 5-5 Isometric Sketching

Using the following isometric grid, sketch the given drawing at twice the scale (2:1). Re-create the line thicknesses shown.

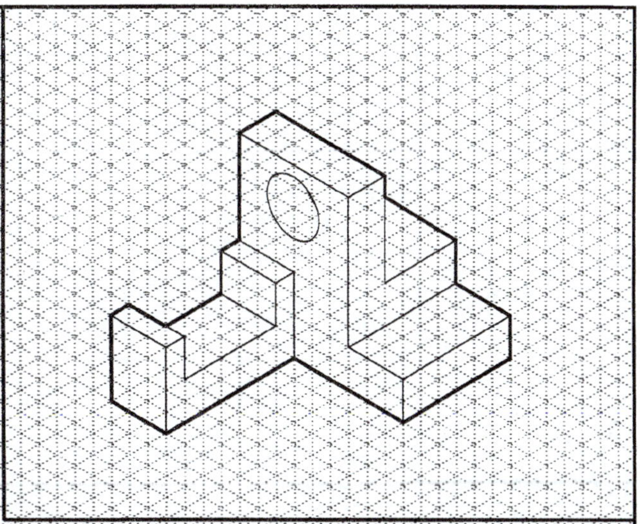

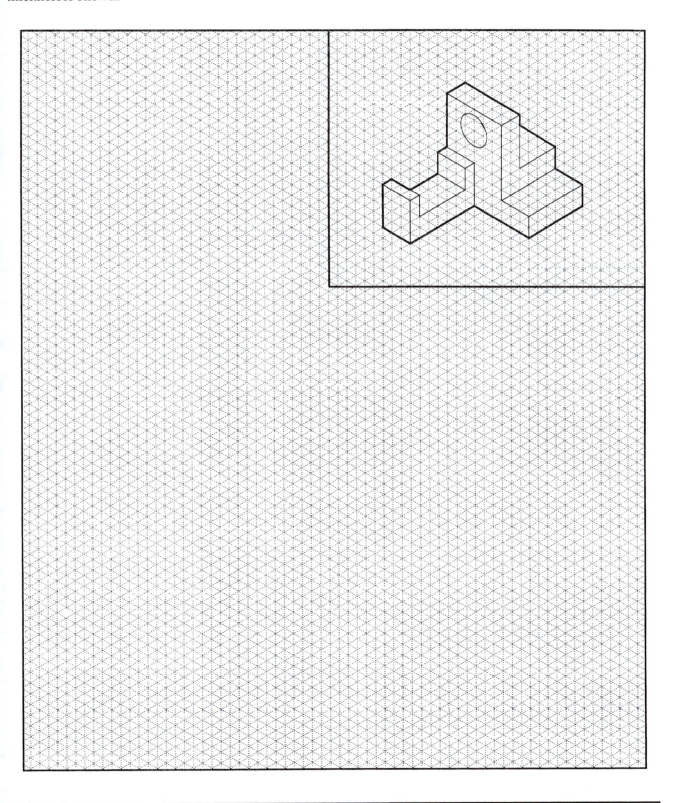

COURSE__________ STUDENT______________________________ DATE__________ PROBLEM 5-5

Problem 5-6 Isometric Sketching

Using the following isometric grid, sketch the given drawing at twice the scale (2:1). Re-create the line thicknesses shown.

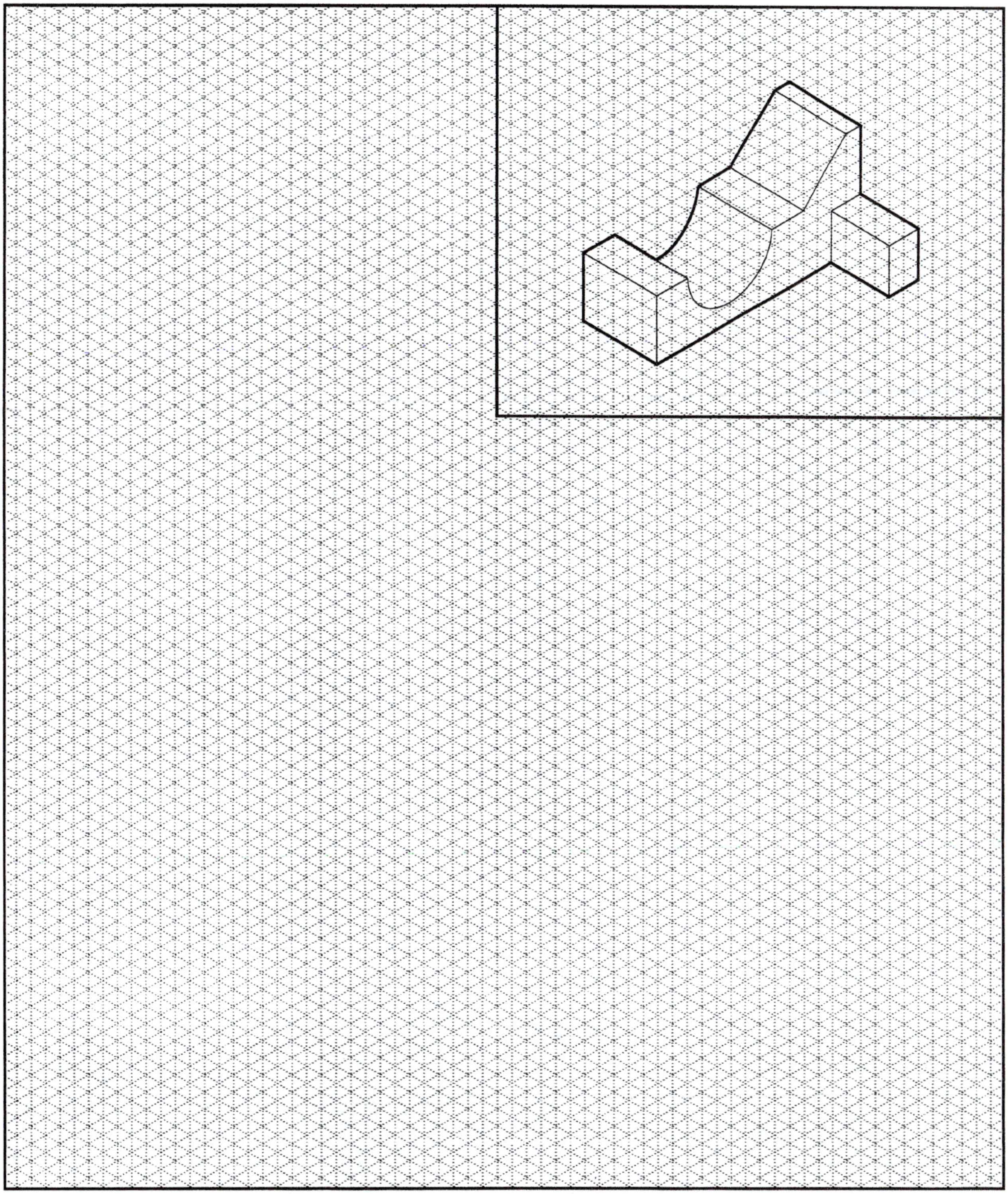

COURSE_________ STUDENT________________________________ DATE_________ PROBLEM 5-6

CHAPTER 6

Lines and Lettering

Problem 6-1 Line Exercise

Review the Supplemental Material for Chapter 6 on the CD that accompanies the textbook. Shown below are six different types of lines drawn according to ANSI standards. Trace over each given line and then reproduce the line three times on the dotted lines shown below each group. At the right side, label the type of line and the standard thickness of the line.

A.

B.

C.

D.

E.

F.

COURSE ________	STUDENT ______________________________	DATE ________	PROBLEM 6-1

Problem 6-2 Section Lines

Review the Supplemental Material for Chapter 6 on the CD that accompanies the textbook. Sketch in the coded section lines of the materials shown below. Do not use a straightedge.

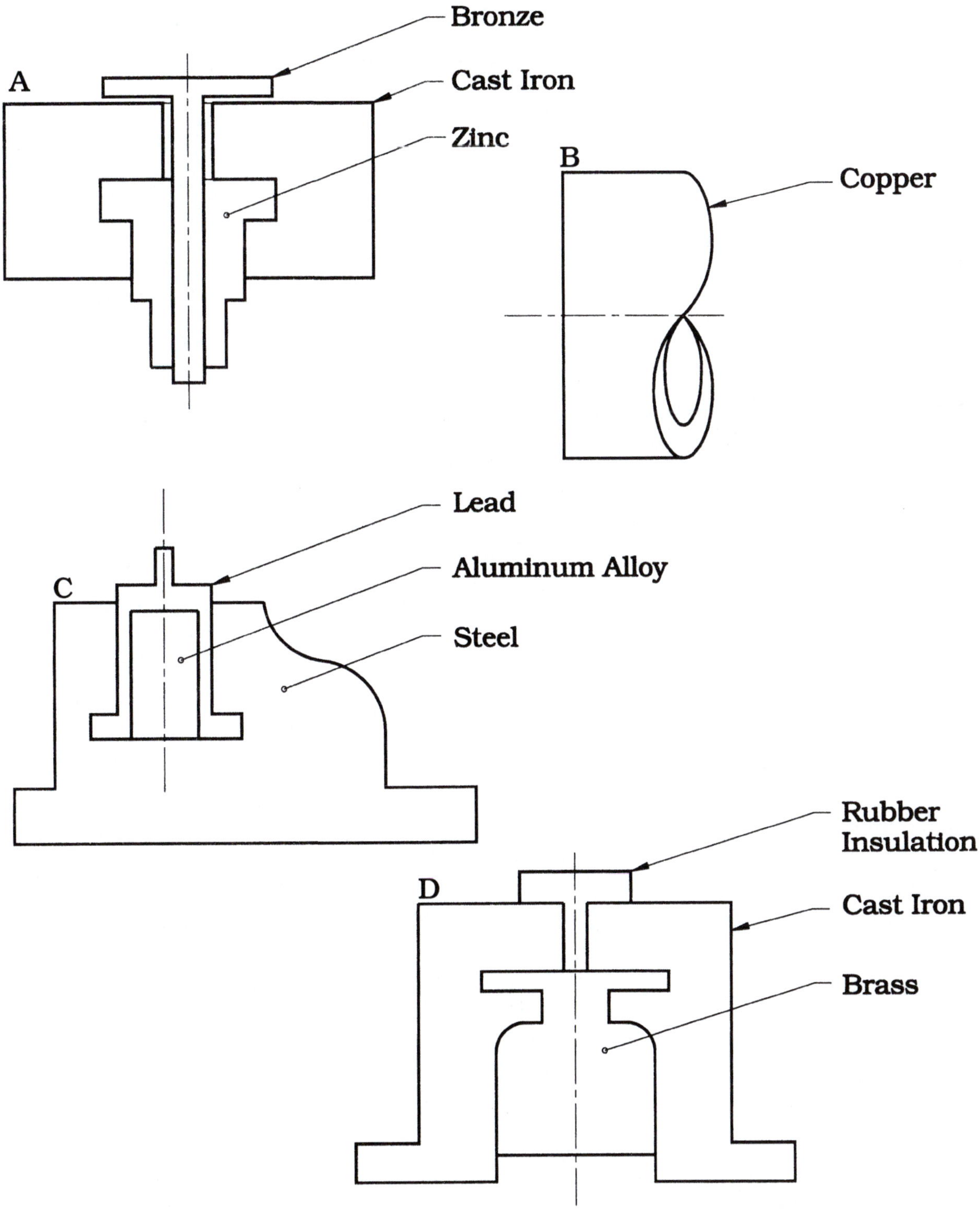

COURSE________ STUDENT____________________ DATE________ PROBLEM 6-2

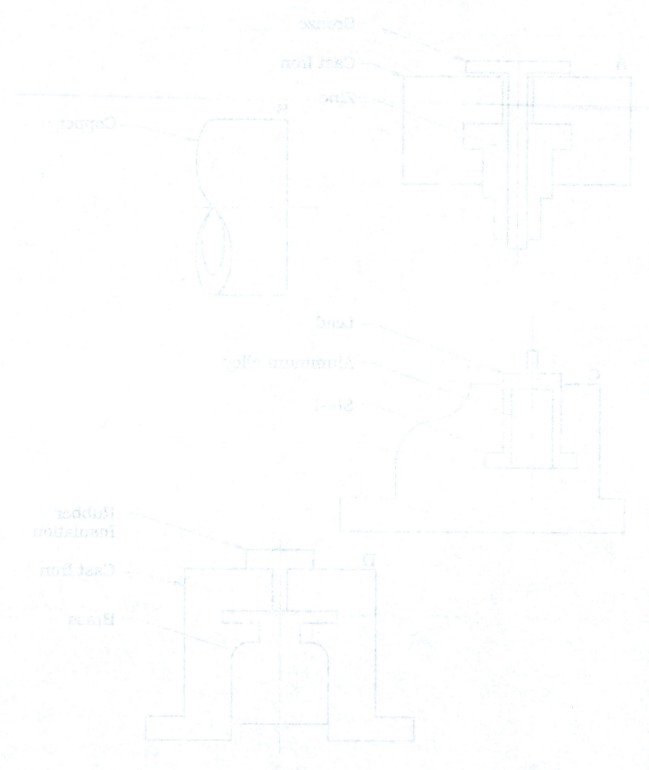

Problem 6-3 Section Lines

Review the Supplemental Material for Chapter 6 on the CD that accompanies the textbook. Show the section lines of the materials below. Do not use a straightedge.

A

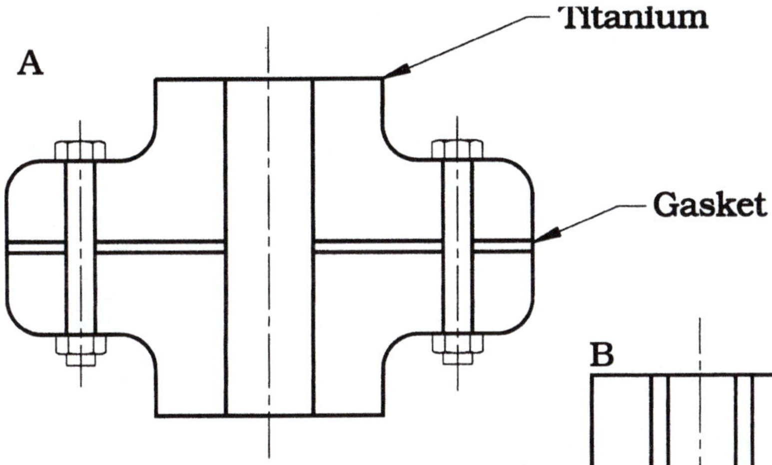

B

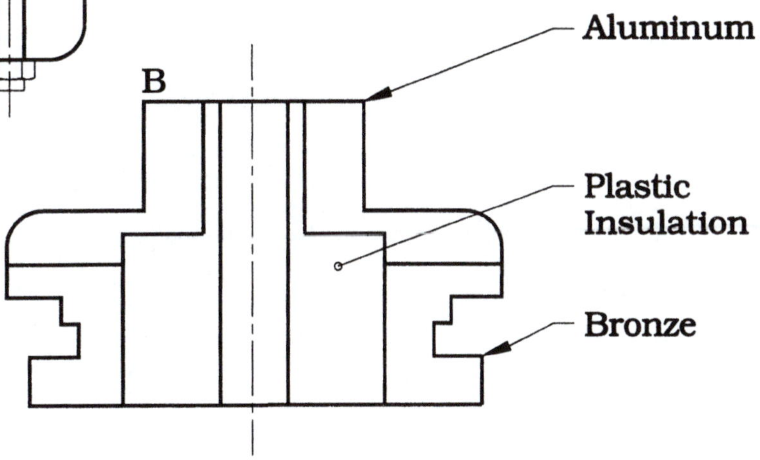

C

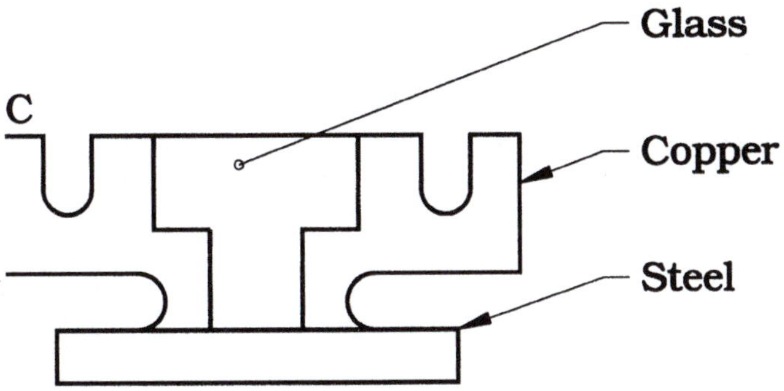

Show break lines of the following items.

D

Metal

E

Copper Tube

COURSE ________	STUDENT ______________________________	DATE ________	PROBLEM 6-3

Problem 6-4 Spur Gears

Spur gears are mounted on parallel shafts to transmit rotary motion. The gear teeth are parallel to the axis of rotation and are cut with an involute profile. Spur gear terms are listed below and are referenced to the figure. Construct guidelines for lettering, and print the terms in the spaces provided using a vertical Gothic style.

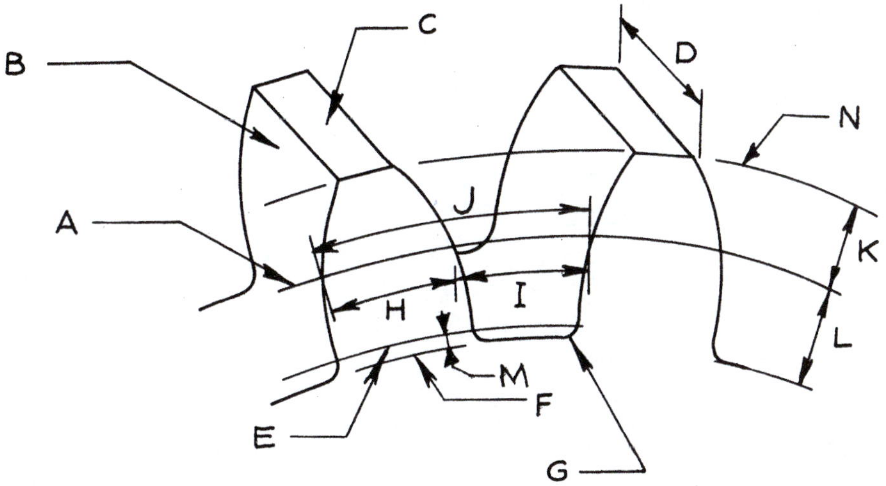

A. PITCH CIRCLE

B. FACE

C. TOP

D. FACE WIDTH

E. CLEARANCE CIRCLE

F. DEDENDUM CIRCLE

G. FILLET

H. TOOTH THICKNESS

I. SPACE

J. CIRCULAR PITCH

K. ADDENDUM

L. DEDENDUM

M. CLEARANCE

N. ADDENDUM CIRCLE

COURSE________ STUDENT____________________ DATE________ PROBLEM 6-4

Problem 6-5 Line Types

Letters identify each different type of line in the example drawing, which shows line types used in technical drawing. From the list, match the type of line to the appropriate letter and print the line type in the space provided. Use guidelines at .125" spacing and vertical Gothic capitals. (Note: Two of the line types occur twice.)

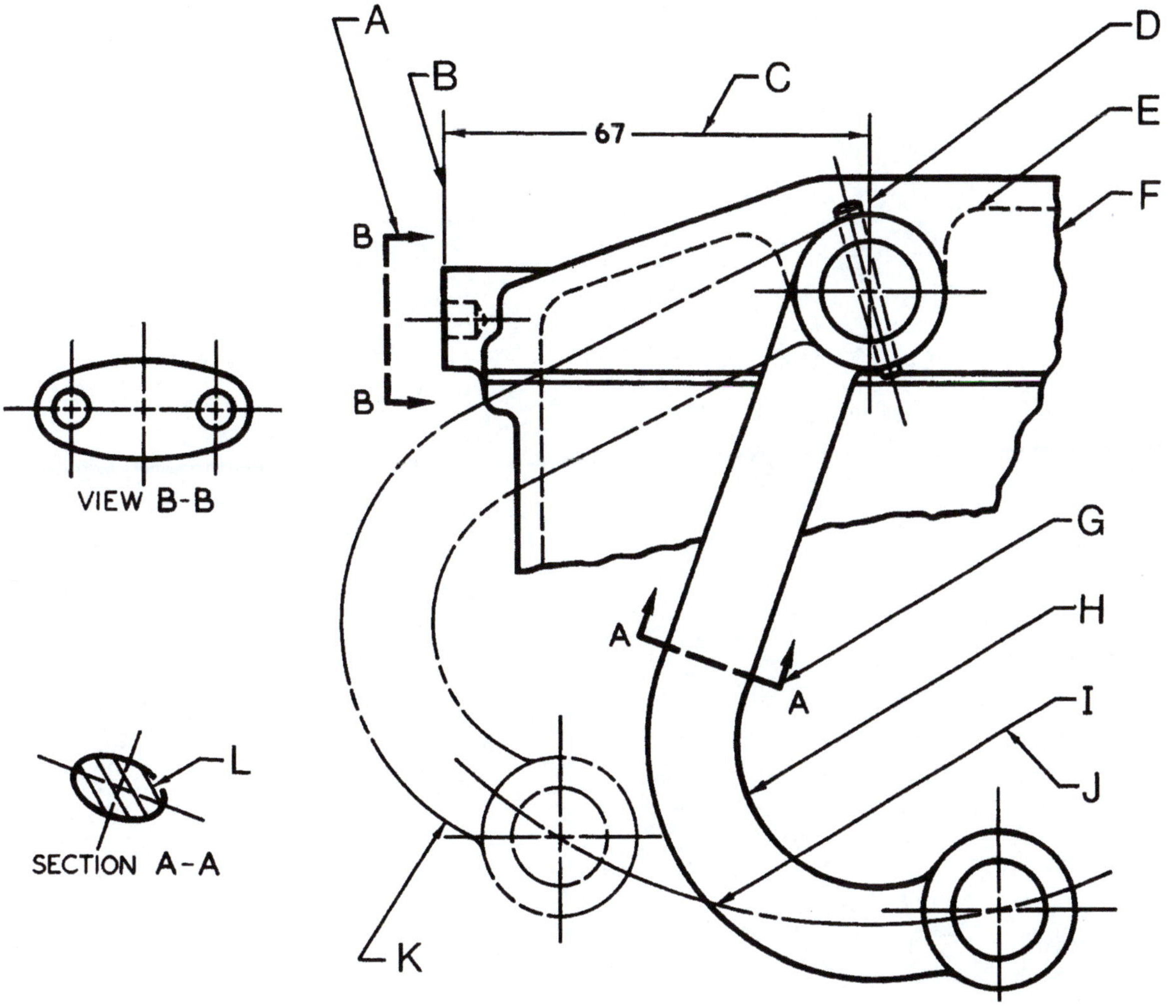

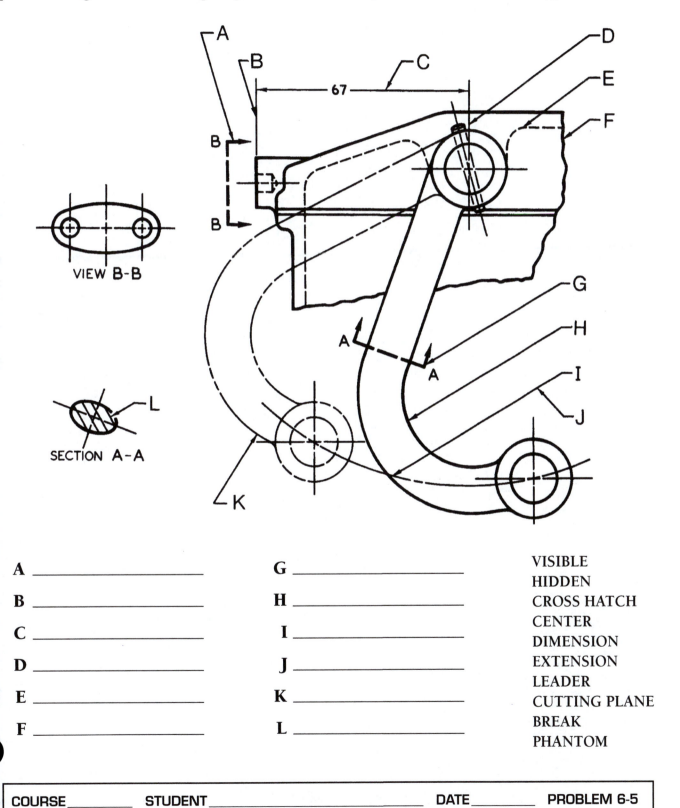

A	______	G	______	VISIBLE
B	______	H	______	HIDDEN
C	______	I	______	CROSS HATCH
D	______	J	______	CENTER
E	______	K	______	DIMENSION
F	______	L	______	EXTENSION
				LEADER
				CUTTING PLANE
				BREAK
				PHANTOM

COURSE ________ STUDENT ____________________ DATE ________ PROBLEM 6-5

Problem 6-6 Lettering

Review the Supplemental Material for Chapter 6 on the CD that accompanies the textbook. Using the lettering techniques outlined, letter the paragraph below using the gridlines provided.

THE TERM SINGLE STROKE COMES FROM THE FACT THAT EACH LETTER IS MADE UP OF A SINGLE STRAIGHT OR CURVED LINE ELEMENT THAT MAKES IT EASY TO DRAW AND CLEAR TO READ. INDUSTRY HAS BECOME ACCUSTOMED TO USING VERTICAL UPPERCASE LETTERS AS STANDARD.

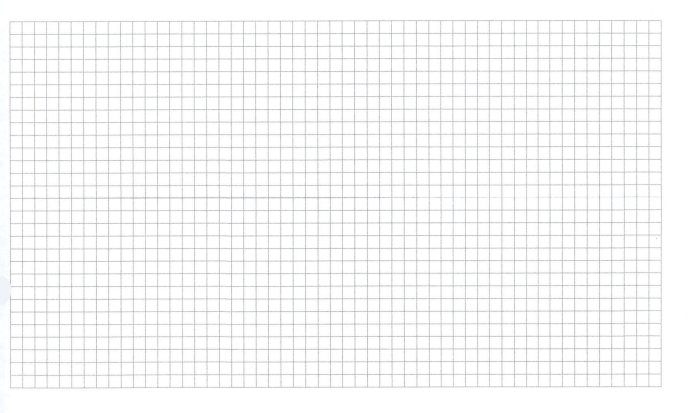

COURSE________ STUDENT____________________ DATE________ PROBLEM 6-6

Problem 6-7 Lettering

Review the Supplemental Material for Chapter 6 on the CD that accompanies the textbook. Using the lettering techniques outlined, letter the following text using the gridlines below.

SOME COMPANIES PREFER INCLINED LETTERING. THE GENERAL SLANT OF INCLINED LETTERS IS 68°. ONE EDGE OF THE AMES LETTERING GUIDE HAS A 68° SLANT, WHICH MAY BE USED TO HELP MAINTAIN THE PROPER ANGLE. STRUCTURAL DRAFTING IS ONE FIELD IN WHICH SLANTED LETTERING MAY BE COMMONLY FOUND.

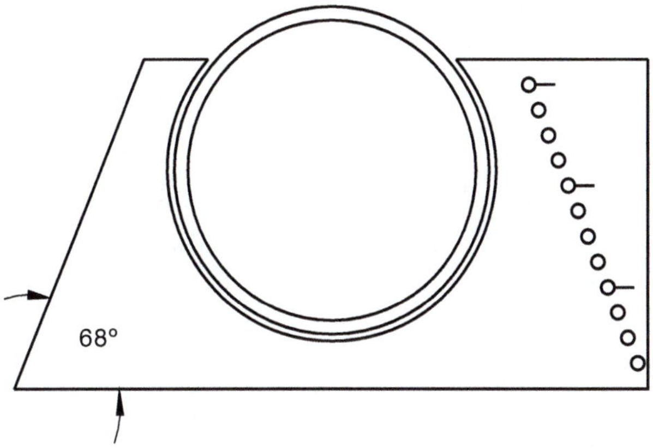

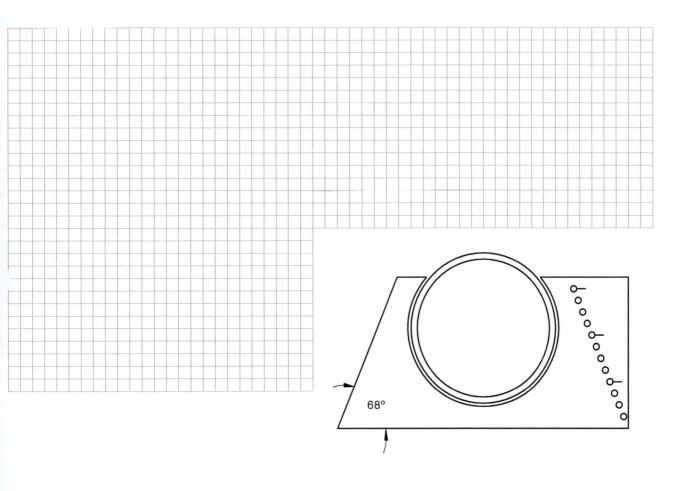

COURSE________ STUDENT______________________________ DATE________ PROBLEM 6-7

Problem 6-8 Lettering

Review the Supplemental Material for Chapter 6 on the CD that accompanies the textbook. Using the lettering techniques outlined, letter the paragraph below using the gridlines provided.

Occasionally, lowercase letters are used, but they are very uncommon in mechanical drafting. Civil or map drafters use lowercase lettering for some practices.

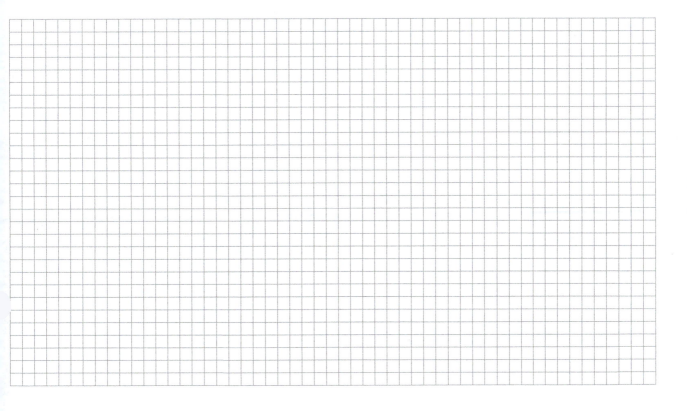

COURSE________ STUDENT________________________ DATE________ PROBLEM 6-8

CHAPTER 7

Drafting Geometry

Problem 7-1 Guided Practice Bisecting an Angle

Step 1. Draw a circle with the center located at the vertex of angle O. The radius of the circle should be about half the distance of one of the sides of the angle.

COURSE ________	STUDENT ______________________________	DATE ________	PROBLEM 7-1

Step 2. Draw a circle (any size radius may be used) with the center located at the intersection of the first circle and one of the sides of the angle (line AO).

Step 3. Draw a circle (use the same radius size as in step 2) with the center located at the intersection of the first circle and the other side of the angle (line BO).

Note: The circles from steps 2 and 3 should intersect; if they don't, readjust the radius of the circles (in steps 2 and 3) so that they do intersect.

COURSE ________	STUDENT ______________________________	DATE ________	PROBLEM 7-1

Step 4. Draw a line from the intersection of the circles in steps 2 and 3 to the vertex bisecting the angle.

COURSE________	STUDENT________________________	DATE________	PROBLEM 7-1

Problem 7-2 Guided Practice Constructing a Triangle

Step 1. From the location where the triangle is to be constructed, use the LINE command to draw a line equal in length to one of the sides of the triangle.

Step 2. Use the CIRCLE command to draw a circle at the end of the first line (B) that has a radius equal to the length of the second line.

COURSE________ STUDENT______________________________ DATE_________ PROBLEM 7-2

Step 3. Use the CIRCLE command to draw a circle that has a radius equal to the radius of the third line at the opposite end of the first line drawn.

Step 4. Use the LINE command to draw a line from one of the end points of the first line drawn to the intersection of the two circles (from steps 2 and 3) to the opposite end point.

Note: Either point of intersection may be used.

COURSE________	STUDENT________________________	DATE________	PROBLEM 7-2

Problem 7-3 Guided Practice Constructing an Equilateral Triangle

A polygon is a multisided geometric figure containing three or more sides and angles (examples: triangles, squares, pentagons, hexagons, and so forth). An equilateral triangle is a third-sided polygon in which all three sides are equal in length. Equilateral triangles can be constructed by drawing a (base) circle with a radius equal to 0.577350269 or [2/3 $\sqrt{}$length^2 – (length ÷ 2)^2] times the length of one of its sides. Using the DIVIDE command, the circle is then divided into three equal segments. The segments are then connected using the PLINE command to form the equilateral triangle. The following is the sequence of steps outlining the procedure described here.

Note: The coefficient was calculated using the formula 2/3 $\sqrt{1^2}$ (1 ÷ 2)^2.

CIRCLE-DIVIDE METHOD (NONISOMETRIC)

Step 1. Use the length of one side of the equilateral triangle to calculate the radius of the base circle that inscribes the triangle by multiplying the length of one of the sides by 0.577350269.

Step 2. Use the CIRCLE command to draw the base circle at the location where the equilateral triangle is to be constructed.

COURSE________	STUDENT________________________	DATE________	PROBLEM 7-3

Step 3. Use the DIVIDE command to split the circle into three equal segments, placing nodes at each division point.

Step 4. Use the PLINE command to connect the nodes produced in step 3 to form the equilateral triangle.

An equilateral triangle can also be constructed by first drawing a line equal in length to one of the sides of the triangle. Next, using the CIRCLE command, draw a circle whose center is positioned at one end point of the first line drawn and whose radius is equal in length to the side of the triangle. The same procedure is repeated, but this time the circle's center is positioned on the opposite end point of the baseline. Finally, the

COURSE ________	STUDENT ________________________	DATE ________	PROBLEM 7-3

triangle is constructed by drawing a line from the intersection of the two circles created to each end point of the baseline. (This is basically the same procedure used to construct a triangle with three sides given.) The following is the sequence of steps outlining this procedure for the equilateral triangle.

LINE–CIRCLE METHOD (NONISOMETRIC)

Step 1. Use the LINE command to draw a line equal in length to one of the sides of the equilateral triangle and at the location where the triangle is to be positioned.

COURSE________	STUDENT______________________________	DATE________	PROBLEM 7-3

Step 2. Draw a circle whose radius is equal in length to one side of the triangle and whose center is located at one end point of the baseline constructed in step 1 (A).

COURSE ________	STUDENT ______________________________	DATE ________	PROBLEM 7-3

Step 3. Draw a circle whose radius is equal in length to one of the sides of the triangle and whose center is located at the opposite end point of the baseline constructed in step 1 (B).

COURSE________ STUDENT______________________________ DATE________ PROBLEM 7-3

Step 4. Use the LINE command to draw a line from one of the end points of the first line drawn to the intersection of the two circles (from steps 2 and 3) to the opposite end point.

COURSE ________	STUDENT ______________________________	DATE ________	PROBLEM 7-3

Problem 7-4 Guided Practice Constructing an Isometric Equilateral Triangle

The main drawback of the POLYGON command is that it is not designed to construct isometric polygons. In order to draw an isometric equilateral triangle, the preceding procedure must be employed with one slight modification. Instead of using the CIRCLE command, the Isocircle option of the ELLIPSE command is used. The rest of the procedure is the same for either of the two methods previously mentioned.

CIRCLE-DIVIDE METHOD (ISOMETRIC)

Step 1. Use the length of one side of the equilateral triangle to calculate the radius of the base circle that inscribes the triangle by multiplying the length of one of the sides by 0.577350269.

Step 2. Use the Isocircle option of the ELLIPSE command to draw the base circle at the location where the equilateral triangle is to be constructed.

COURSE________	STUDENT____________________	DATE________	PROBLEM 7-4

Step 3. Use the DIVIDE command to split the circle into three equal segments, placing nodes at each division point.

Step 4. Use the PLINE command to connect the nodes produced in step 3 to form the equilateral triangle.

COURSE ________	STUDENT ______________________________	DATE ________	PROBLEM 7-4

LINE–CIRCLE METHOD (ISOMETRIC)

Step 1. Use the LINE command to draw a line equal in length to one of the sides of the equilateral triangle at the location where the triangle is to be positioned.

Step 2. Use the Isocircle option of the ELLIPSE command to draw a circle whose radius is equal in length to one of the sides of the triangle and whose center is located at one of the end points of the baseline constructed in step 1 (A).

COURSE_________ STUDENT_____________________________ DATE_________ PROBLEM 7-4

Step 3. Draw a circle whose radius is equal in length to one of the sides of the triangle and whose center is located at the opposite end point of the baseline constructed in step 1 (B).

COURSE ________	STUDENT ______________________________	DATE ________	PROBLEM 7-4

Step 4. Use the LINE command to draw a line from one of the end points of the first line drawn to the intersection of the two circles (from steps 2 and 3) to the opposite end point.

COURSE______	STUDENT____________________	DATE______	PROBLEM 7-4

Problem 7-5 Guided Practice Constructing an Isometric Square

A square is a four-sided polygon with sides that are equal in length. It can be constructed by creating a circle whose radius is equal to 0.707106781 (Radius = $\{\sqrt{\text{length}^2 + \text{length}^2}\}/2$) times the length of one of the sides of the square. The circle is then divided into four equal segments, and each segment is connected using the PLINE command. The following steps are involved in constructing a square using CAD:

Step 1. Use the CIRCLE command to construct a circle whose radius is equal to 0.707106781 times the length of one side of the square.

COURSE ________	STUDENT ________________________	DATE ________	PROBLEM 7-5

Step 2. Use the DIVIDE command to divide the circle into four equal segments.

COURSE________ STUDENT________________________ DATE________ PROBLEM 7-5

Step 3. Use the PLINE command to connect the nodes that mark the division points on the circle.

Note: To construct an isometric square, the procedure described previously is employed, except the technician facilitates the Isocircle option of the ELLIPSE command instead of using the CIRCLE command.

COURSE________ STUDENT______________________________ DATE________ PROBLEM 7-5

Problem 7-6 Guided Practice Constructing a Hexagon Using AutoCAD

A **hexagon** is a six-sided polygon with sides that are equal in length. When a hexagon is inscribed in a circle, the length of one of its sides is equal to the radius of the inscribed circle. Therefore, a hexagon can be constructed by drawing a circle whose radius is equal to the length of one of the sides of the hexagon and dividing the circumference of the circle into six equal segments. The following steps are used to construct a hexagon:

Step 1. Use the CIRCLE command to construct a circle whose radius is equal in length to one of the sides of the hexagon.

COURSE________	STUDENT________________________________	DATE________	PROBLEM 7-6

Step 2. Use the DIVIDE command to divide the circle into six equal segments.

COURSE ________ STUDENT ______________________________ DATE ________ PROBLEM 7-6

Step 3. Use the PLINE command to connect the nodes created by the DIVIDE command to form the hexagon.

Again, the procedure for constructing an isometric hexagon is almost the same as for a standard hexagon. The only difference is that the Isocircle option of the ELLIPSE command is used in place of the CIRCLE command.

COURSE________ STUDENT______________________________ DATE________ PROBLEM 7-6

Problem 7-7 Geometric Construction

Solve these three problems graphically: (1) Find the center of the circle. (2) Draw an arc to pass through C, D, and E. (3) Construct the largest circle within triangle F-G-H. (Hint: Use bisectors to locate the center of the triangle.) Do not erase construction lines.

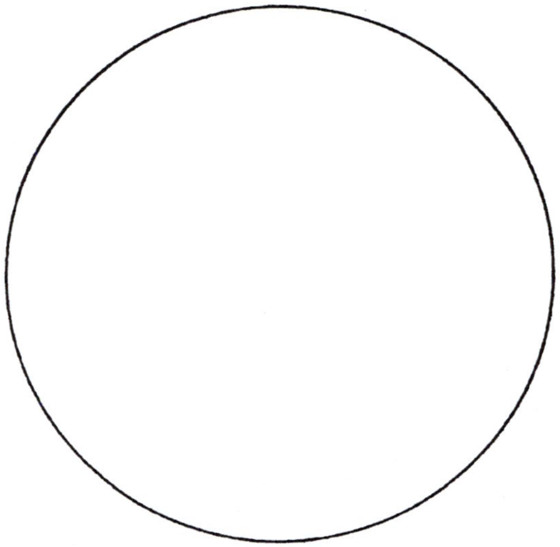

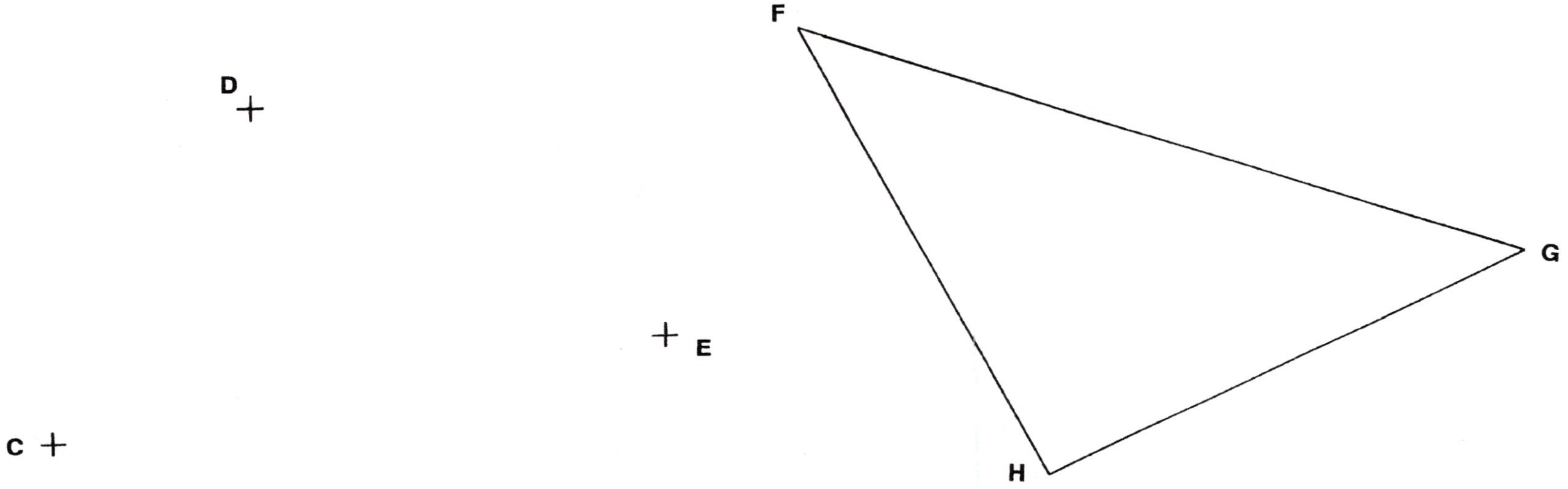

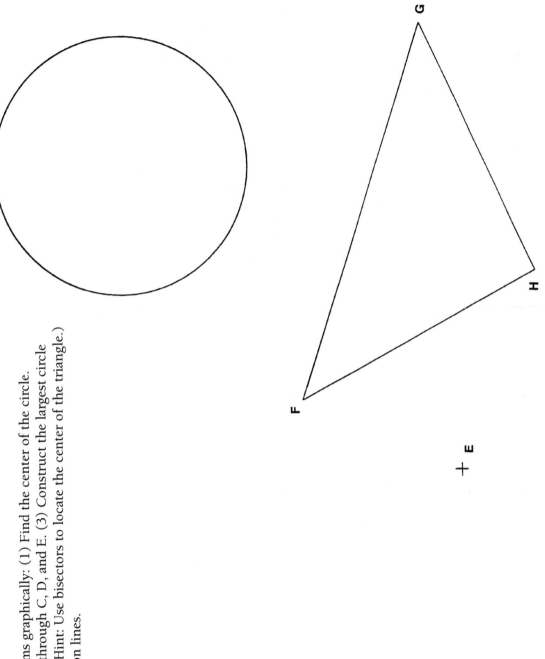

COURSE________ STUDENT____________________________ DATE________ PROBLEM 7-7

Problem 7-8 Perpendicular

Review the Supplemental Material for Chapter 6 on the CD that accompanies the textbook and then solve these three problems graphically: (1) Construct a perpendicular to line M-N at 1.5" from point N. (2) Divide line C-D into seven equal segments. (3) Locate a point X at three quarters of the distance from point F to point G. Do not erase construction lines.

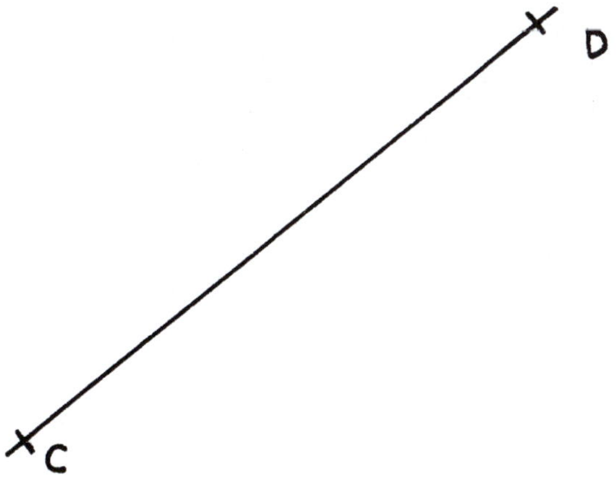

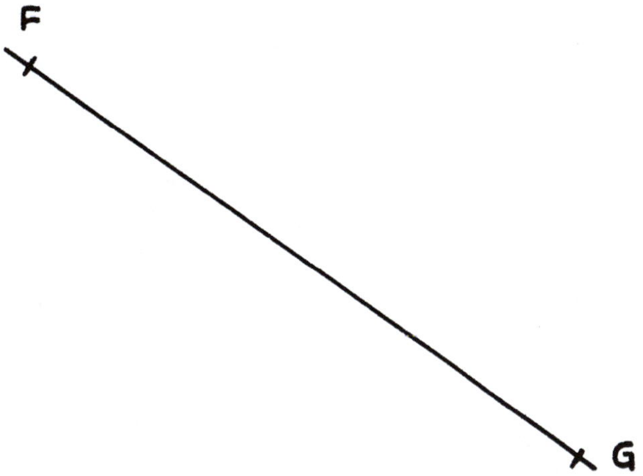

COURSE________ STUDENT____________________________ DATE________ PROBLEM 7-8

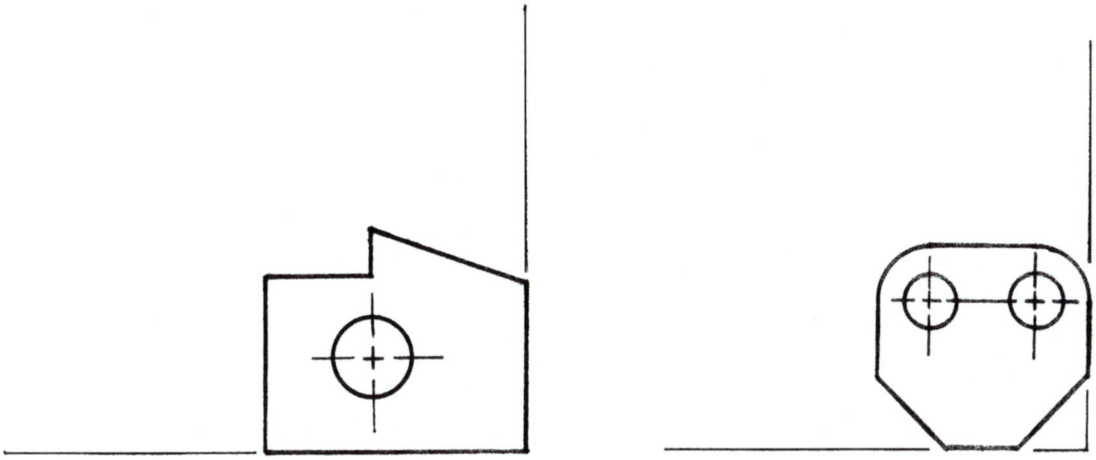

Problem 7-9 Transfer Shapes

Solve these two problems graphically: (1) Transfer the shape with corners X and Y to the new position X-Y. (2) Double the size of the two figures shown at right. Do not erase construction lines.

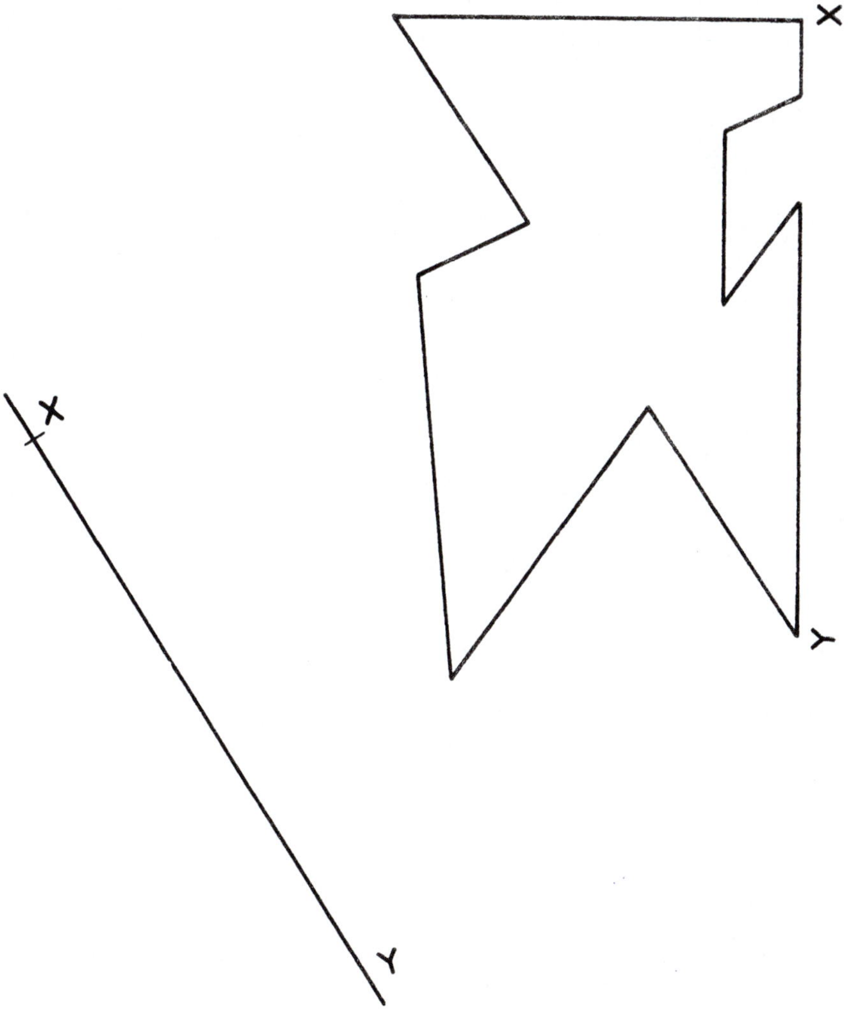

COURSE________ STUDENT________________________ DATE________ PROBLEM 7-9

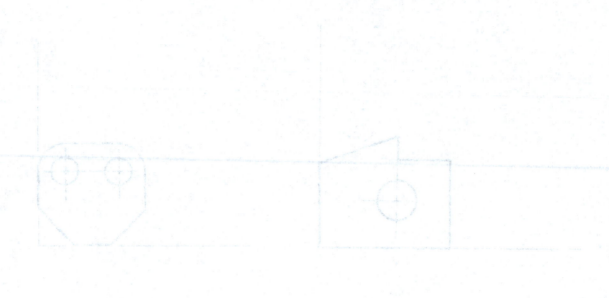

Problem 7-10 Geometric Construction

Solve these six problems graphically as indicated in the spaces. Do not erase construction lines.

A. Construct a pentagon within a 1.5" circle.

B. Construct one hexagon with 36 mm across the flats.

C. Construct an octagon with 1.8" across the flats.

D. Construct an arc tangent to the angle at a radius of 14 mm.

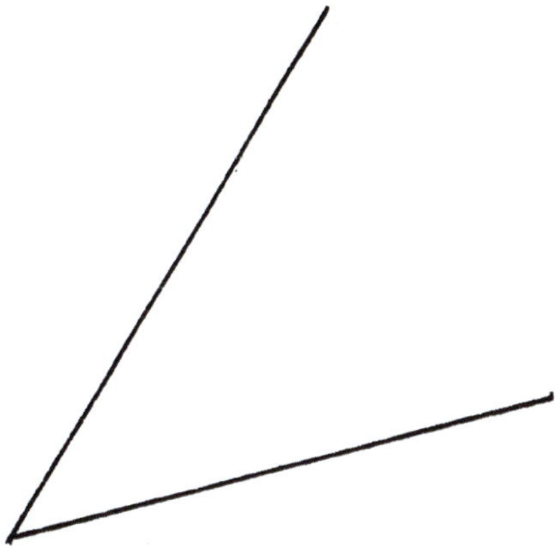

E. Construct an arc tangent to the circle and line at a radius of 18 mm.

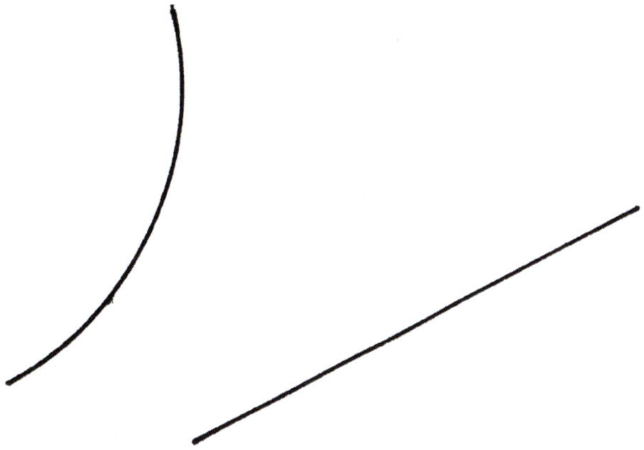

F. Construct an arc tangent to the circles at a radius of 1".

COURSE________ STUDENT____________________ DATE________ PROBLEM 7-10

Problem 7-11 Geometric Construction

Answer the questions below using your best lettering.

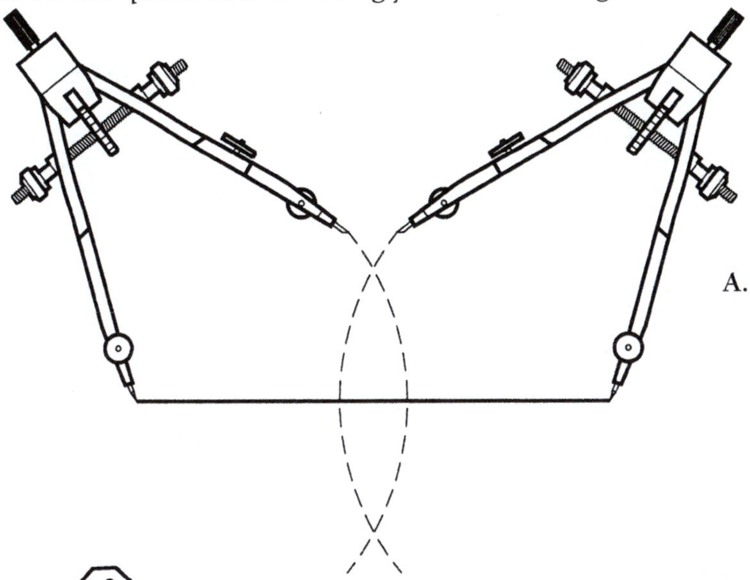

A. What type of geometric construction is this?

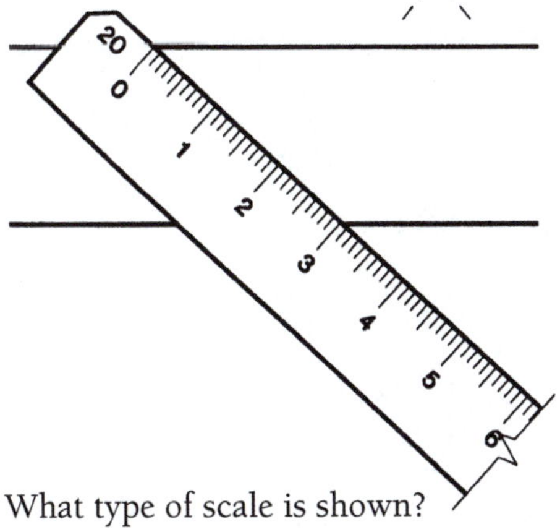

B. Given the space below, paraphrase (describe in your own words) how you would divide the space into six equal parts.

C. What type of scale is shown?

D. Is a rhombus or a trapezoid a type of parallelogram?

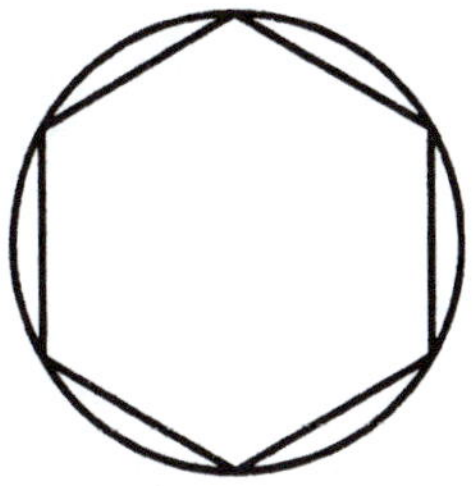

E. Which type of triangle has one interior angle greater than 90 degrees?

F. Is the polygon shown at the left a regular polygon or an irregular polygon?

G. What specific type of polygon is it?

H. Is the circle inscribed or circumscribed?

COURSE ________ STUDENT ________________ DATE ________ PROBLEM 7-11

Problem 7-12 Geometric Construction

Answer the questions below using your best lettering.

A. How many times does a line that is tangent to a circle cross the circle?

B. From the point of tangency of a line with a circle to the center of the circle is called the ________________ of the circle.

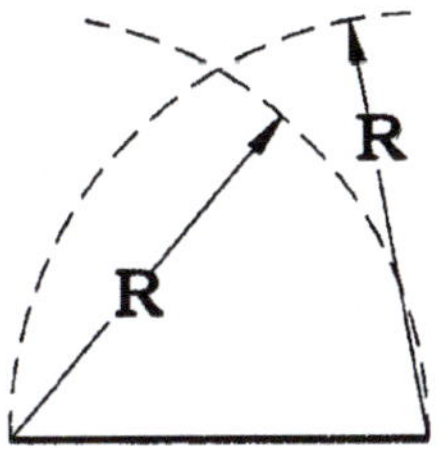

C. The geometric construction shown at the left is an example of the construction of what type of shape?

D. Describe in your own words how you would construct an eight-sided regular polygon.

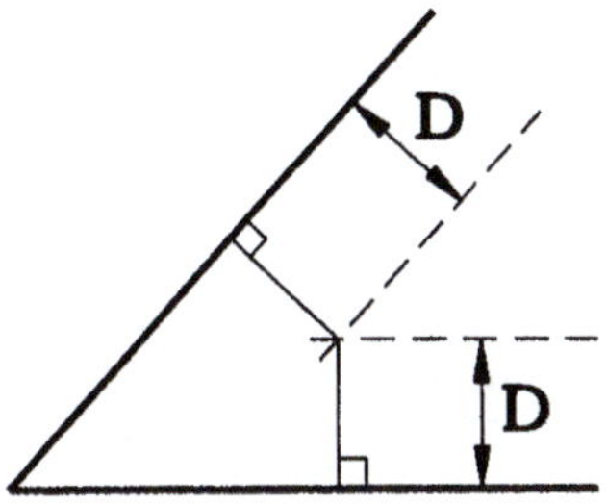

E. The geometric construction shown at the left is an example of the construction of what type of shape? Describe why this is set up in this manner.

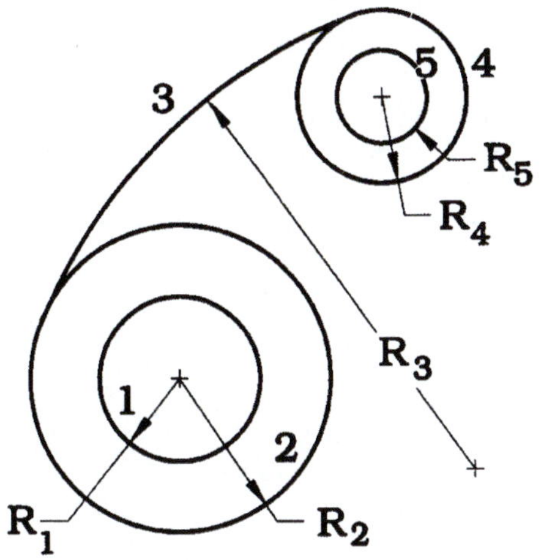

F. Study the example shown at the left. You are given circles 1, 2, 4, and 5. You are also given the radius of arc 3. Using geometric construction, how would you construct arc 3 so that it is tangent to circle 2 and circle 4?

COURSE	STUDENT	DATE	PROBLEM 7-12

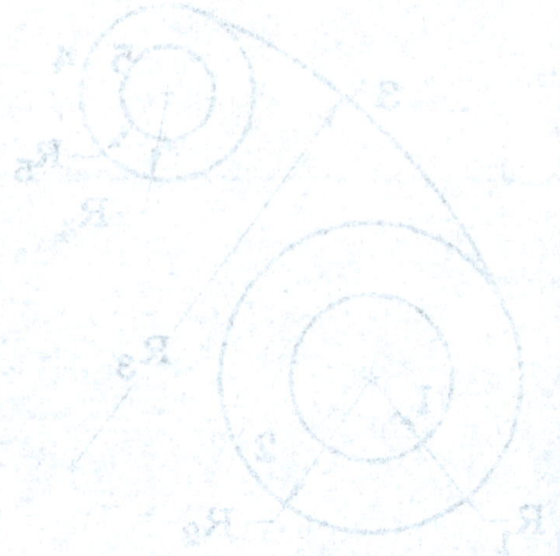

Problem 7-13 Geometric Construction

Perform the following:

A. Construct an octagon within the square to the right following these instructions: Draw the two diagonals of the square. At *each* corner of the square, strike an arc (draw the arc) with a radius equal to one-half the square's diagonal length. Let these arcs be drawn through the sides adjoining the arc center (i.e., through the two sides adjoining the corner of the square). With straight lines, connect the intersections of the arcs and the sides of the square.

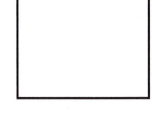

B. Construct a pentagon within the circle to the right following these instructions: Draw two chords in the circle. (A chord is a line that intersects the circle twice.) Draw perpendicular bisectors of each chord. The center of the circle is the point at which these two bisectors intersect. Draw the diameter of the circle horizontally and vertically through the center point. Label the horizontal diameter line A on the left, B on the right, and C in the middle. Label the vertical diameter line D at the top. Bisect line BC (which is actually the radius) and label the new point E. Draw an arc from point D through the horizontal diameter using point E as the center, and distance DE as the radius. Label the intersection of the arc and the horizontal diameter as point F. Draw another arc, this time from point F through the side of the circle to the left using point D as the center and distance DF as the radius. Label this new point on the circle as point G. Using the previous distance DF as a chord, strike several arcs through the circle in a counterclockwise direction, first using point G as the center of the arc, then using each new intersection as the center for the next arc. When you have struck five arcs and are back at point D, connect the five intersections with straight lines.

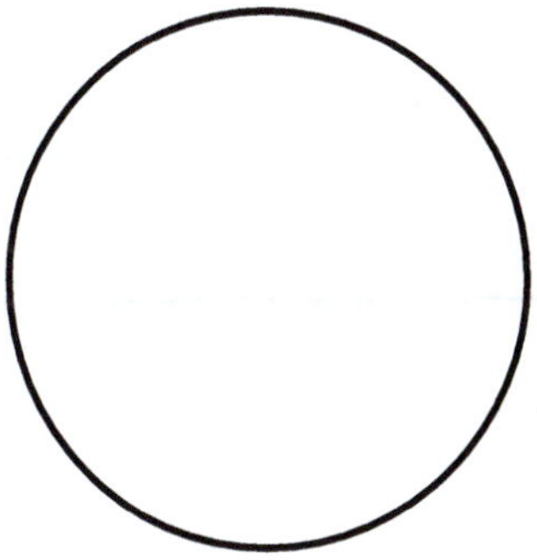

C. Construct an inscribed circle and a circumscribed circle using the polygon to the right and by following these instructions: Draw perpendicular bisectors of line AB and line BC. The point at which these bisectors intersect is the center of both the inscribed circle and the circumscribed circle. Label this point H. Using point H as the center and distance AH as the radius, draw a circle around the polygon. What type of circle is this?

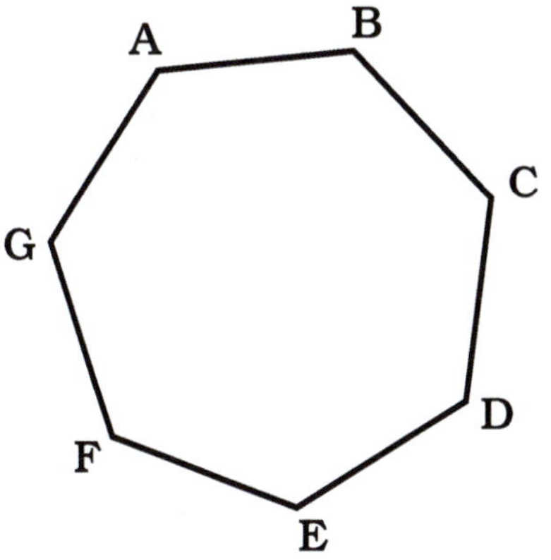

____

____

Now draw another circle, also using point H as the center, but this time using the distance from H to the middle of line AB as the radius. What type of circle is this?

____

____

COURSE________ **STUDENT**____________________ **DATE**________ **PROBLEM 7-13**

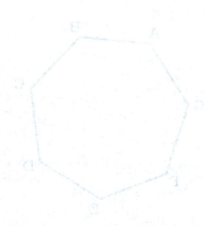

SECTION 3

Drawing Views and Annotations

Problem 8-1 Orthographic Questions

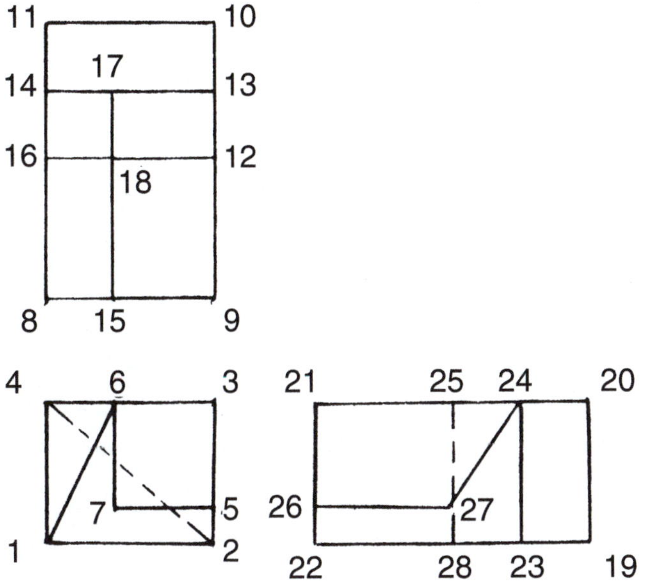

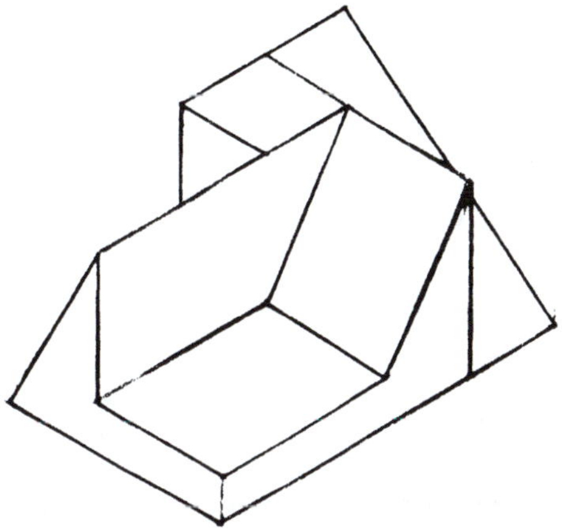

Use numbers to describe surfaces and answer the following questions (1 and 2 are examples):

1. Surface 1-2-5-7-6	in the F view is	8-15-9	in the T view
2. Surface 15-9-12-18	in the T view is	26-27	in the S view
3. Surface 18-12-13-17	in the T view is	________	in the F view
4. Surface 18-12-13-17	in the T view is	________	in the S view
5. Surface 4-2	in the F view is	________	in the T view
6. Surface 1-6	in the F view is	________	in the S view
7. Surface 1-6	in the F view is	________	in the T view
8. Surface 16-18-17-14	in the T view is	________	in the S view
9. Surface 25-28	in the S view is	________	in the F view
10. Surface 25-28	in the S view is	________	in the T view
11. Surface 6-7	in the F view is	________	in the T view
12. Surface 6-7	in the F view is	________	in the S view
13. Surface 23-24	in the S view is	________	in the T view
14. Surface 23-24	in the S view is	________	in the F view
15. Surface 4-2	in the F view is	________	in the S view

COURSE ________ STUDENT ______________________________ DATE ________ PROBLEM 8-1

Problem 8-2 Isometric Matching

Match the isometric drawings of objects to the appropriate three-view drawings of the objects. Note: Arrows point toward the front views.

Label the grid below as follows:

	A	B	C	D
1				
2				
3				
4				

Record your answers.

Example: **A-1** matches B-3
A-2 matches ________
B-1 matches ________
B-2 matches ________
C-1 matches ________
C-2 matches ________

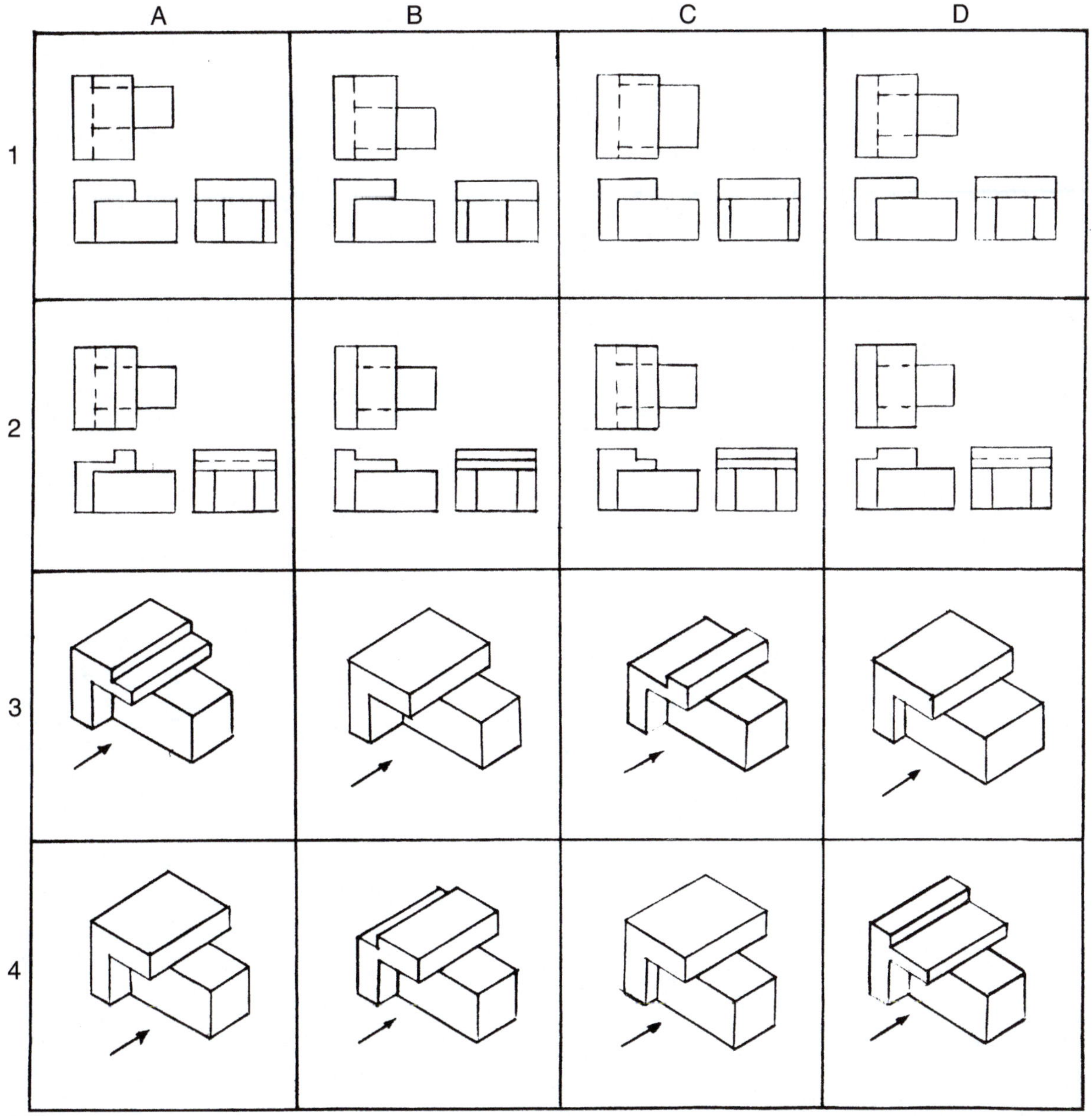

COURSE________ STUDENT________________________ DATE________ PROBLEM 8-2

roblem 8-3 Add the Missing Lines

A

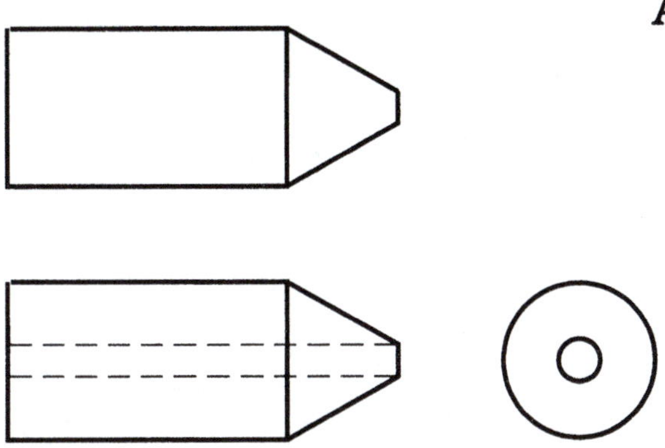

B

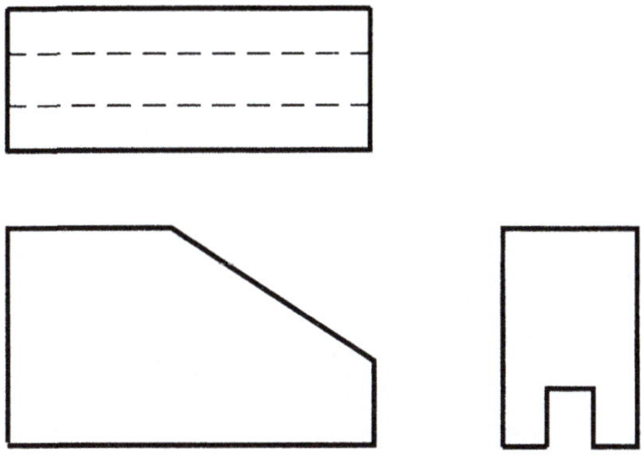

C

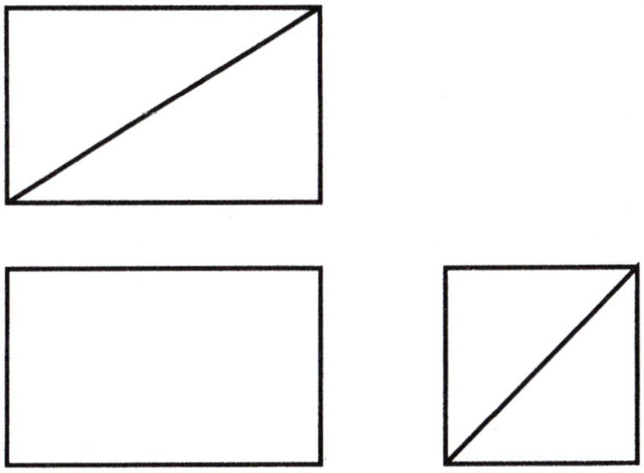

D

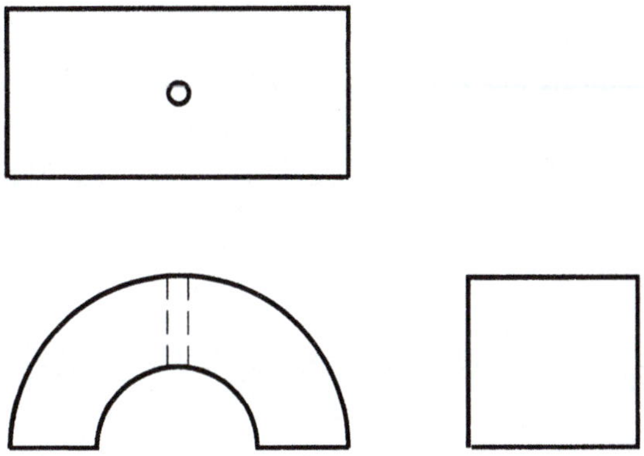

E

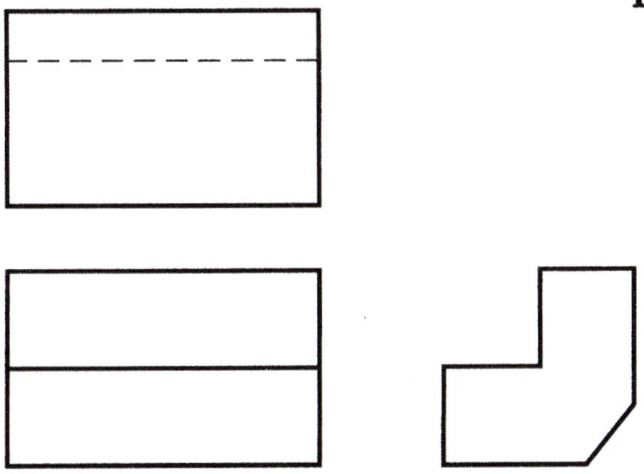

F

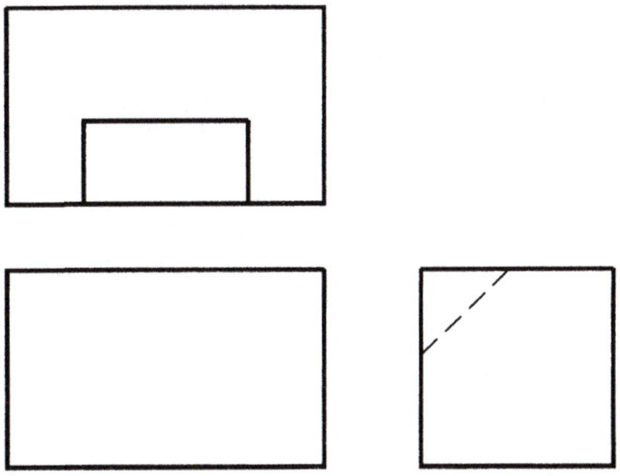

COURSE_________ STUDENT______________________________ DATE________ PROBLEM 8-3

Problem 8-4 Add the Missing Lines

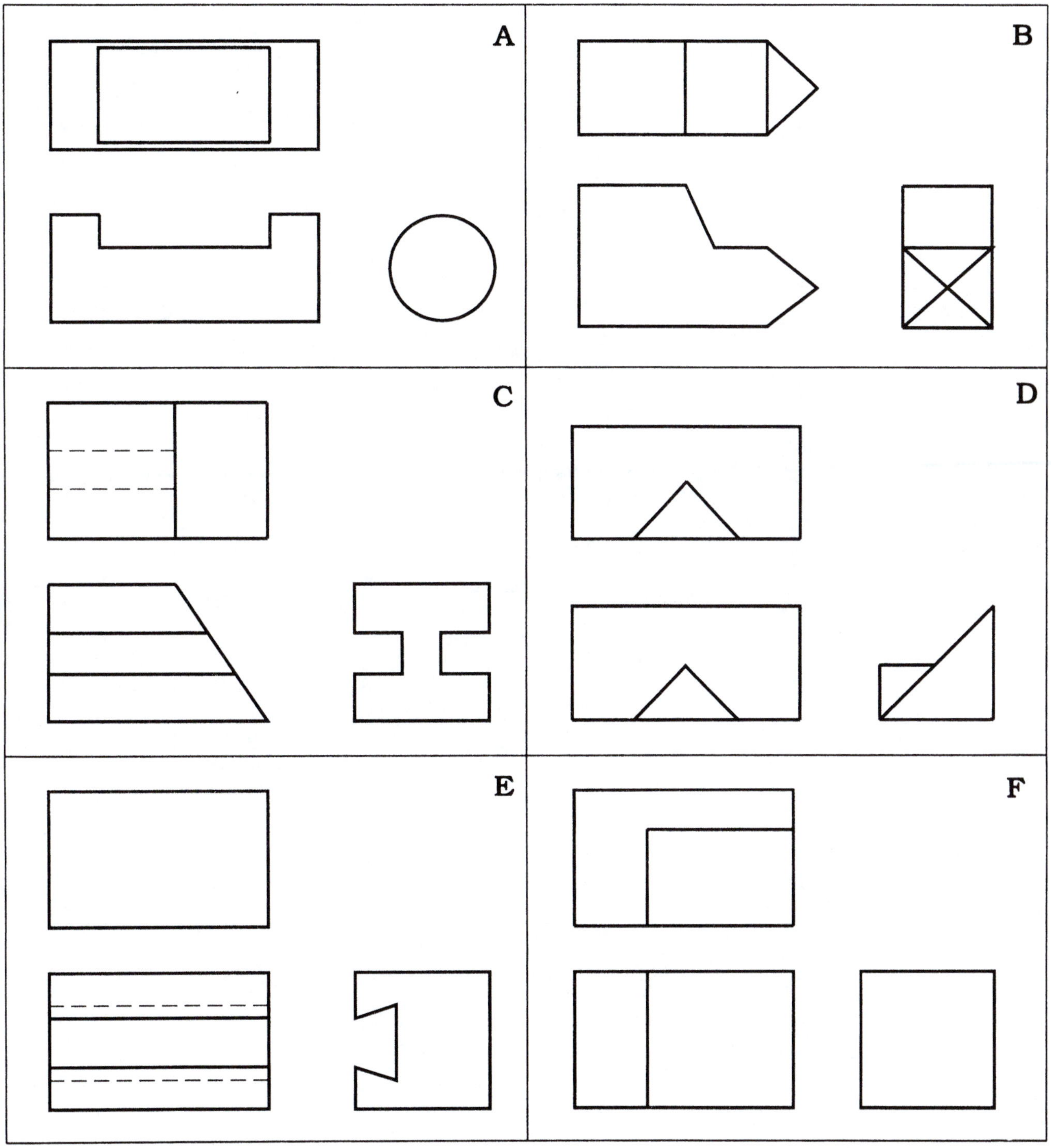

COURSE __________ STUDENT ______________________________ DATE __________ PROBLEM 8-4

Problem 8-5 Complete the View

A

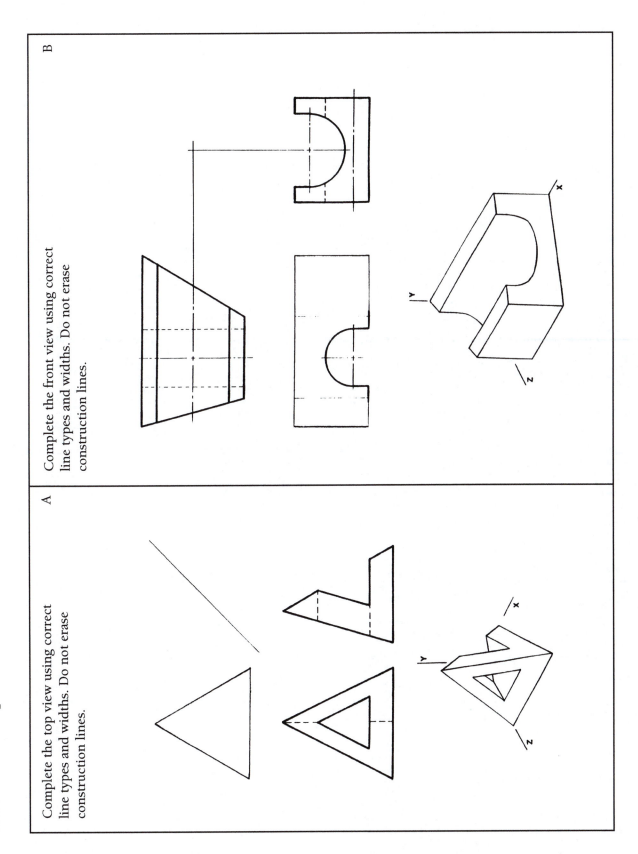

Complete the top view using correct line types and widths. Do not erase construction lines.

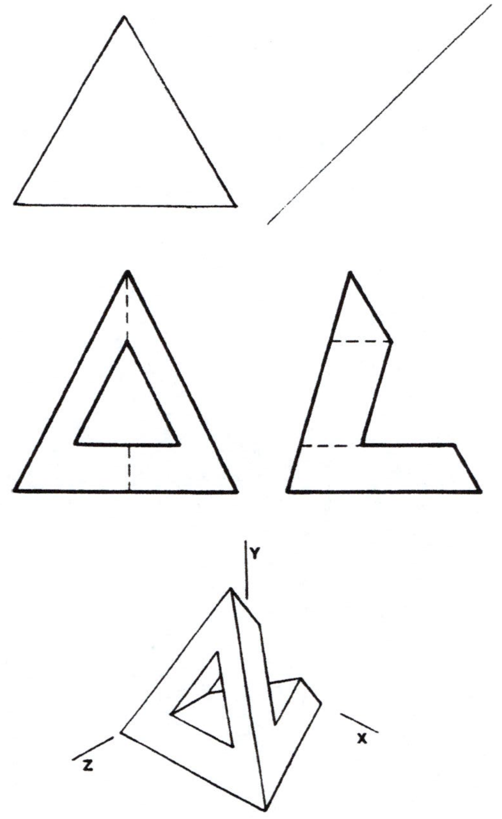

B

Complete the front view using correct line types and widths. Do not erase construction lines.

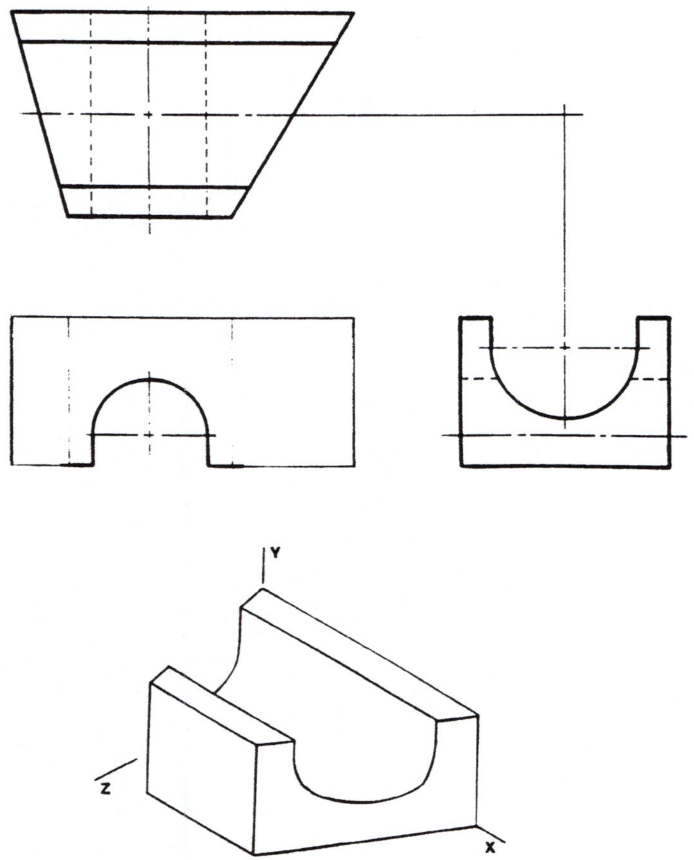

COURSE________ STUDENT________________________ DATE________ PROBLEM 8-5

Problem 8-6 Complete the View

A

Complete the front view using correct line types and widths. Do not erase construction lines.

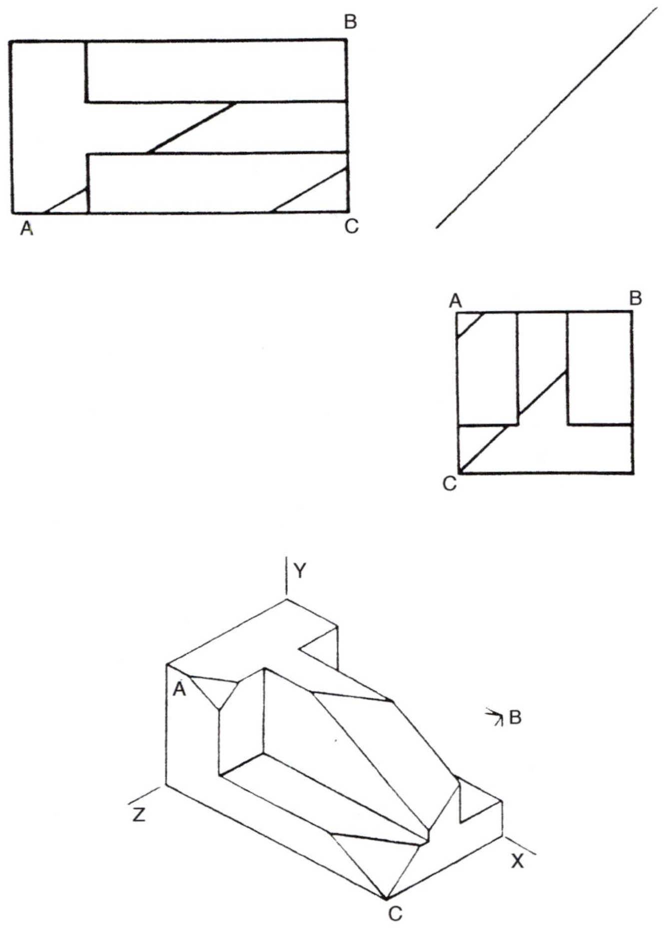

B

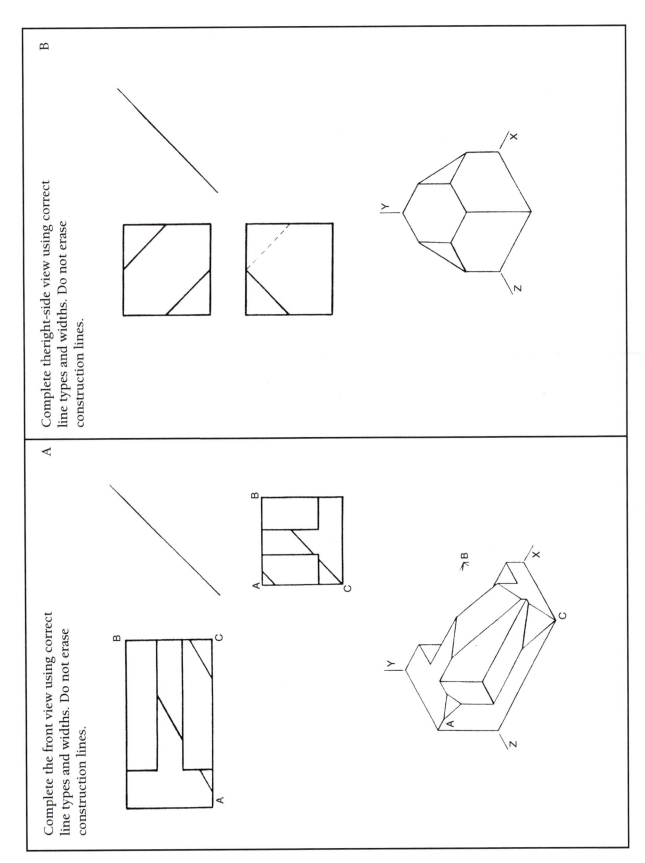

Complete theright-side view using correct line types and widths. Do not erase construction lines.

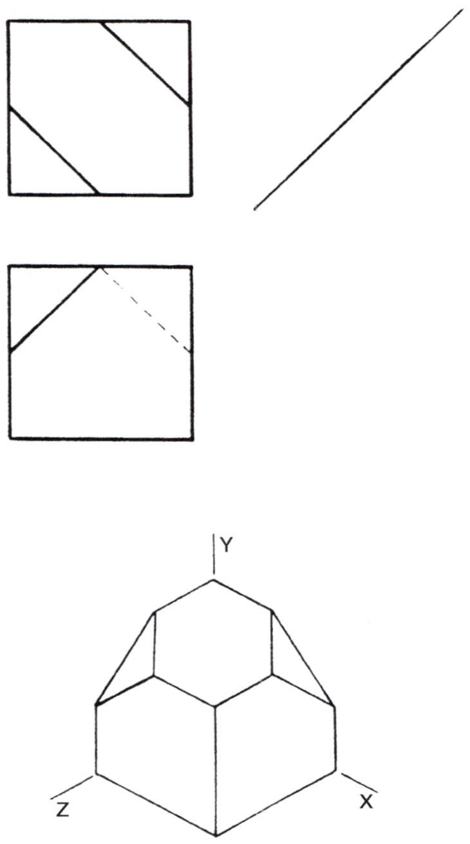

COURSE________ STUDENT____________________________ DATE________ PROBLEM 8-6

Problem 8-7 Complete the View

A

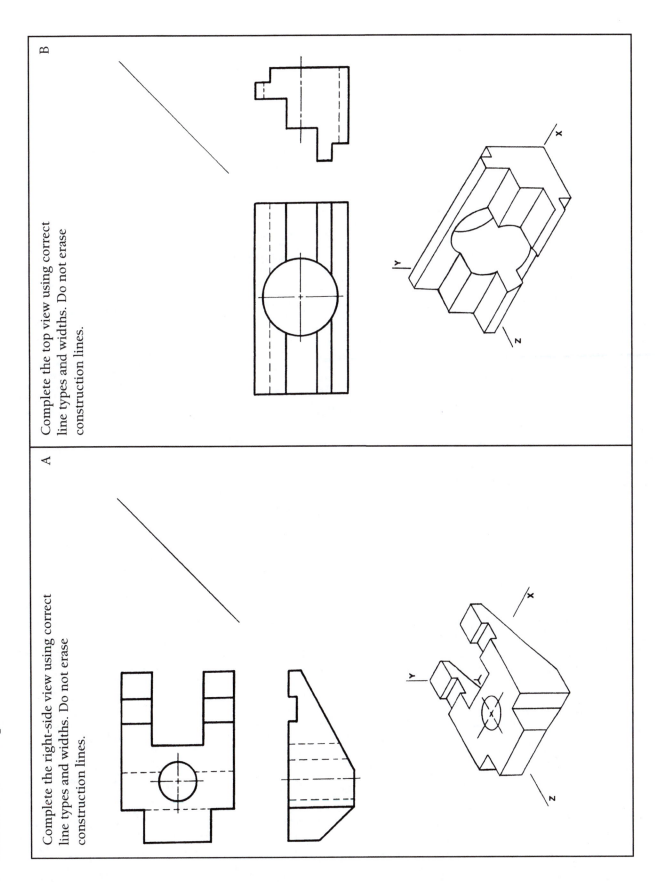

Complete the right-side view using correct line types and widths. Do not erase construction lines.

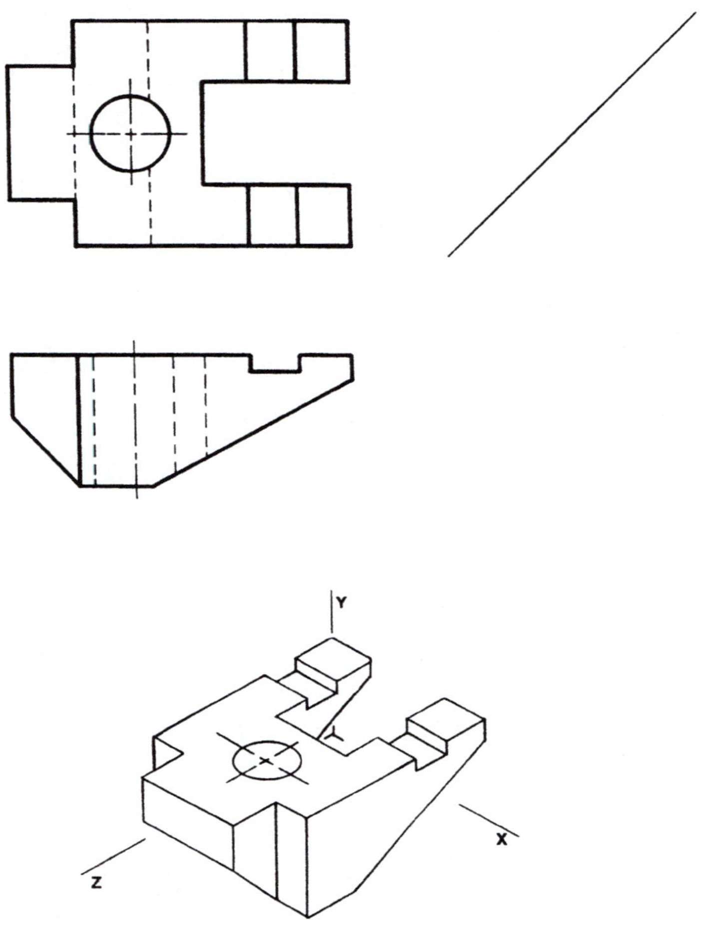

B

Complete the top view using correct line types and widths. Do not erase construction lines.

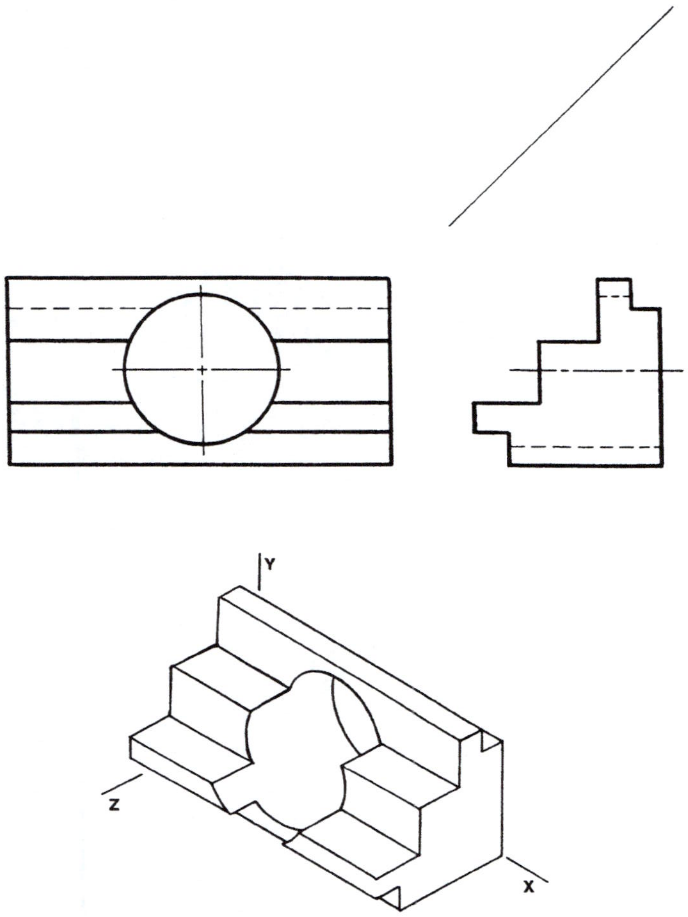

COURSE________ STUDENT________________________ DATE________ PROBLEM 8-7

Problem 8-8 Orthographic

Sketch the six orthographic views of the isometric object.

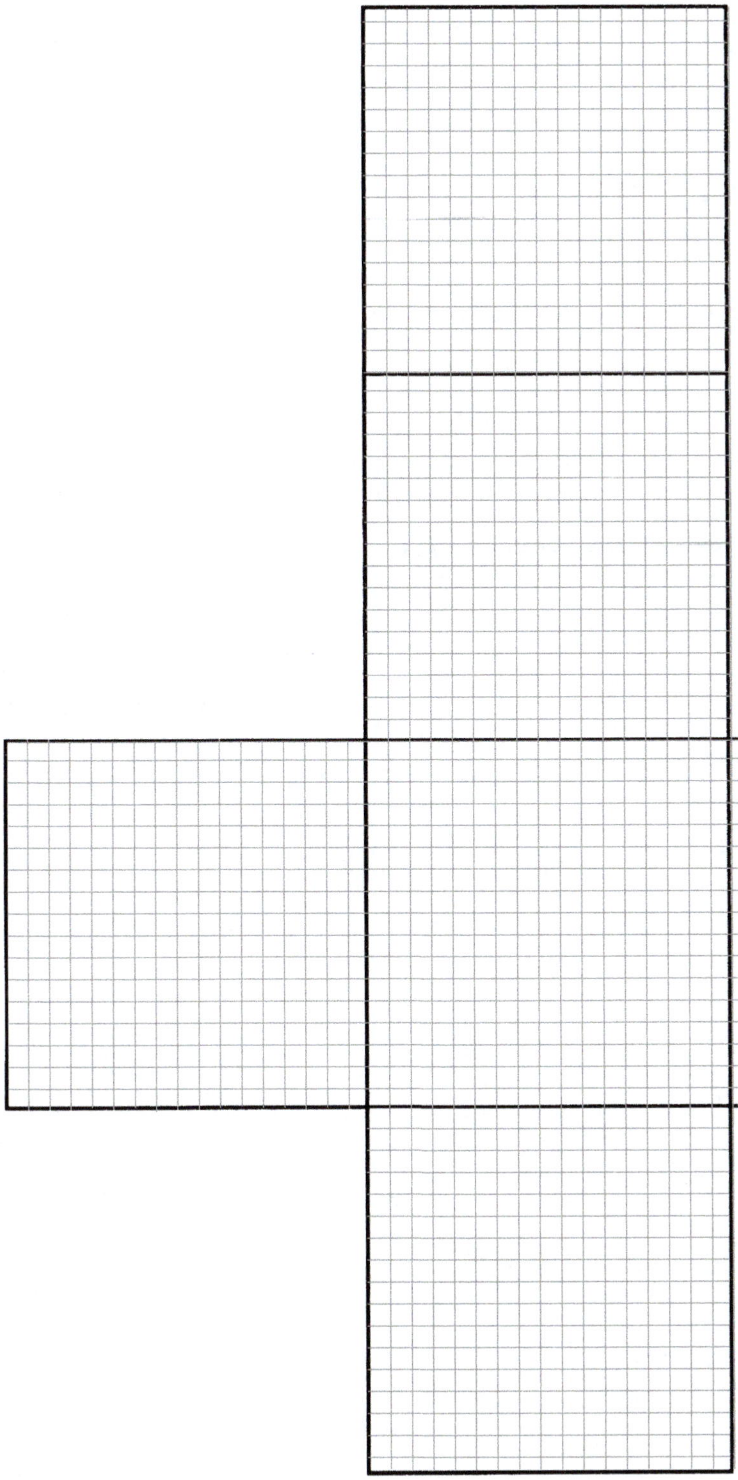

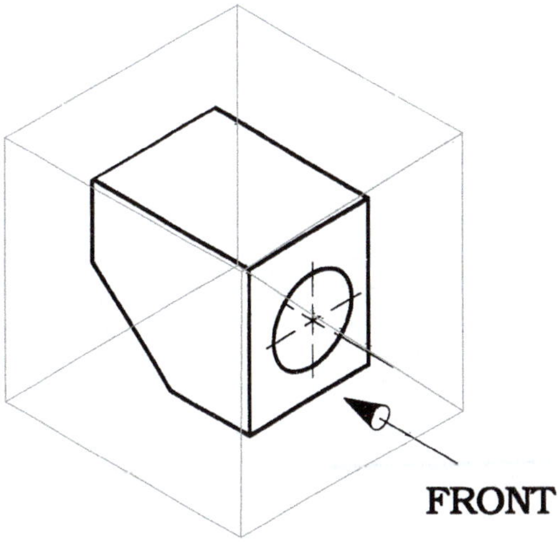

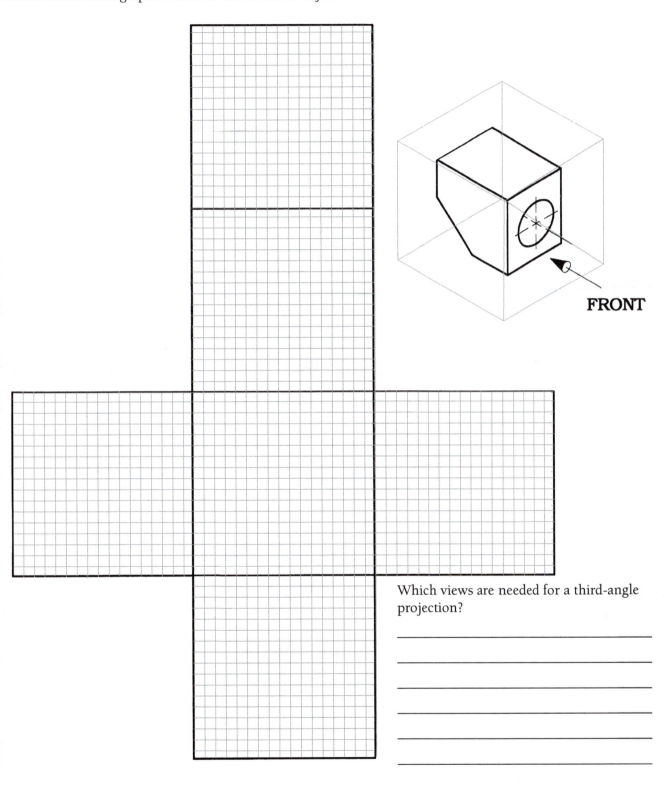

Which views are needed for a third-angle projection?

____

____

____

____

____

____

COURSE ________	STUDENT ______________________	DATE ________	PROBLEM 8-8

Problem 8-9 Sketching

Sketch a third-angle projection using the isometric drawing as a guide. Sketching at twice the scale (2:1), use ANSI standards.

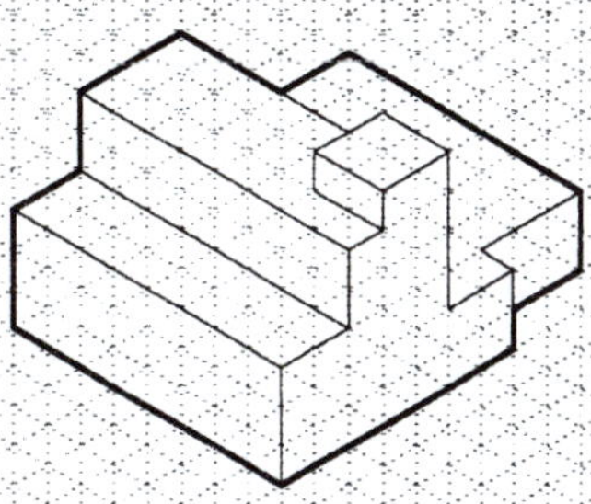

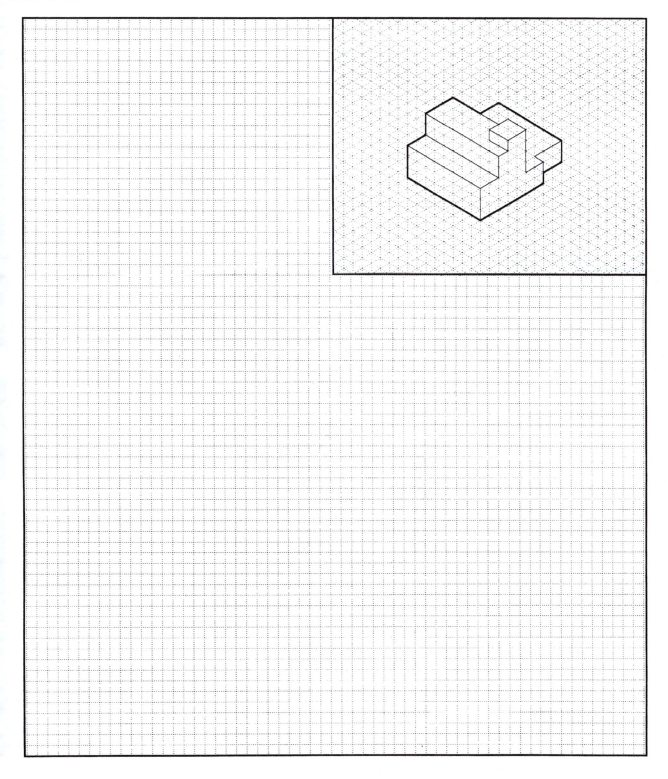

COURSE________ STUDENT____________________ DATE________ PROBLEM 8-9

Problem 8-10 Orthographic Sketching

Using the grid below, sketch the orthographic views of the given drawing. Sketching at twice the scale (2:1), re-create the line thicknesses shown.

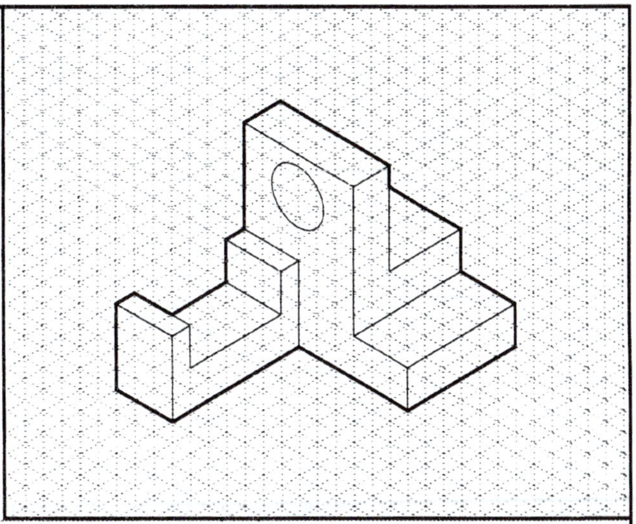

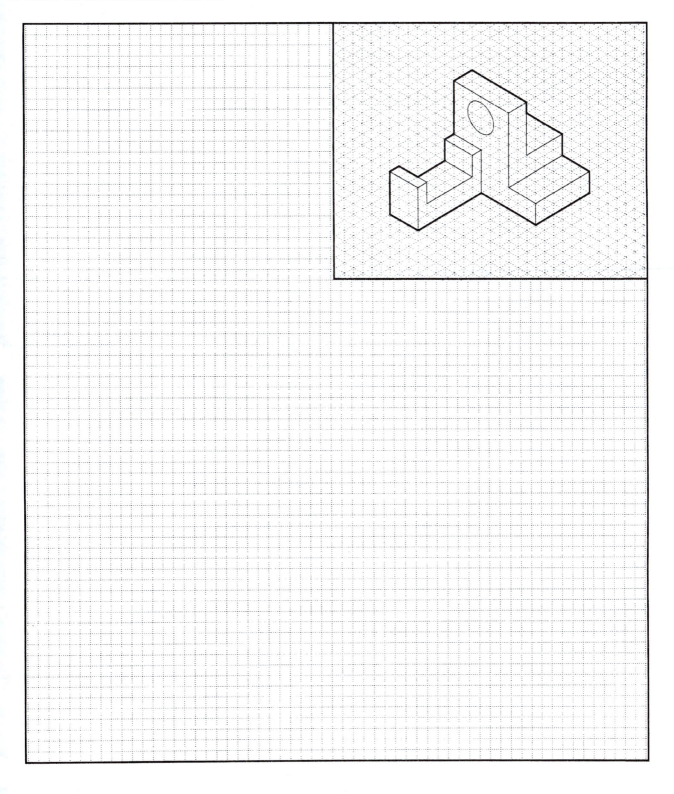

COURSE________	STUDENT________________________	DATE________	PROBLEM 8-10

Problem 8-11 Orthographic Sketching

Using the grid below, sketch the orthographic views of the given drawing. Sketching at twice the scale (2:1), re-create the line thicknesses shown.

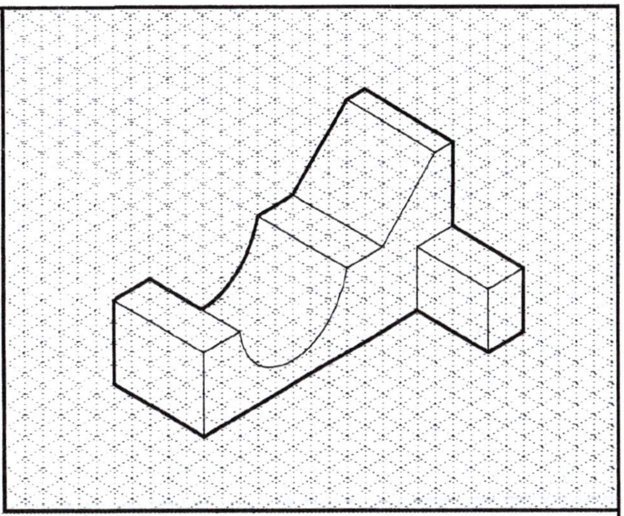

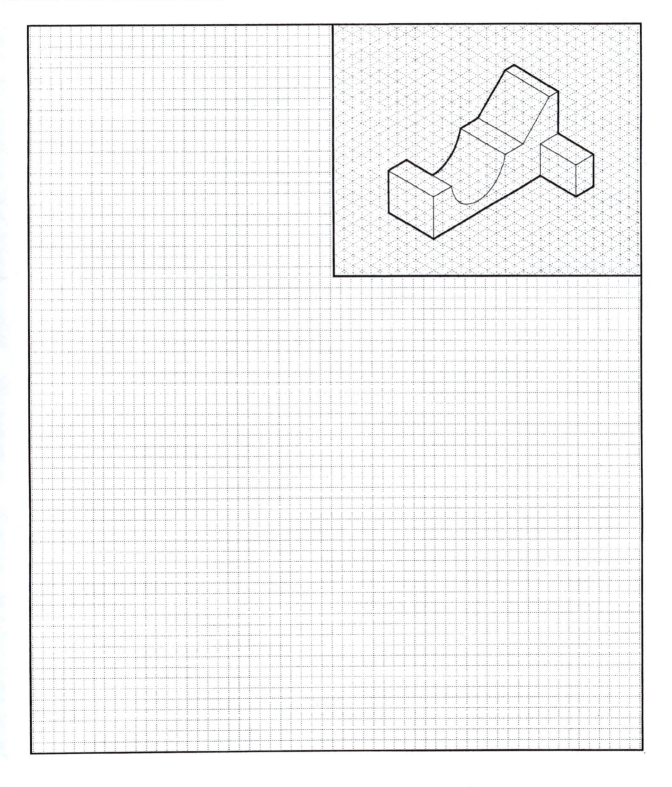

COURSE________ STUDENT______________________________ DATE________ PROBLEM 8-11

Problem 8-12 Complete the Missing Views

For the object shown, complete the front, top, and side views on the projection surfaces. Label the corners as needed. Show all the necessary projection lines to complete the front view. Show a few of the projection lines for the top and side views.

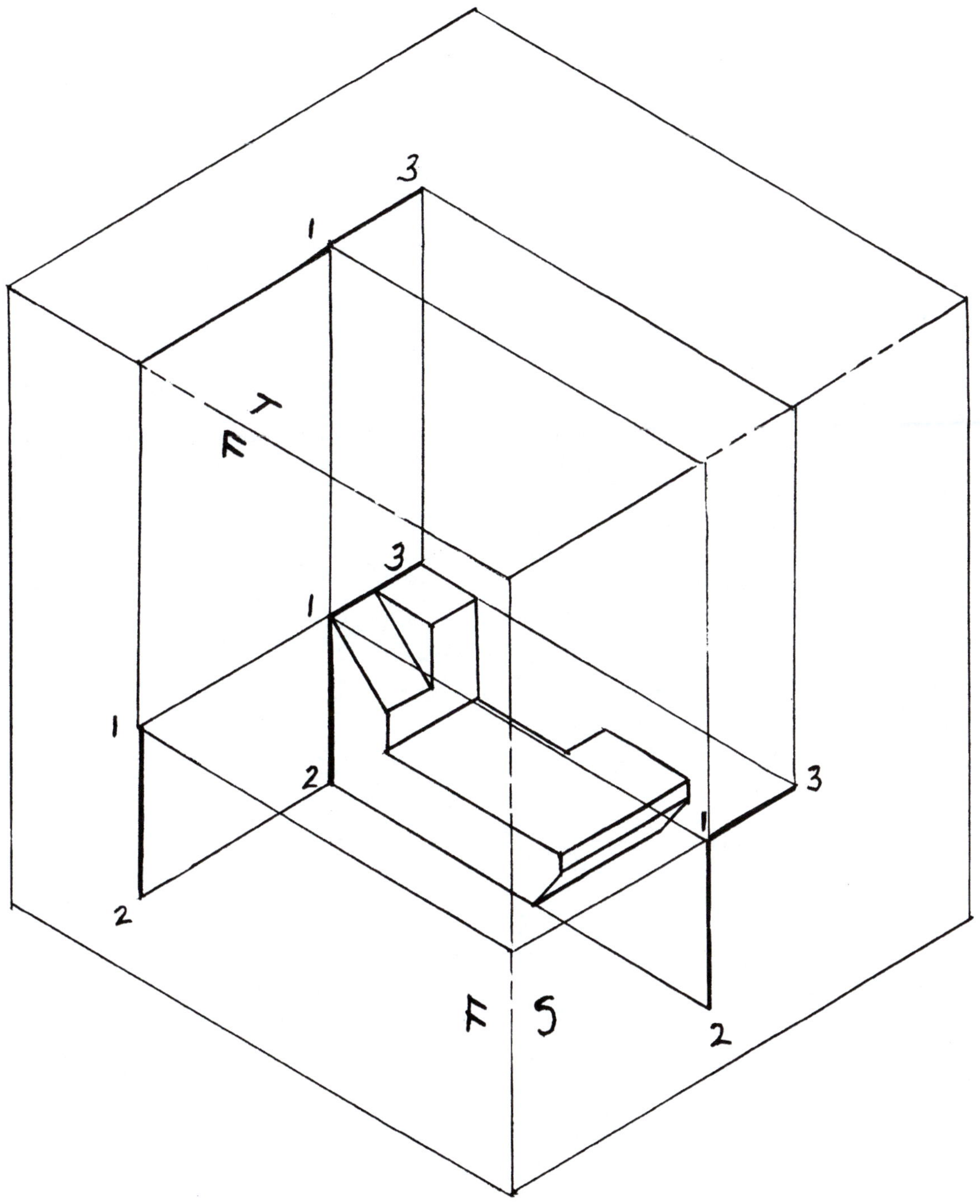

COURSE______	STUDENT______	DATE______	PROBLEM 8-12

Problem 8-13 Hand Sketching

Using standard lines, sketch three freehand views of the objects given in the spaces indicated. Leave three spaces between views.

A

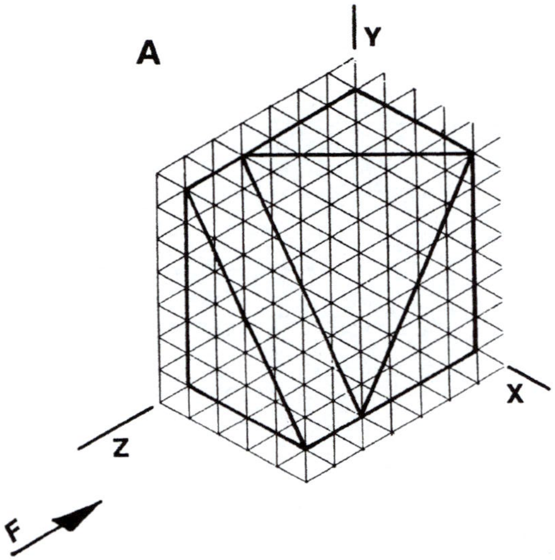

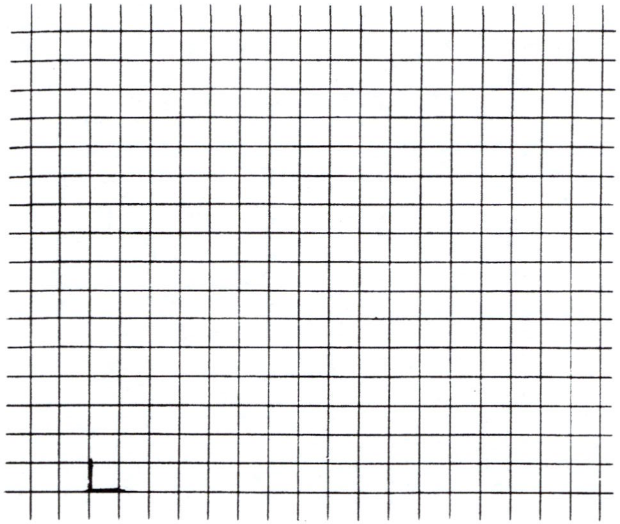

B

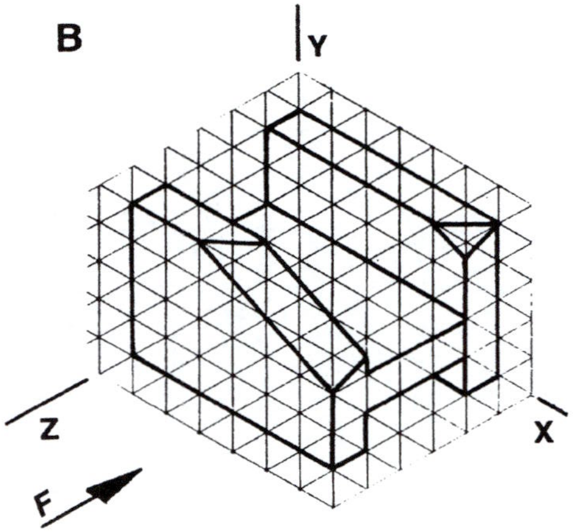

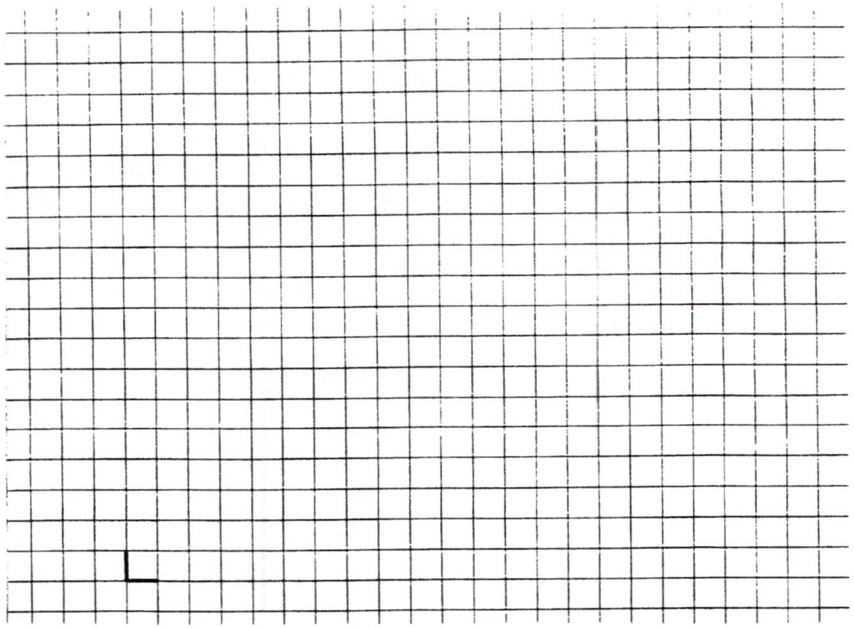

COURSE________ STUDENT______________________________ DATE________ PROBLEM 8-13

Problem 8-14 Hand Sketching

Using standard lines, sketch three freehand views of the objects given in the spaces indicated. Leave three spaces between views.

A

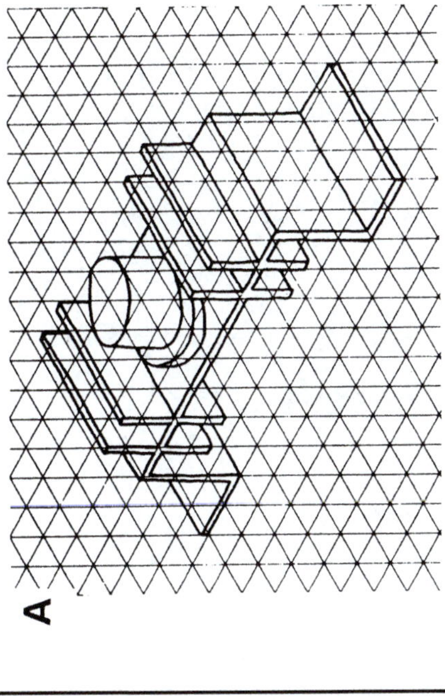

B

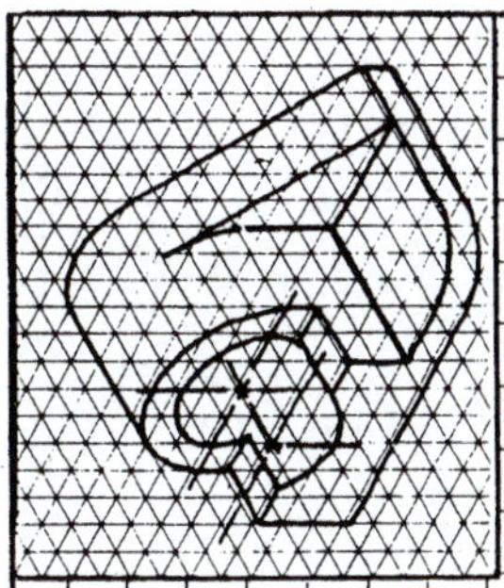

C

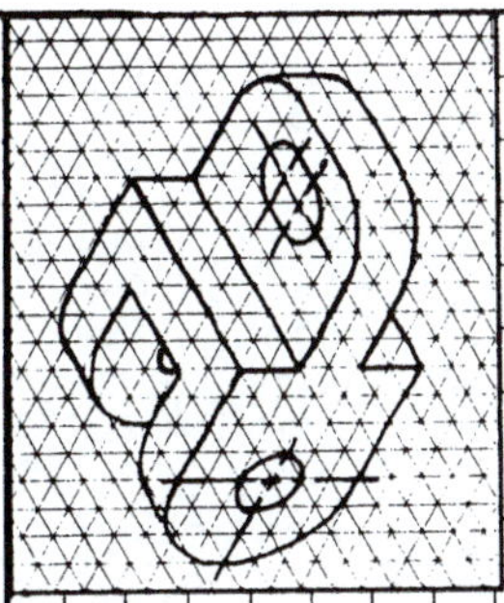

COURSE ________ STUDENT ________________________ DATE ________ PROBLEM 8-14

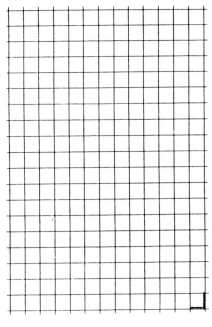

Problem 8-15 Orthographic Projection

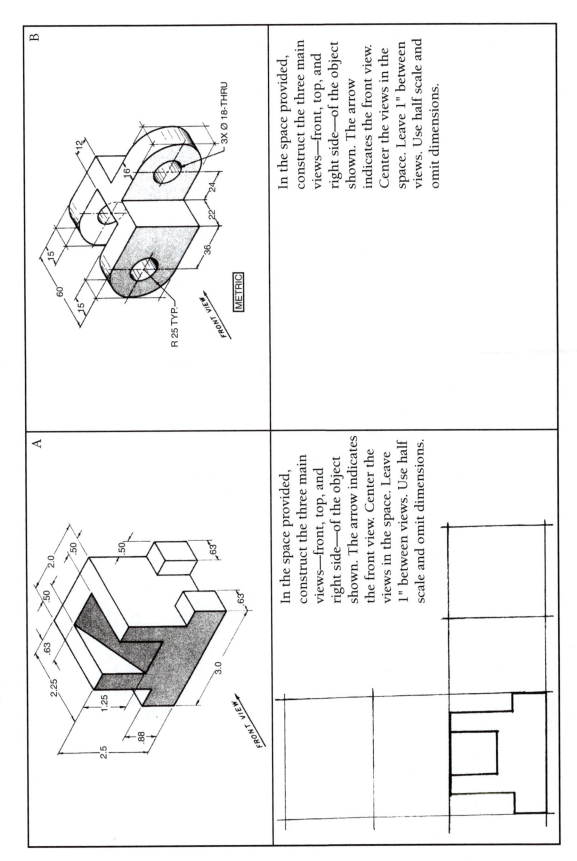

A

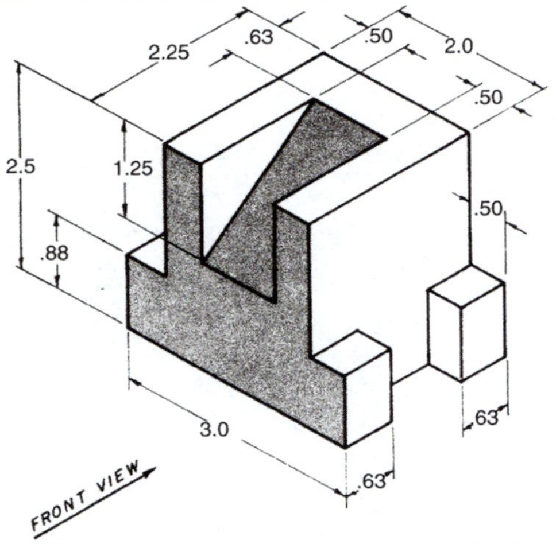

In the space provided, construct the three main views—front, top, and right side—of the object shown. The arrow indicates the front view. Center the views in the space. Leave 1" between views. Use half scale and omit dimensions.

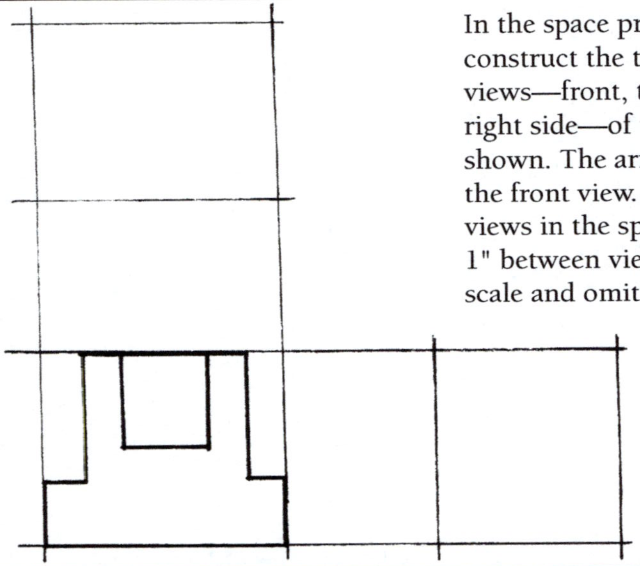

B

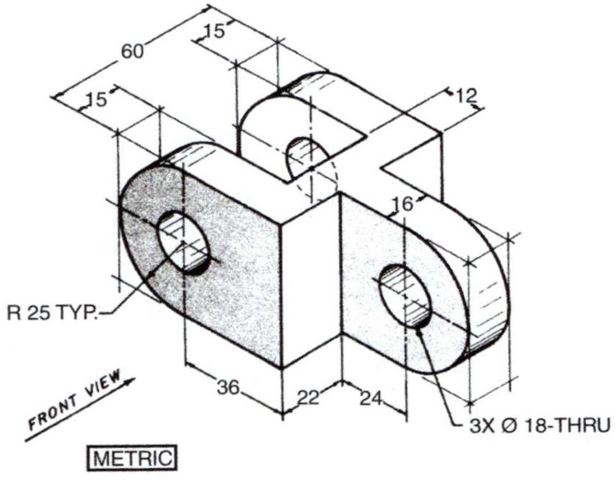

In the space provided, construct the three main views—front, top, and right side—of the object shown. The arrow indicates the front view. Center the views in the space. Leave 1" between views. Use half scale and omit dimensions.

COURSE________ STUDENT________ DATE________ PROBLEM 8-15

Problem 8-16 Orthographic Projection

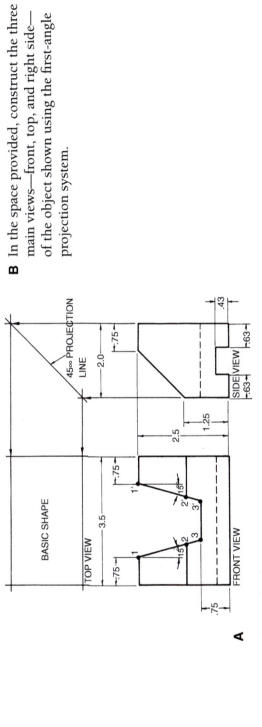

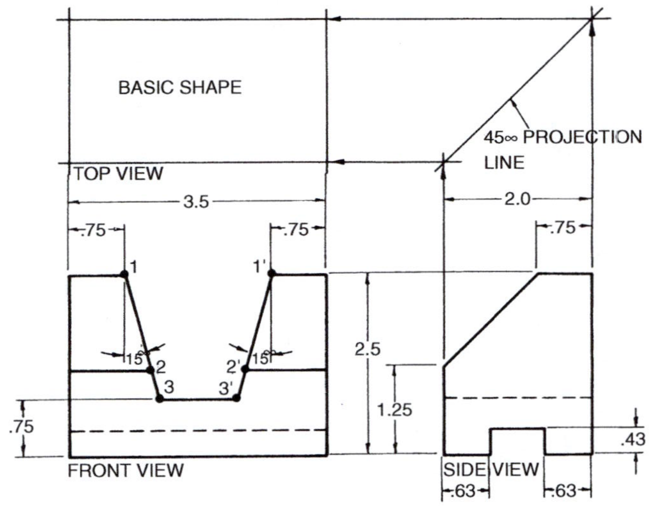

A

In the space provided, construct the three main views—front, top, and right side—of the object shown using the third-angle projection system.

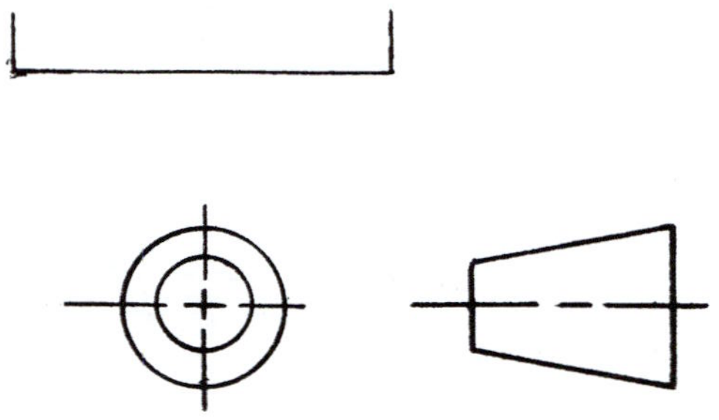

B In the space provided, construct the three main views—front, top, and right side—of the object shown using the first-angle projection system.

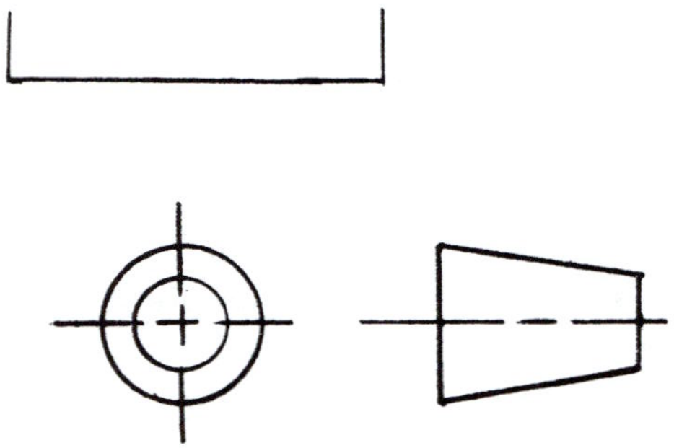

COURSE________ STUDENT________________________ DATE________ PROBLEM 8-16

Problem 8-17 Orthographic Projection

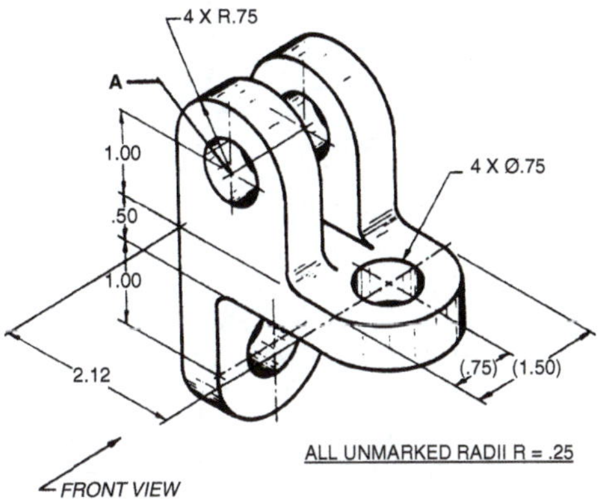

A

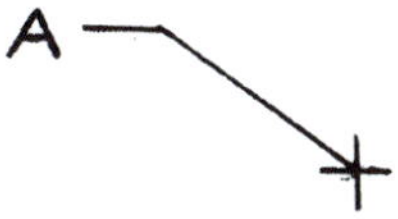

In the space provided, construct the three main views—front, top, and right side—of the object shown. Locate the front view as indicated. Leave 1" between views. Use half scale and omit dimensions.

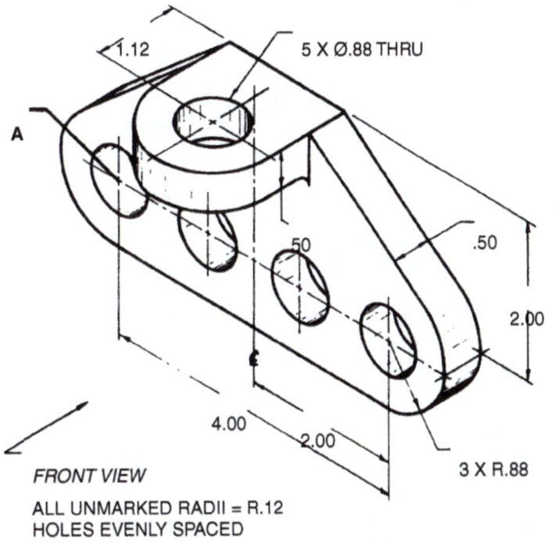

B

In the space provided construct the three main views—front, top, and right side—of the object shown. Locate the front view as indicated. Leave 1" between views. Use half scale and omit dimensions.

COURSE________	STUDENT______________________________	DATE________	PROBLEM 8-17

Problem 8-18 Orthographic Projection

In the space indicated, construct four views—front, top, right side, and left side—of the object shown. Leave 1" between views and center the views in the space provided. Use half scale and omit dimensions.

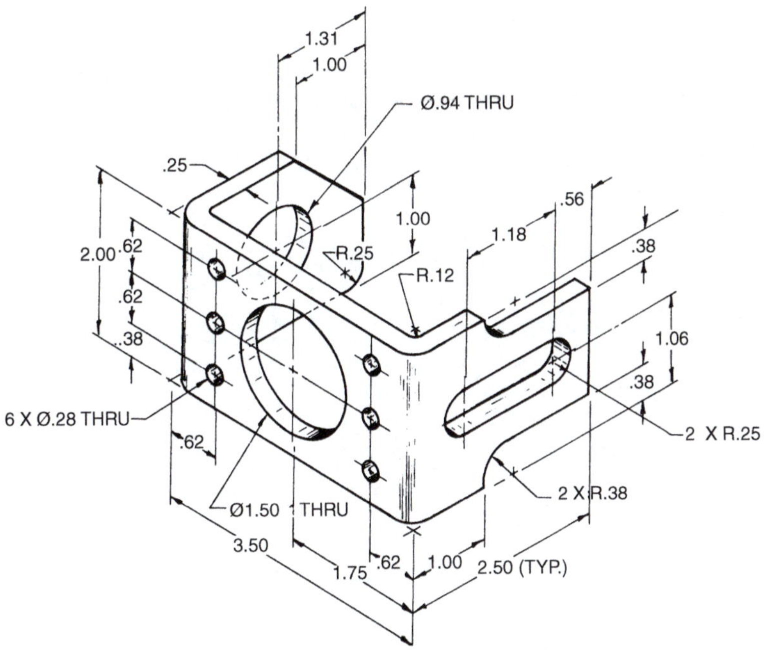

COURSE ________	STUDENT ________	DATE ________	PROBLEM 8-18

Problem 8-19 Orthographic Projection

In the space provided, construct the three main views—front, top, and right side—of the object shown. The curve of the curved neck shows in the front view. Center the views in the space. Leave 1" (25 mm) between views. Use half scale and omit dimensions.

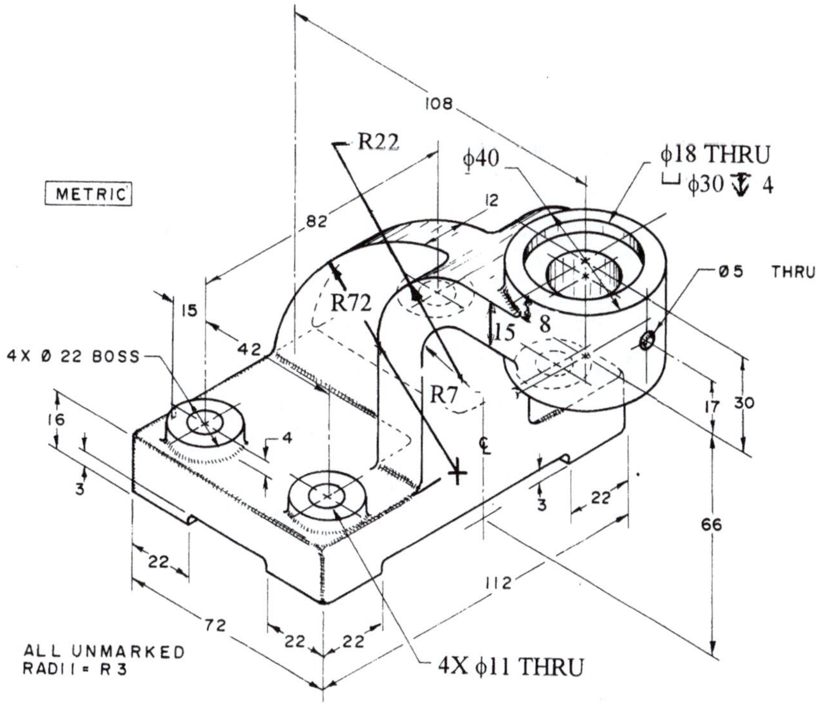

COURSE ________	STUDENT ______________________________	DATE ________	PROBLEM 8-19

Problem 8-20 Orthographic Projection

In the space provided, construct the three main views—front, top, and right side—of the object shown. The curve of the curved neck shows in the front view. Center the views in the space. Leave 1" (25 mm) between views. Use half scale and omit dimensions.

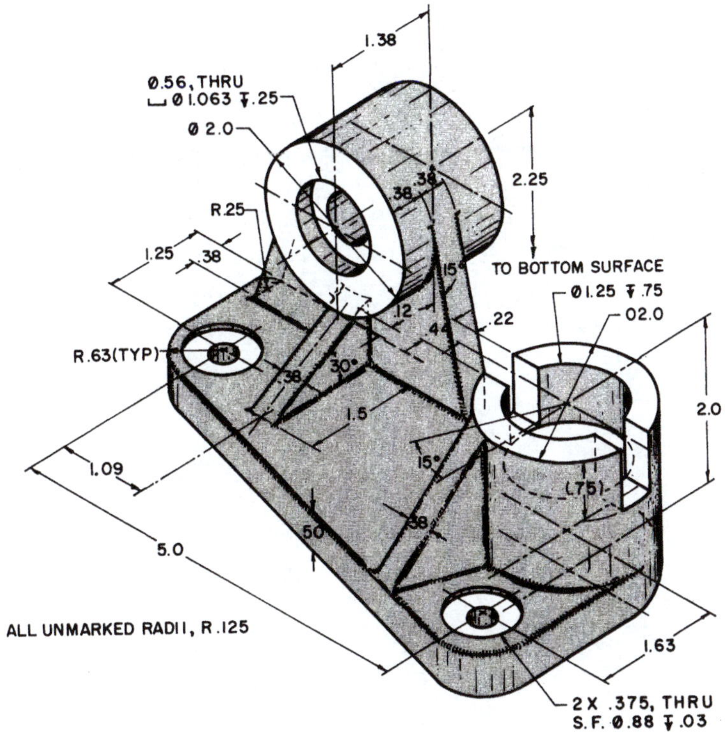

COURSE ________	STUDENT ________________________	DATE ________	PROBLEM 8-20

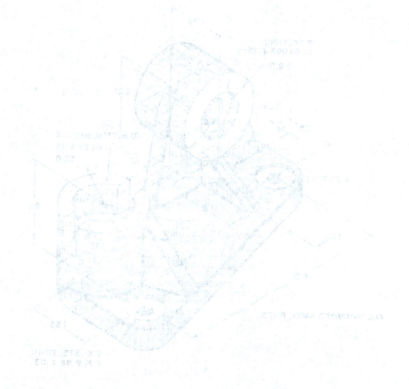

CHAPTER 9

Auxiliary Views

Problem 9-1 Auxiliary Views

Refer to the views shown and answer the following TRUE–FALSE questions.

__________ 1. To be able to draw an auxiliary view of a plane surface, the plane surface must first appear as an edge.

__________ 2. To prepare an auxiliary view of a given plane that appears as an edge, projection lines must be perpendicular to the given plane.

__________ 3. All the lines in a sectional drawing (except for specific conventions such as bolts and shafts) are in true size because the section cutting plane shows as an edge.

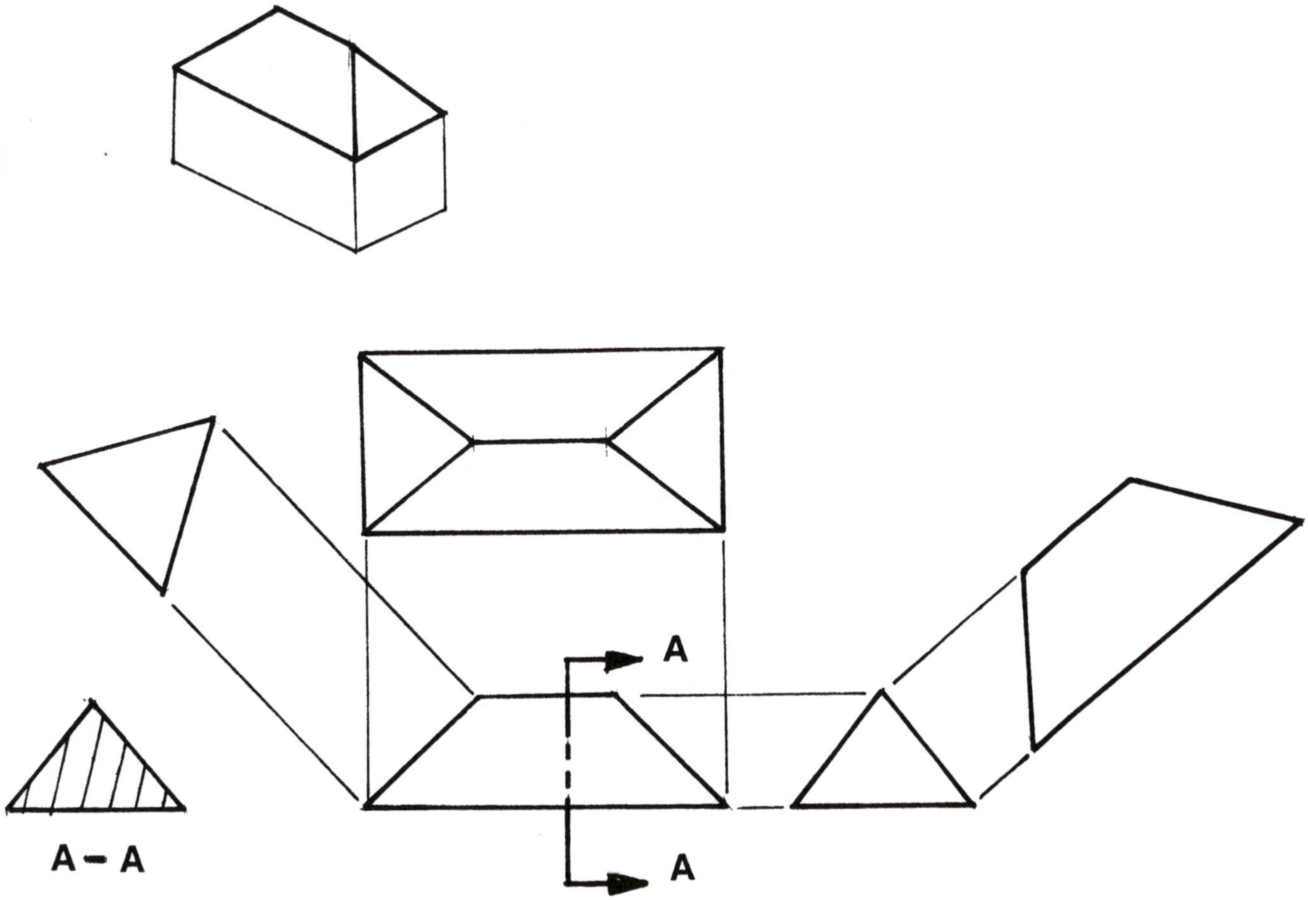

COURSE__________	STUDENT______________________________	DATE__________	PROBLEM 9-1

Problem 9-2 Auxiliary Views

The following figures show possible auxiliary views of the object at the bottom of the page. Circle the view(s) that are not true auxiliary views.

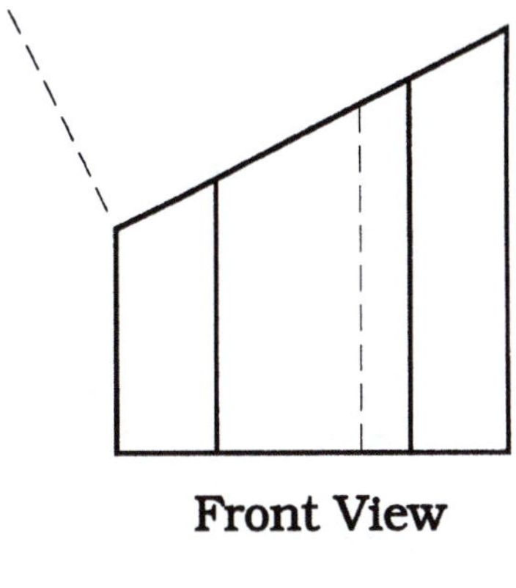

Front View

COURSE ________	STUDENT ____________________	DATE ________	PROBLEM 9-2

Problem 9-3 Auxiliary Views

Add the missing lines.

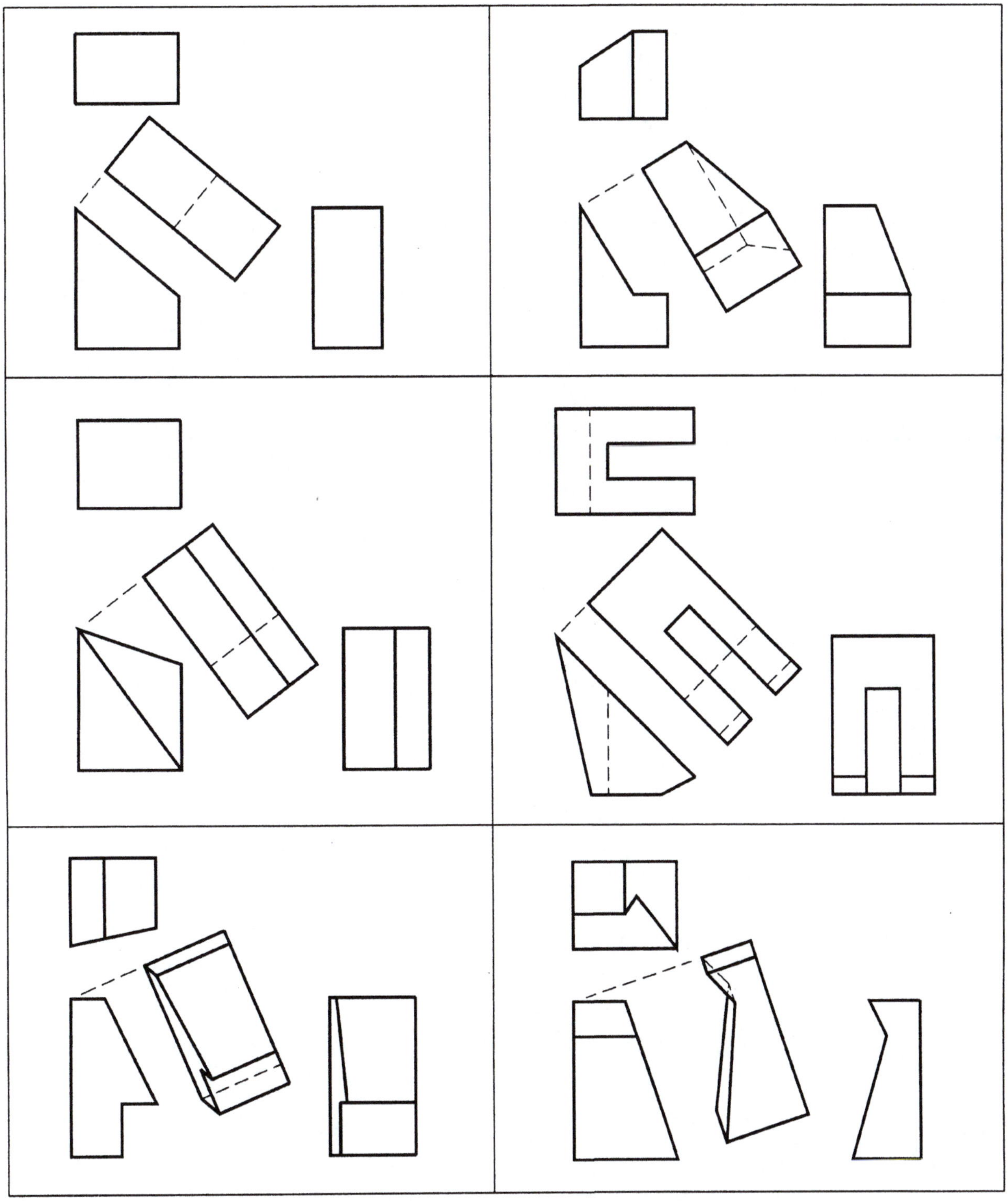

COURSE ________ STUDENT ______________________________ DATE ________ PROBLEM 9-3

Problem 9-4 Auxiliary Views

Complete the auxiliary views.

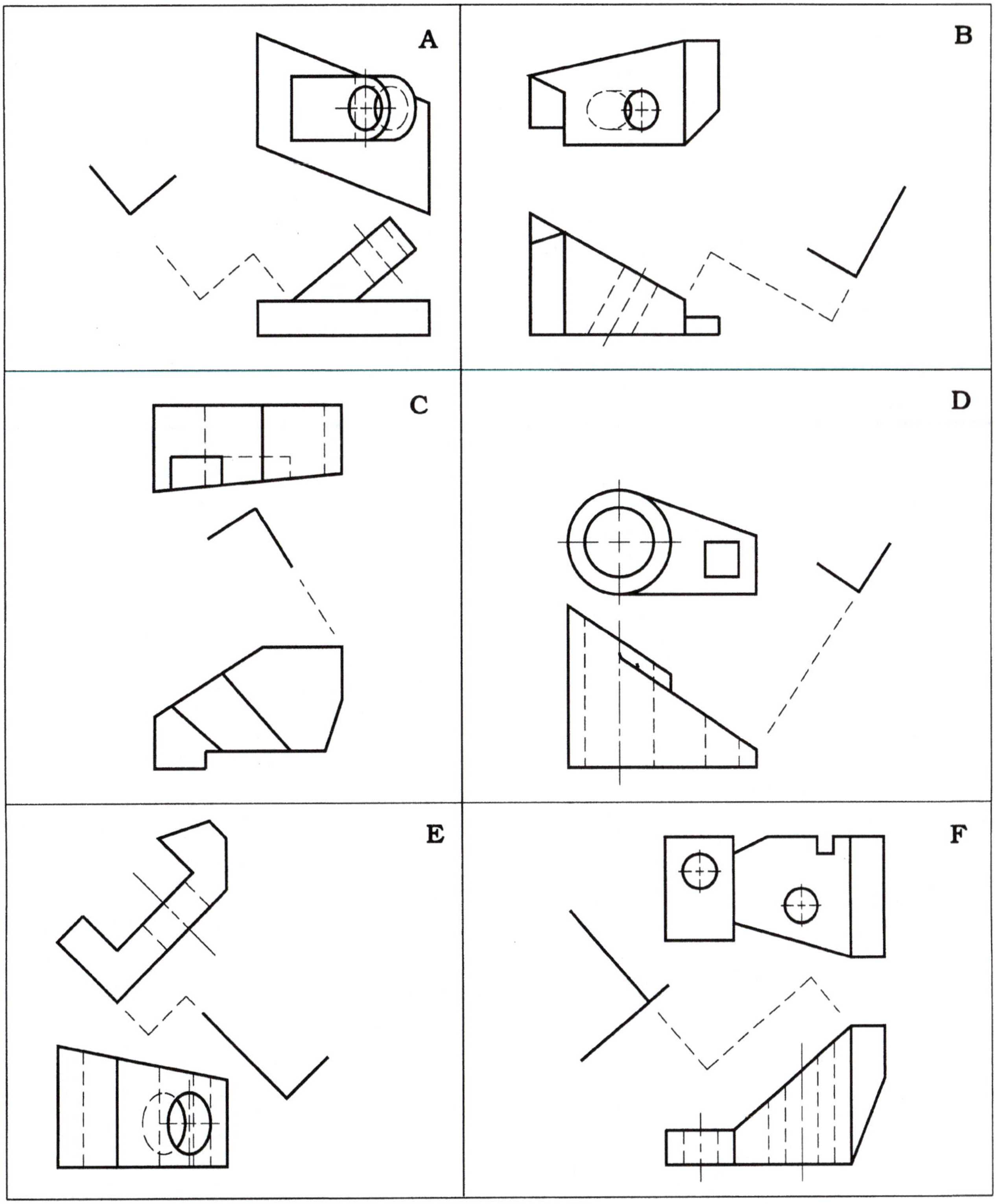

COURSE	STUDENT	DATE	PROBLEM 9-4

Problem 9-4 Auxiliary Views

Complete the auxiliary views.

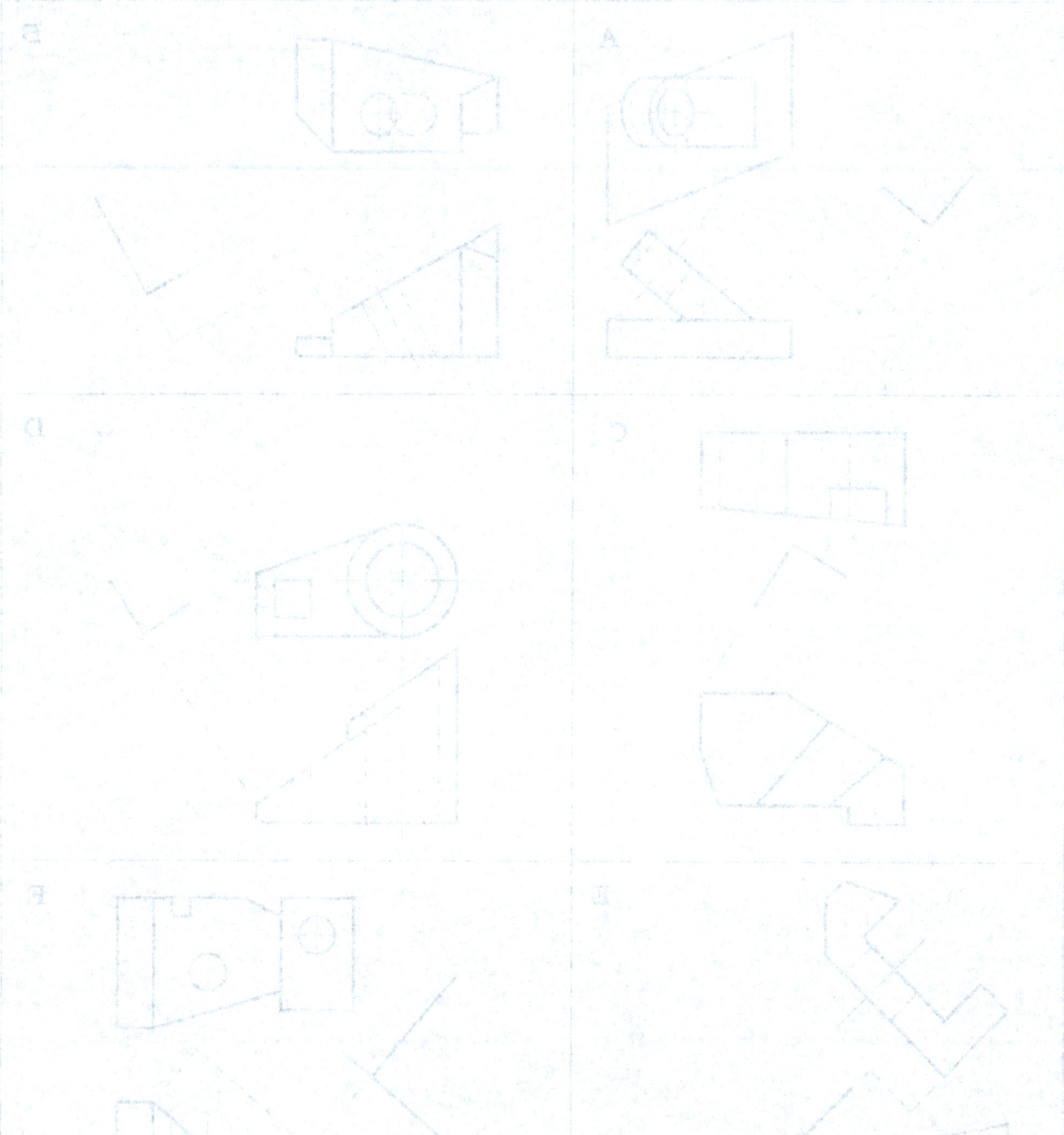

COURSE	STUDENT	DATE	PROBLEM 9-4

Problem 9-5 Auxiliary Views

Sketch the auxiliary views.

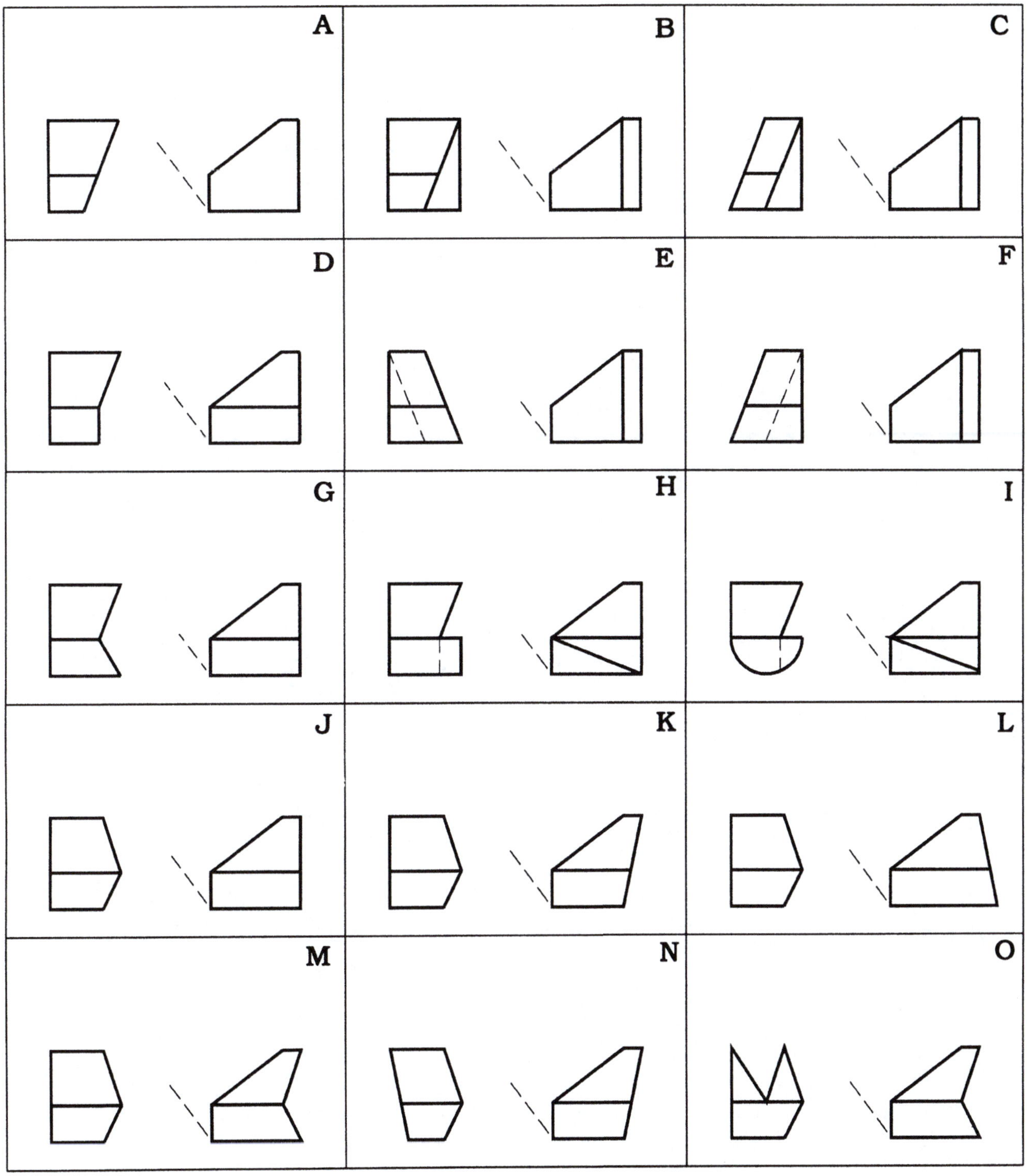

COURSE ________ STUDENT ______________________ DATE ________ PROBLEM 9-5

Problem 9-6 Auxiliary Views

Sketch the auxiliary view of the object below. Be sure to include all construction lines.

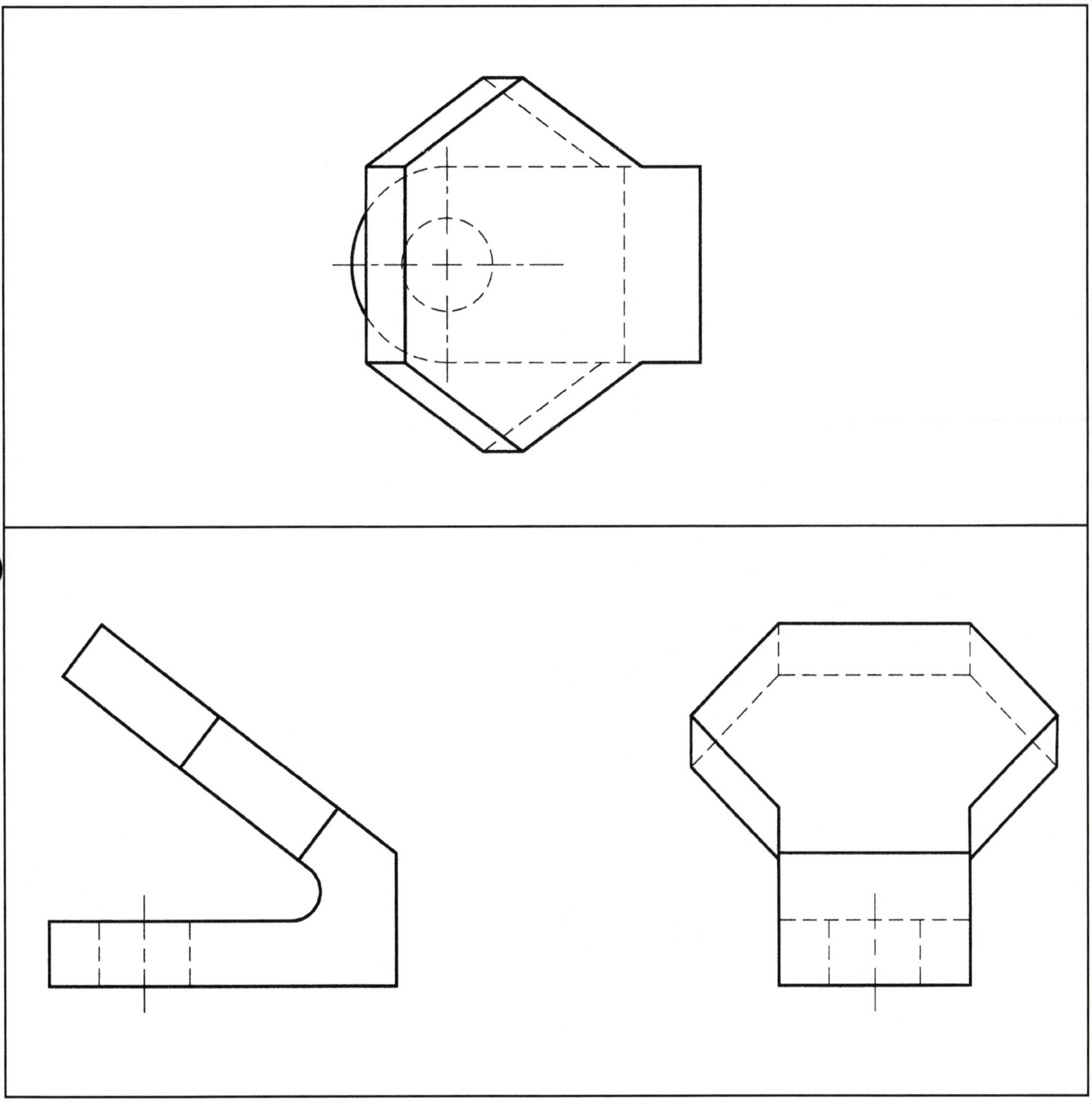

COURSE ________	STUDENT ________________________	DATE ________	PROBLEM 9-6

roblem 9-7 Auxiliary Views

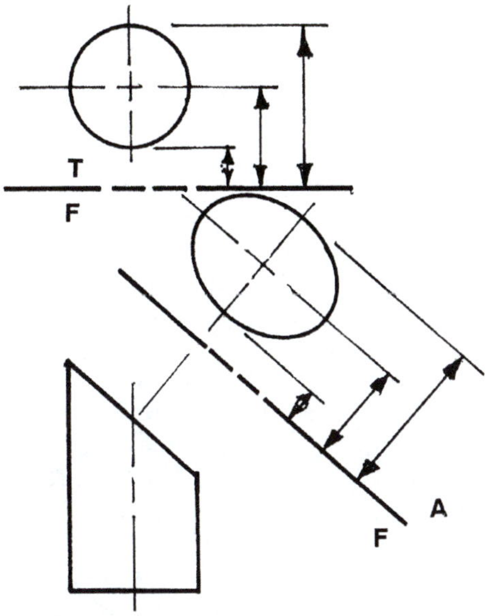

EXAMPLE: FOLD-LINE METHOD

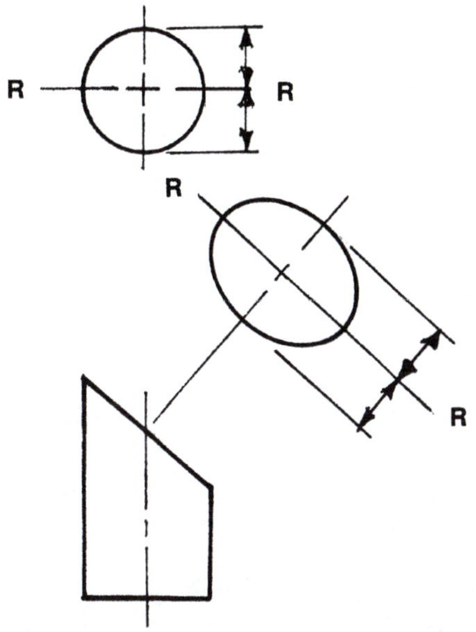

EXAMPLE: REFERENCE-LINE METHOD

Construct an auxiliary view of the inclined portion of the object using fold line A/S. Construct an auxiliary view of the inclined portion of the object using reference line R-R. Do not erase construction lines. Use correct lines. Use a Z break.

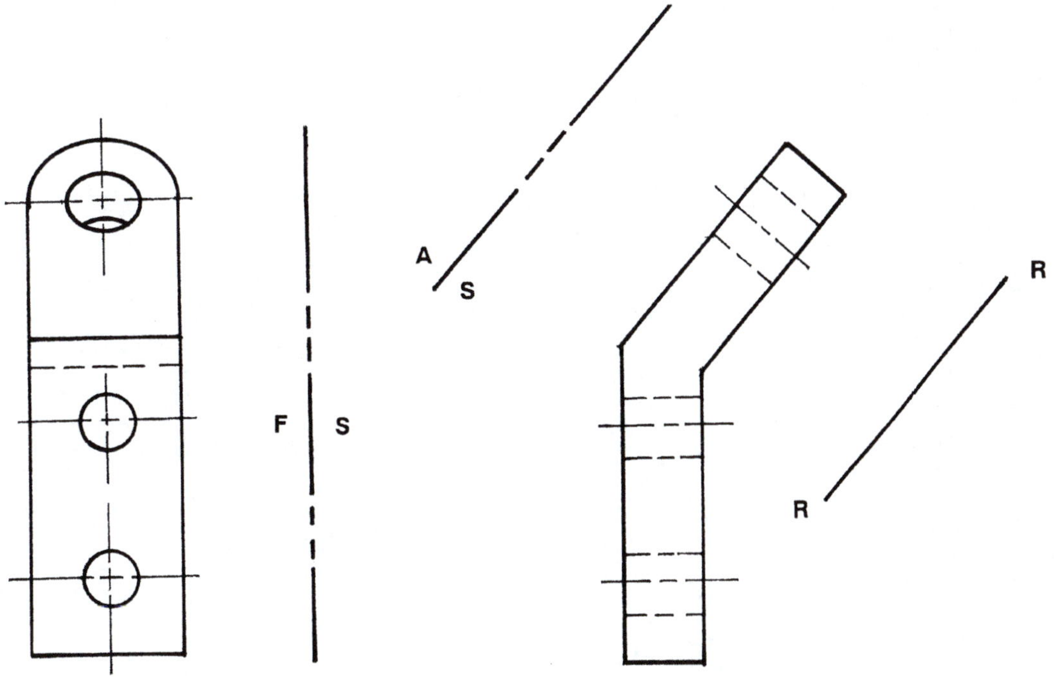

COURSE__________ STUDENT______________________________ DATE__________ PROBLEM 9-7

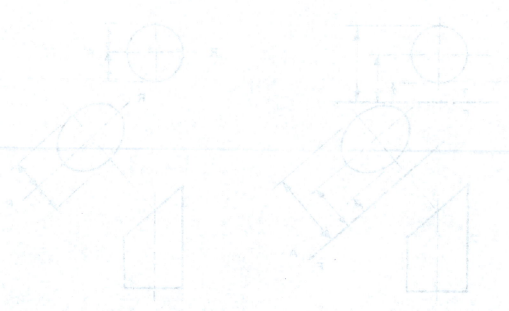

Problem 9-8 Auxiliary Views

Prepare auxiliary views of the inclined planes using the reference-plane method. Do not erase construction lines. Use correct line thickness.

A

B

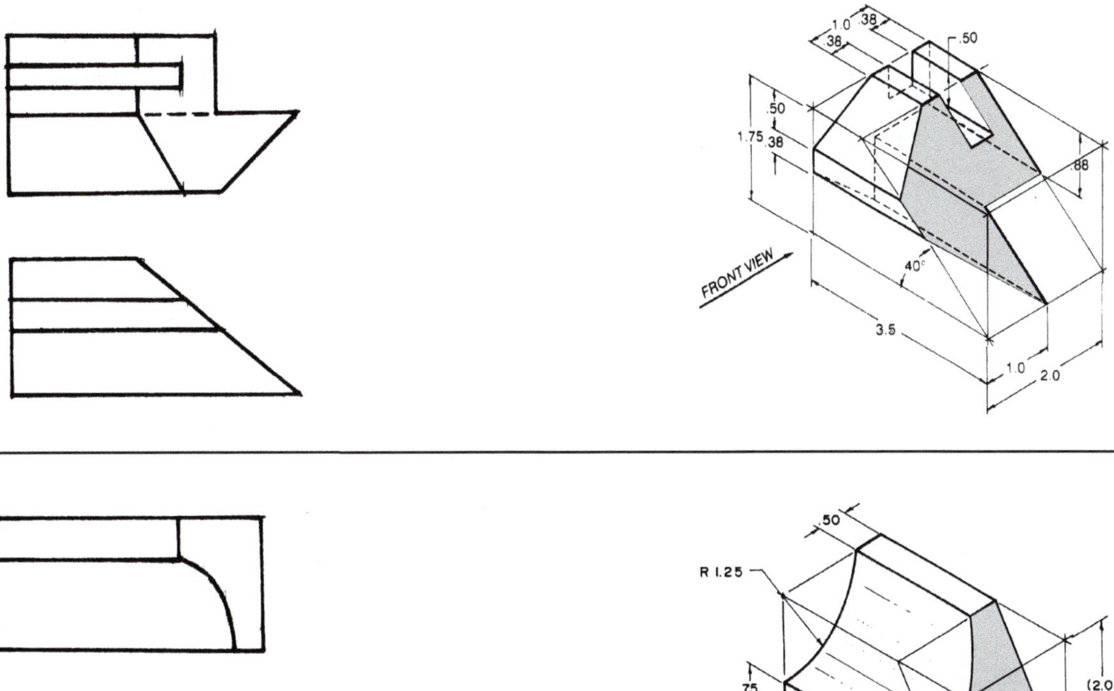

C

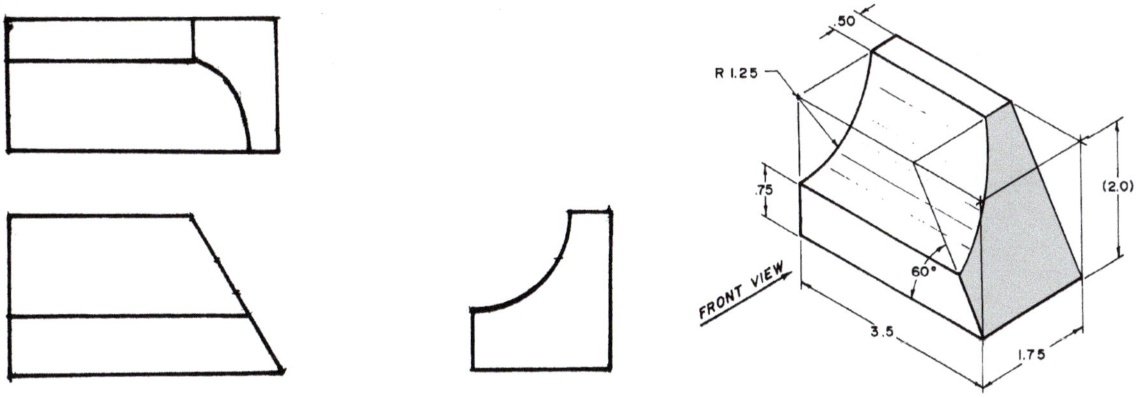

COURSE________ STUDENT________________________ DATE________ PROBLEM 9-8

Problem 9-9 Auxiliary Views

Prepare auxiliary views of the inclined planes using the fold-line method. Do not erase construction lines. Use correct line thickness.

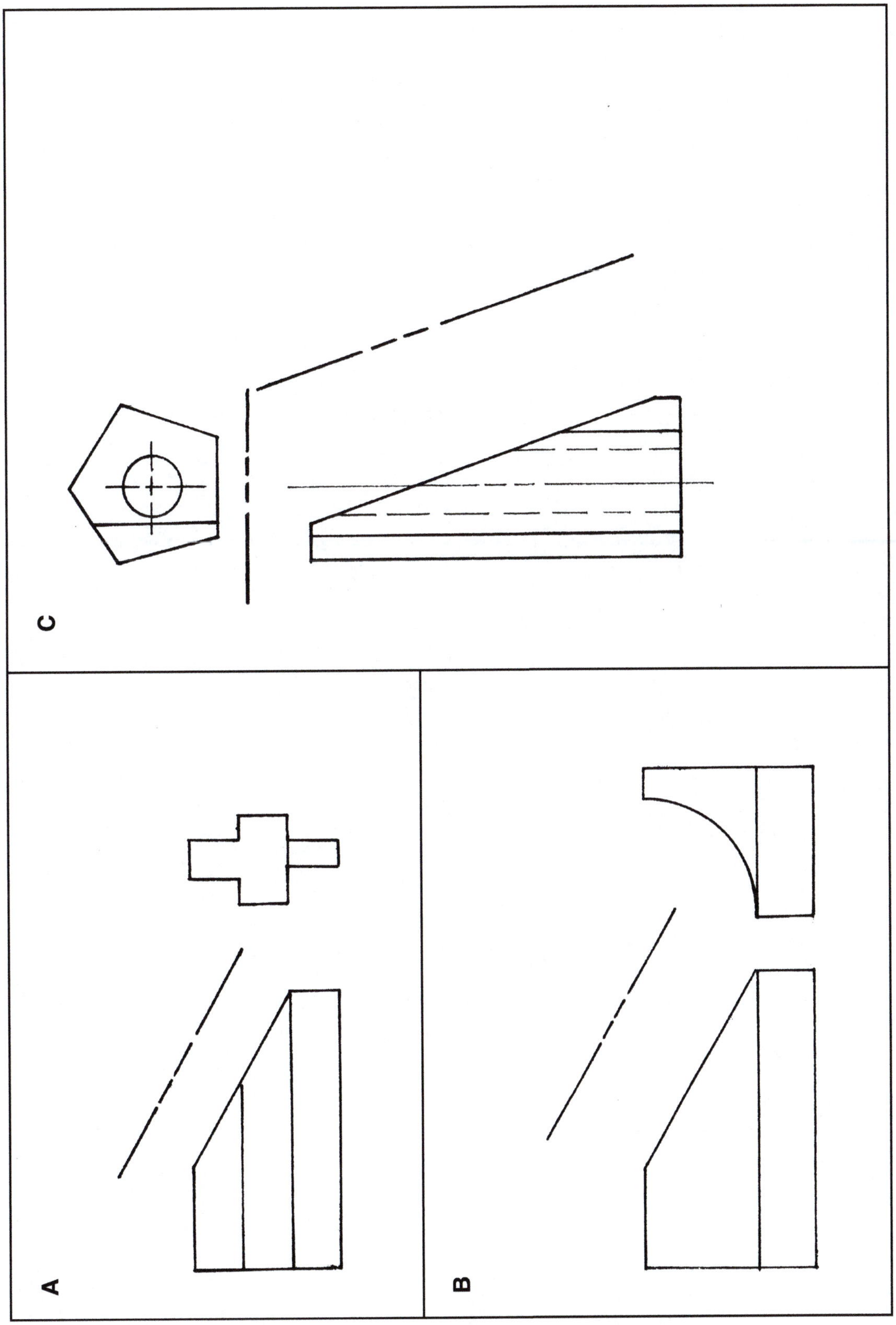

COURSE__________ STUDENT_______________________________ DATE__________ PROBLEM 9-9

Problem 9-10 Auxiliary Views

A

Prepare an auxiliary view of the inclined plane using the fold-line method.

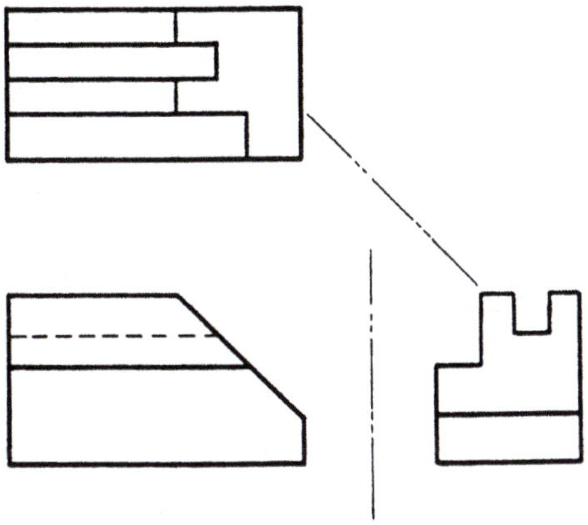

B

Complete the front view. Do not erase construction lines. Use correct line thickness.

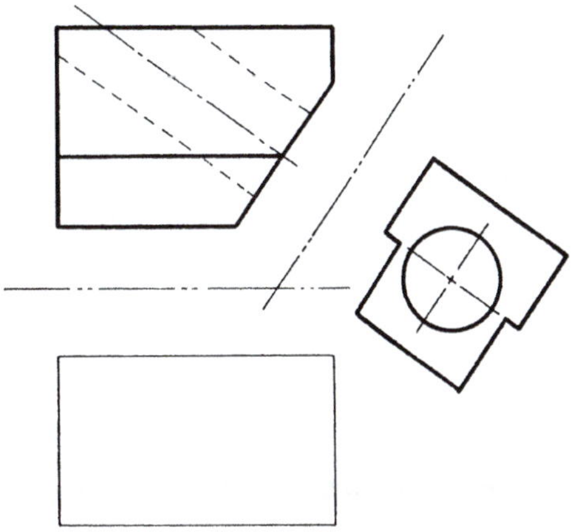

C

Complete the front and right-side views. Do not erase construction lines. Use correct line thickness.

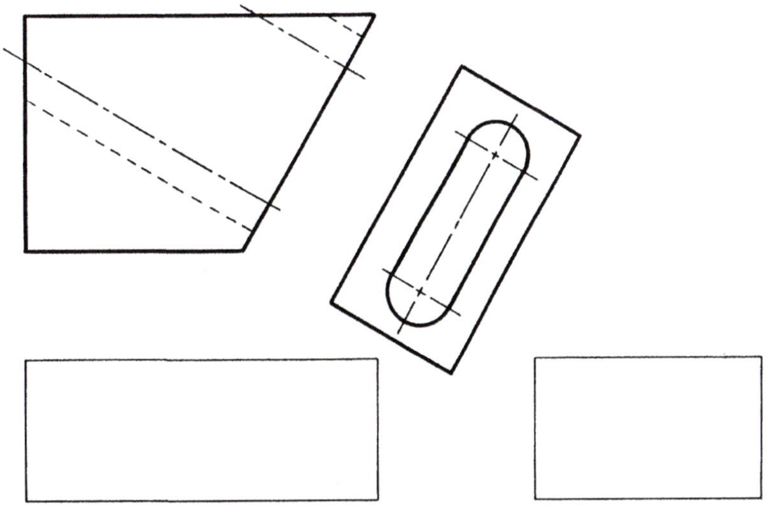

COURSE________ **STUDENT**____________________ **DATE**________ **PROBLEM 9-10**

Problem 9-11 Auxiliary Views

A

Prepare an auxiliary view of the inclined plane using the reference-plane method. Do not erase construction lines. Use correct line thickness.

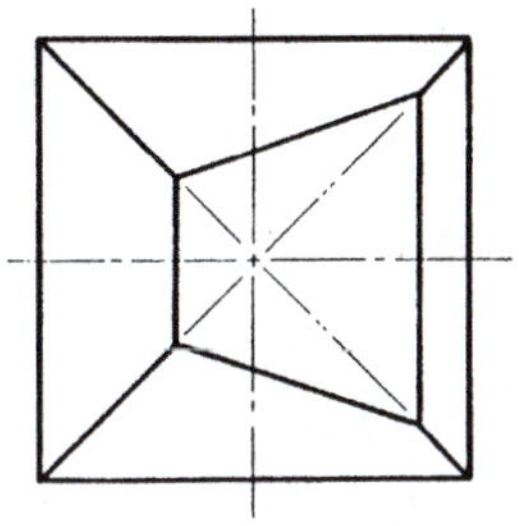

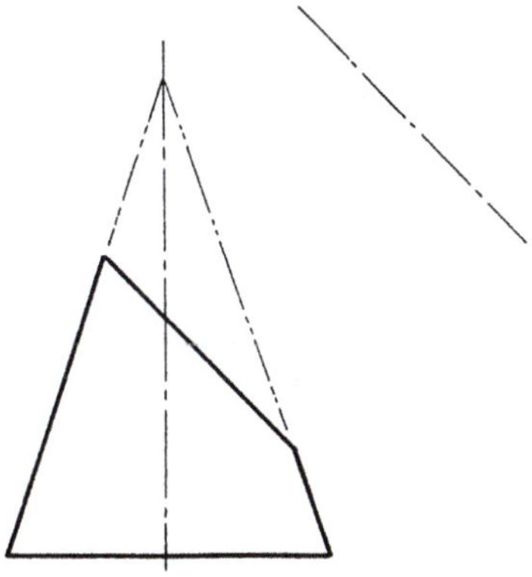

B

Prepare an auxiliary view of the inclined plane using the reference-plane method. Do not erase construction lines. Use correct line thickness.

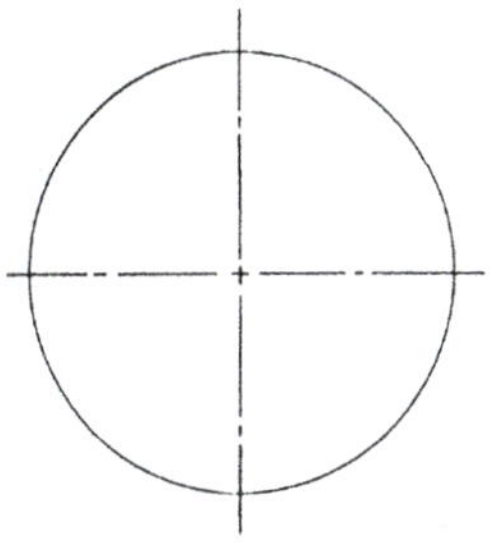

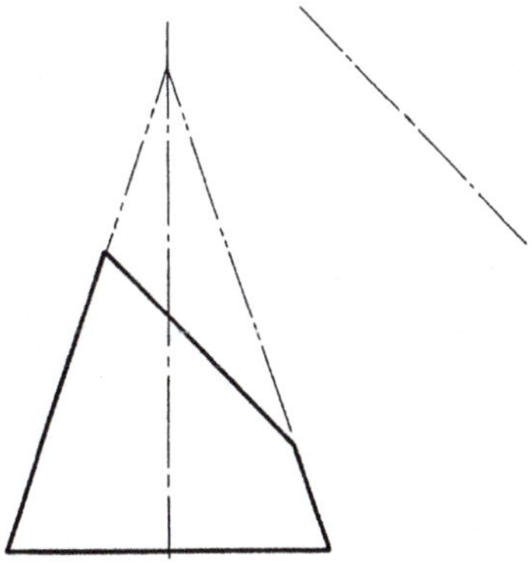

C Prepare an auxiliary view of the inclined plane using the reference-plane method and complete the front view.

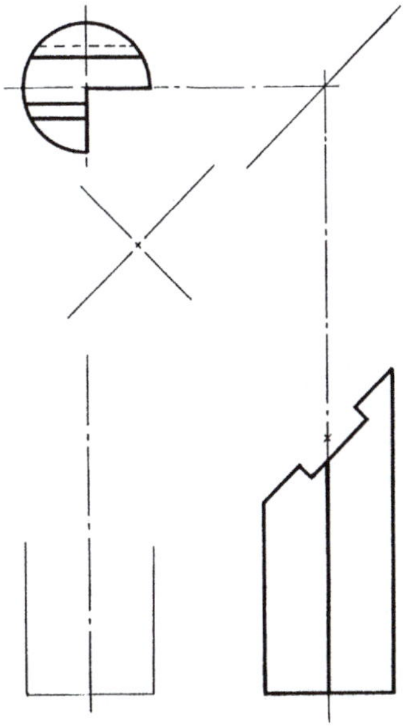

D Prepare an auxiliary view of the inclined plane using the reference-plane method and complete the right-side view.

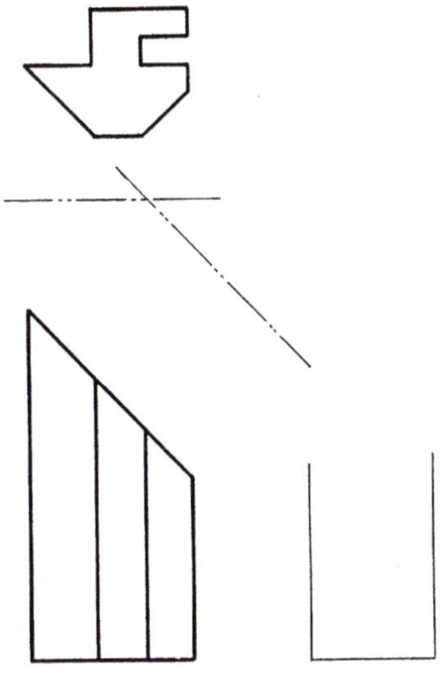

COURSE________ STUDENT____________________ DATE________ PROBLEM 9-11

Problem 9-12 Auxiliary Views

A

Prepare an auxiliary view of the inclined plane using the reference-plane method and complete the right-side view. Do not erase construction lines. Use correct line thickness.

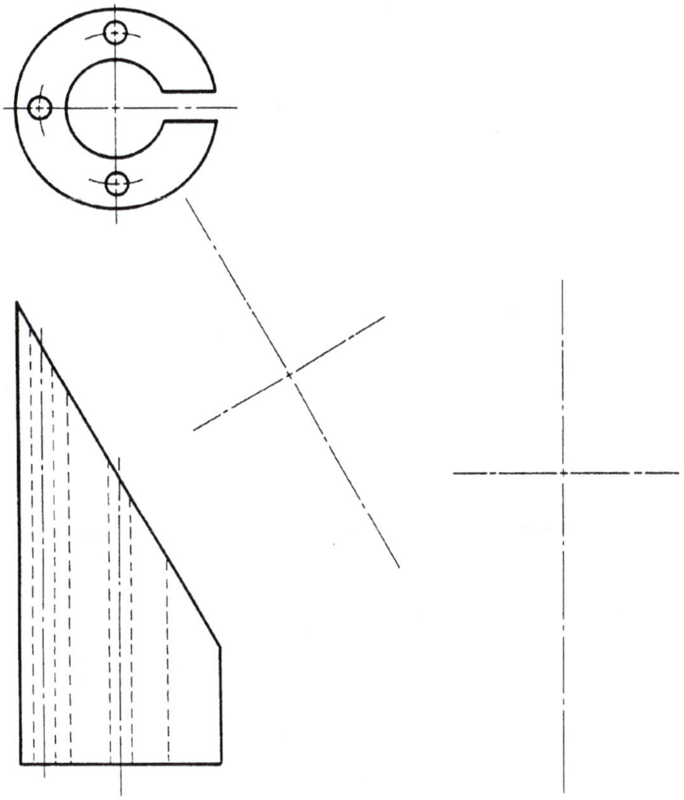

B

Prepare an auxiliary view of the inclined planes using the reference-plane or the fold-line method. Complete the right-side view. Do not erase construction lines. Use correct line thickness.

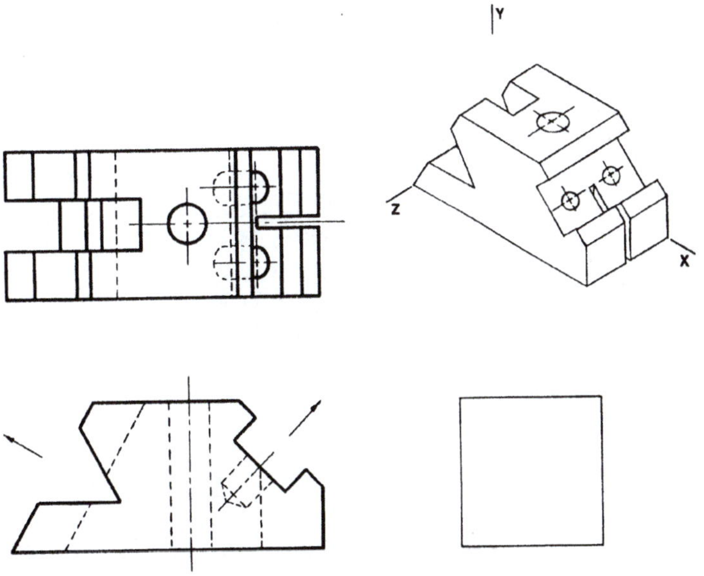

COURSE_________ STUDENT________________________ DATE_________ PROBLEM 9-12

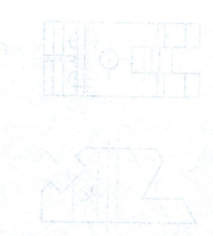

Problem 9-13 Auxiliary Views

A

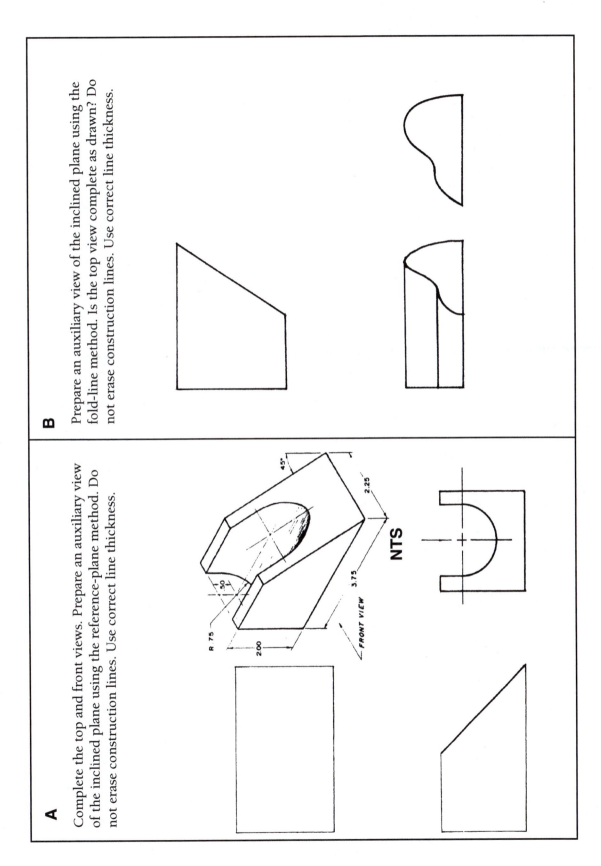

Complete the top and front views. Prepare an auxiliary view of the inclined plane using the reference-plane method. Do not erase construction lines. Use correct line thickness.

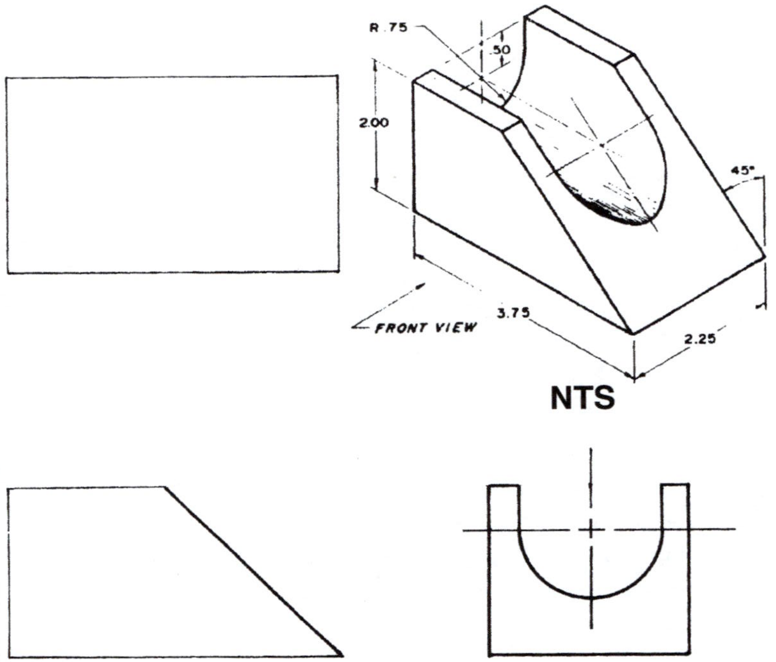

B

Prepare an auxiliary view of the inclined plane using the fold-line method. Is the top view complete as drawn? Do not erase construction lines. Use correct line thickness.

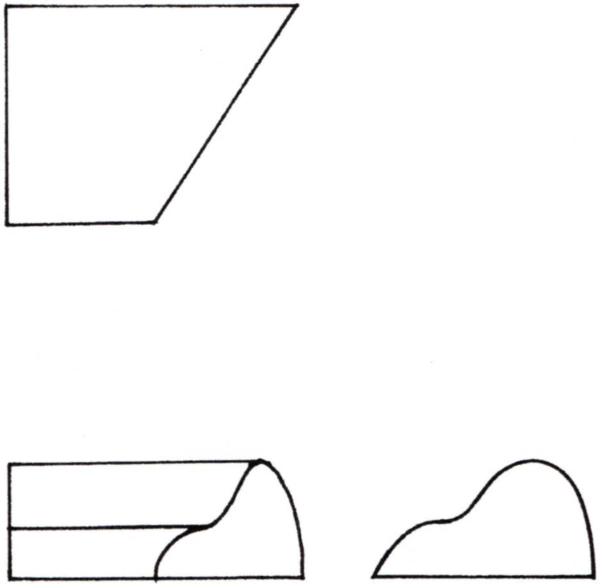

COURSE ________ STUDENT ________ DATE ________ PROBLEM 9-13

Problem 9-14 Auxiliary Views

A

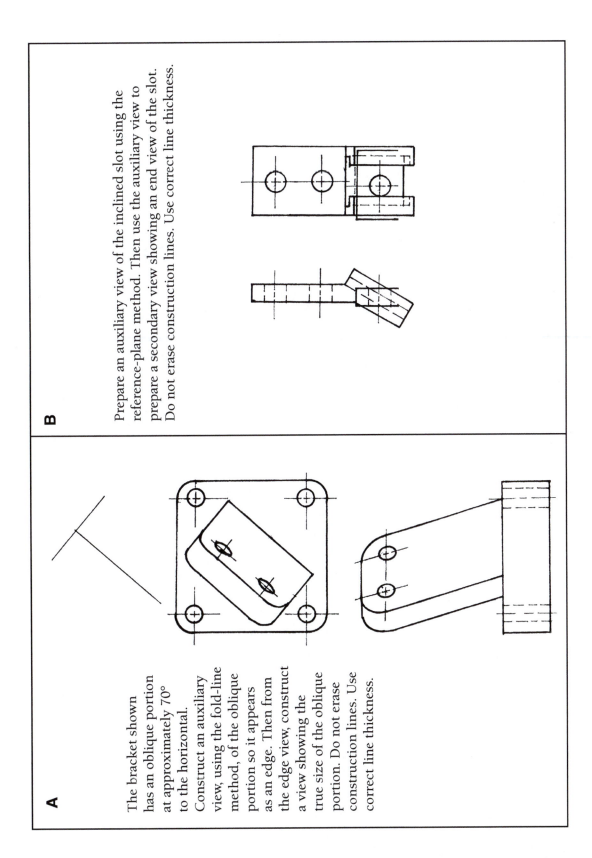

The bracket shown has an oblique portion at approximately 70° to the horizontal. Construct an auxiliary view, using the fold-line method, of the oblique portion so it appears as an edge. Then from the edge view, construct a view showing the true size of the oblique portion. Do not erase construction lines. Use correct line thickness.

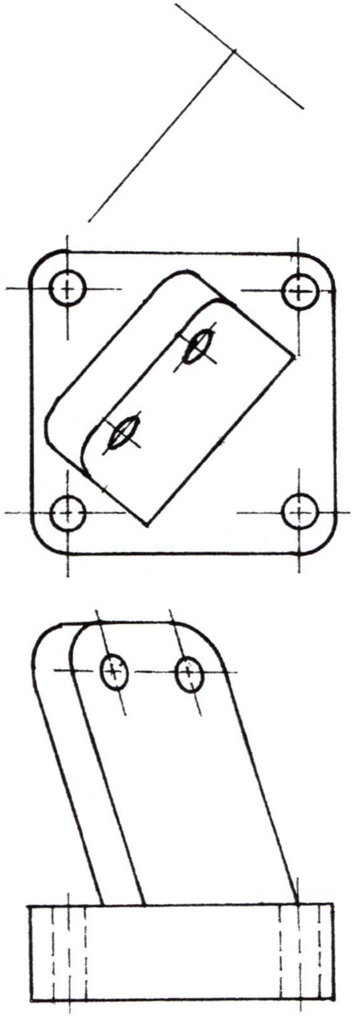

B

Prepare an auxiliary view of the inclined slot using the reference-plane method. Then use the auxiliary view to prepare a secondary view showing an end view of the slot. Do not erase construction lines. Use correct line thickness.

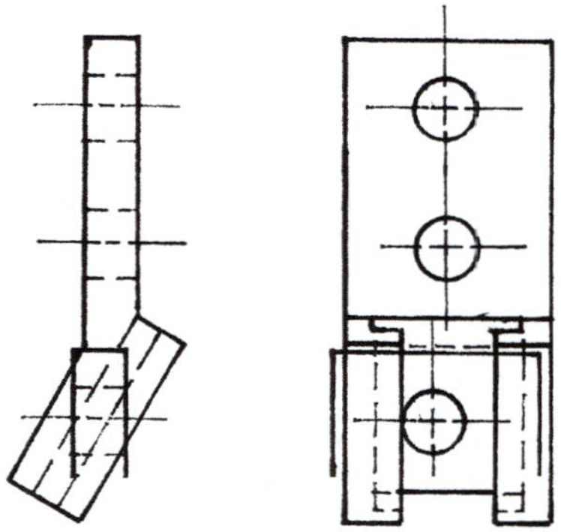

COURSE__________ STUDENT________________________ DATE__________ PROBLEM 9-14

roblem 9-15 Auxiliary Views

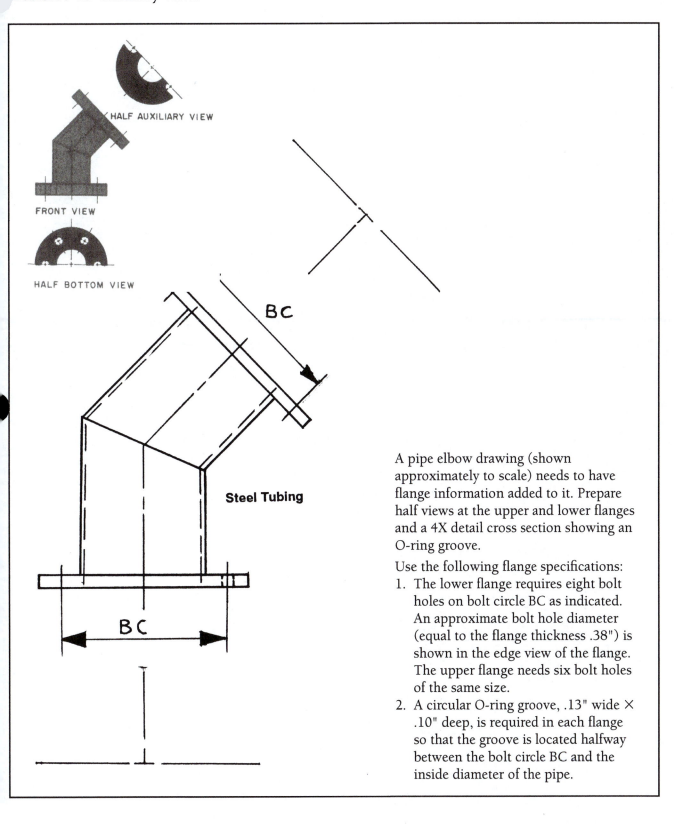

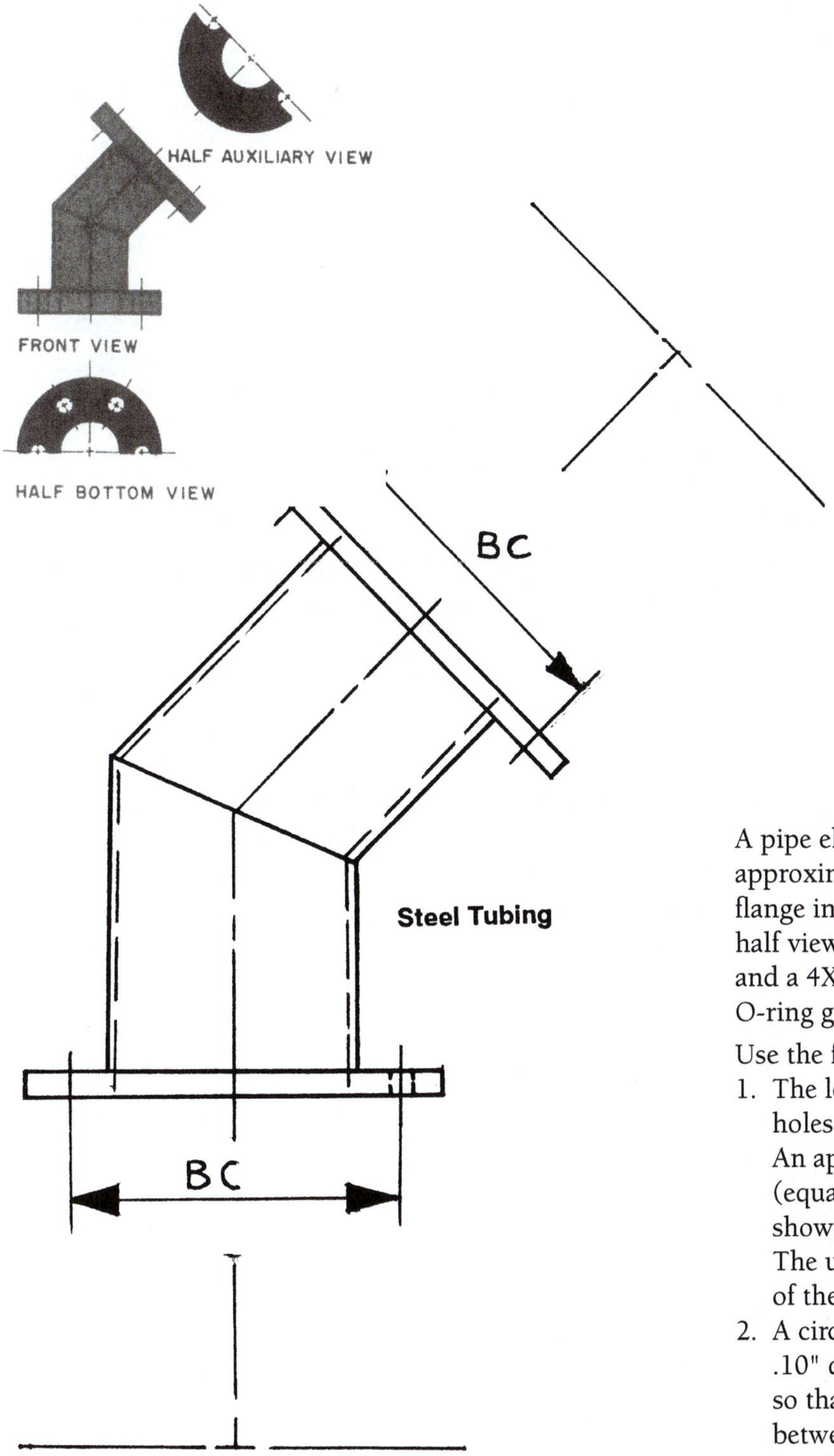

A pipe elbow drawing (shown approximately to scale) needs to have flange information added to it. Prepare half views at the upper and lower flanges and a 4X detail cross section showing an O-ring groove.

Use the following flange specifications:

1. The lower flange requires eight bolt holes on bolt circle BC as indicated. An approximate bolt hole diameter (equal to the flange thickness .38") is shown in the edge view of the flange. The upper flange needs six bolt holes of the same size.
2. A circular O-ring groove, .13" wide × .10" deep, is required in each flange so that the groove is located halfway between the bolt circle BC and the inside diameter of the pipe.

COURSE________ STUDENT____________________ DATE________ PROBLEM 9-15

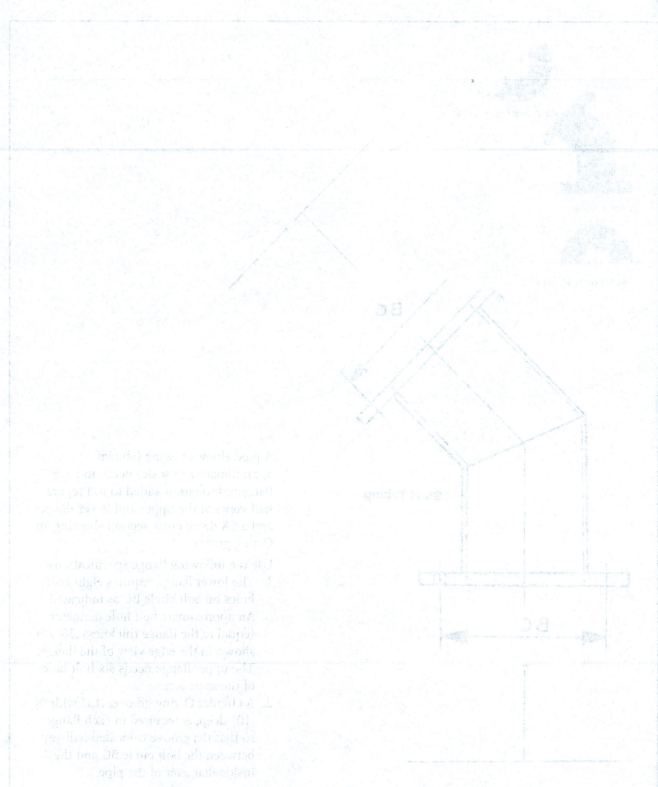

Problem 9-16 Auxiliary Views

Given the following orthographic views, use AutoCAD to construct the indicated auxiliary views of the following parts.

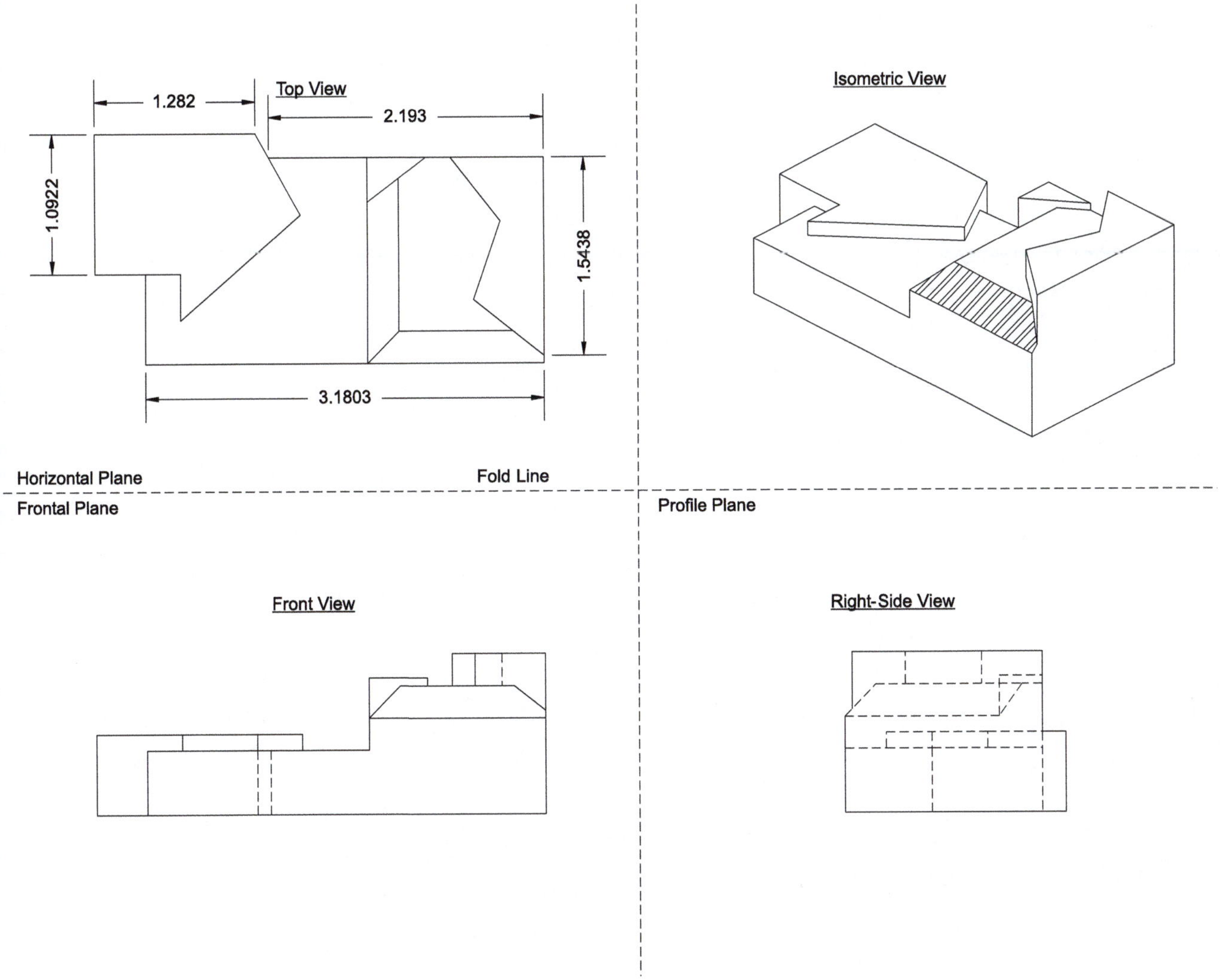

COURSE__________ STUDENT______________________________ DATE__________ PROBLEM 9-16

Problem 9-17 Auxiliary Views

Given the following orthographic views, use AutoCAD to construct the indicated auxiliary views of the following parts.

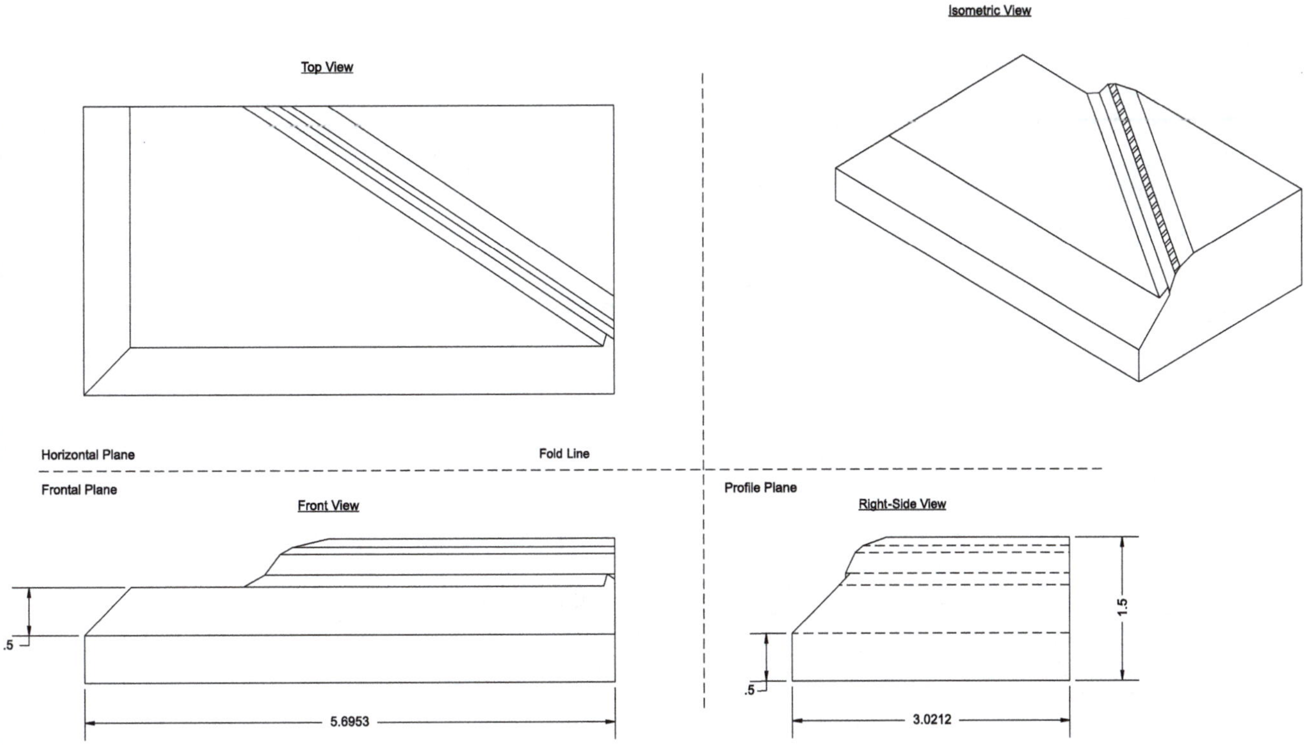

COURSE ________	STUDENT ____________________	DATE ________	PROBLEM 9-17

Problem 9-18 Auxiliary Views

Given the following orthographic views, use AutoCAD to construct the indicated auxiliary views of the following parts.

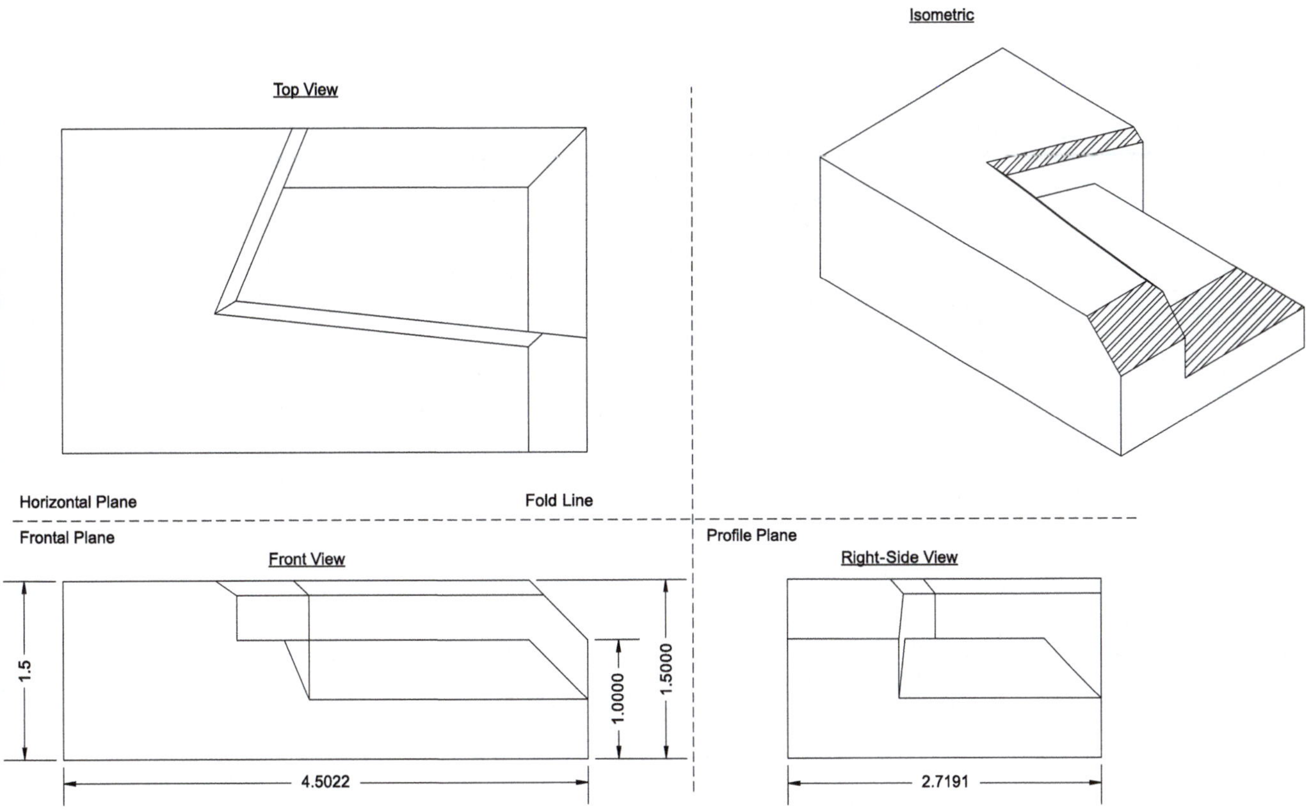

COURSE ______	STUDENT ______	DATE ______	PROBLEM 9-18

Problem 9-19 Auxiliary Views

Given the following orthographic views, use AutoCAD to construct the indicated auxiliary views of the following parts.

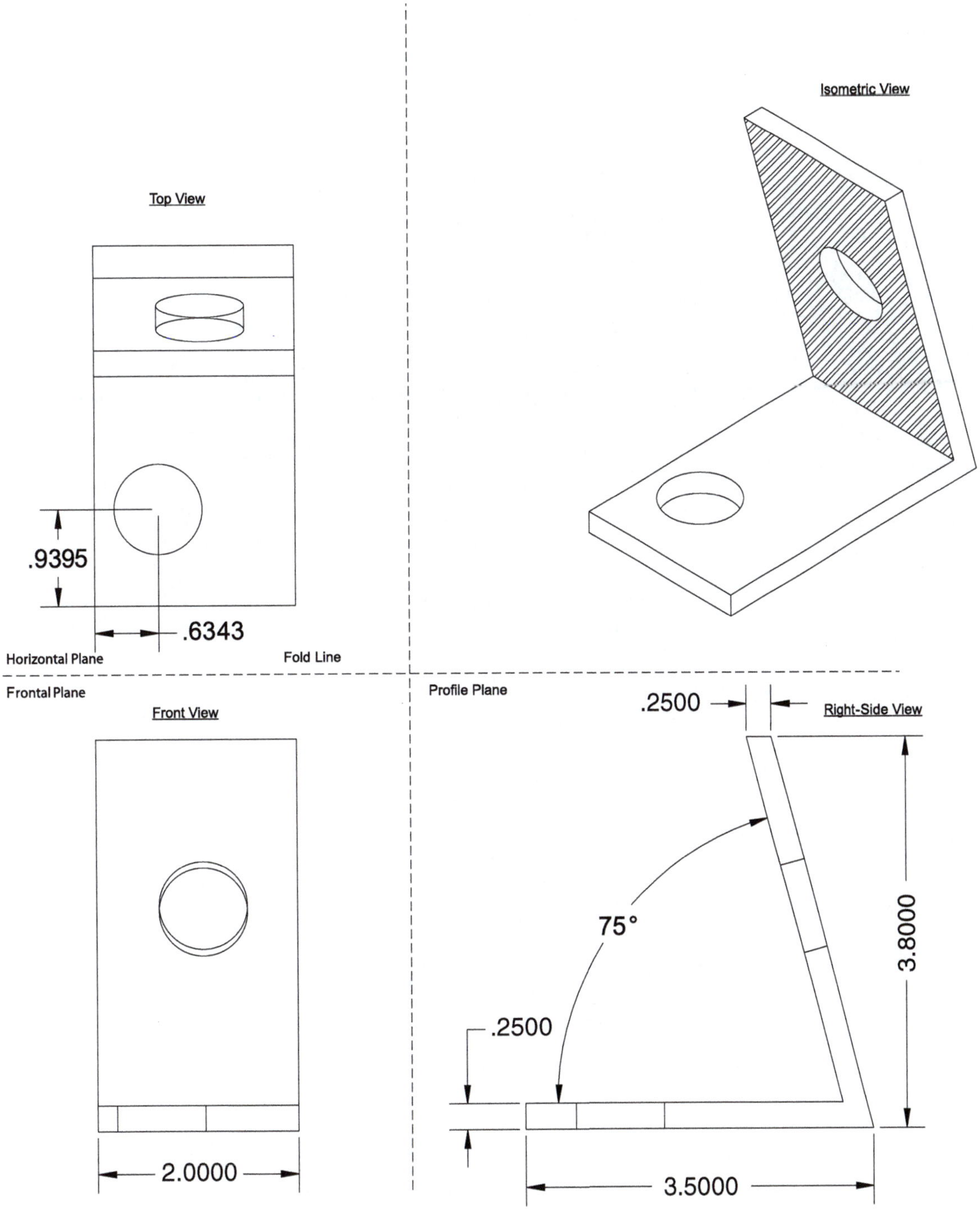

COURSE ________	STUDENT ________________	DATE ________	PROBLEM 9-19

Problem 9-20 Auxiliary Views

Given the following orthographic views, use AutoCAD to construct the indicated auxiliary views of the following parts.

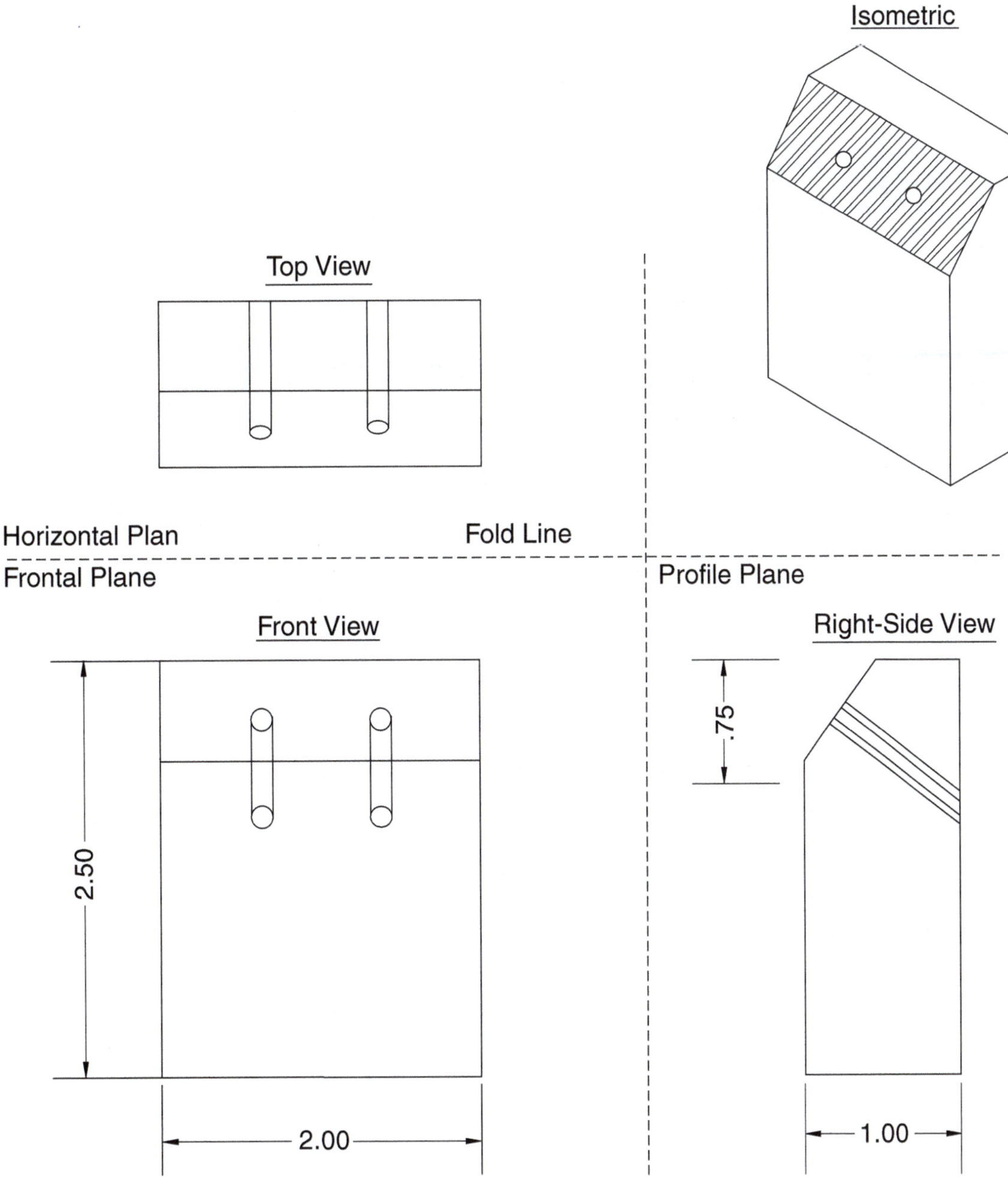

COURSE ______	STUDENT ______	DATE ______	PROBLEM 9-20

Problem 9-21 Auxiliary Views

Given the following orthographic views, use AutoCAD to construct the indicated auxiliary views of the following parts.

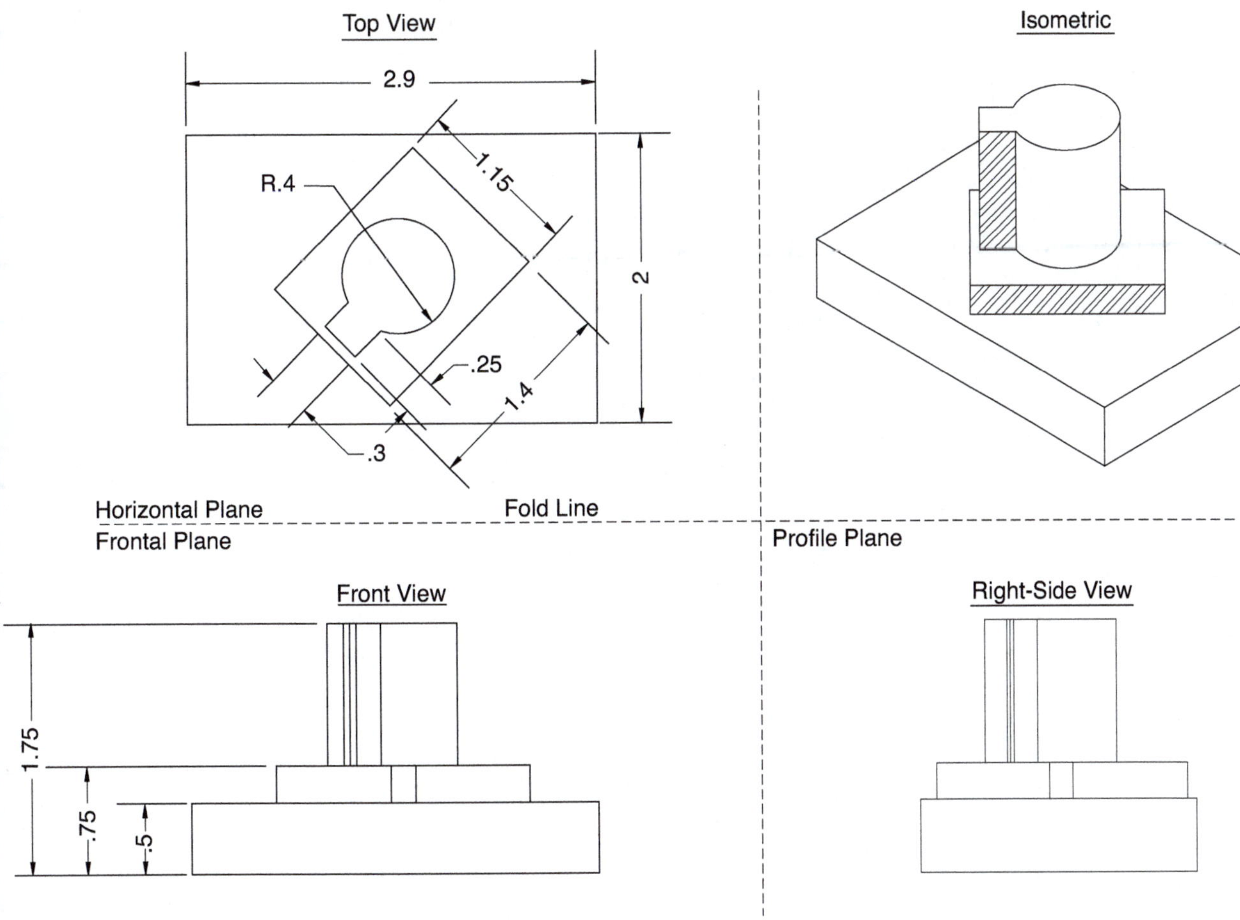

COURSE ________	STUDENT ____________________	DATE ________	PROBLEM 9-21

CHAPTER 10

Dimensioning and Tolerancing

Problem 10-1 Dimensioning

Circle the preferred method of dimensioning.

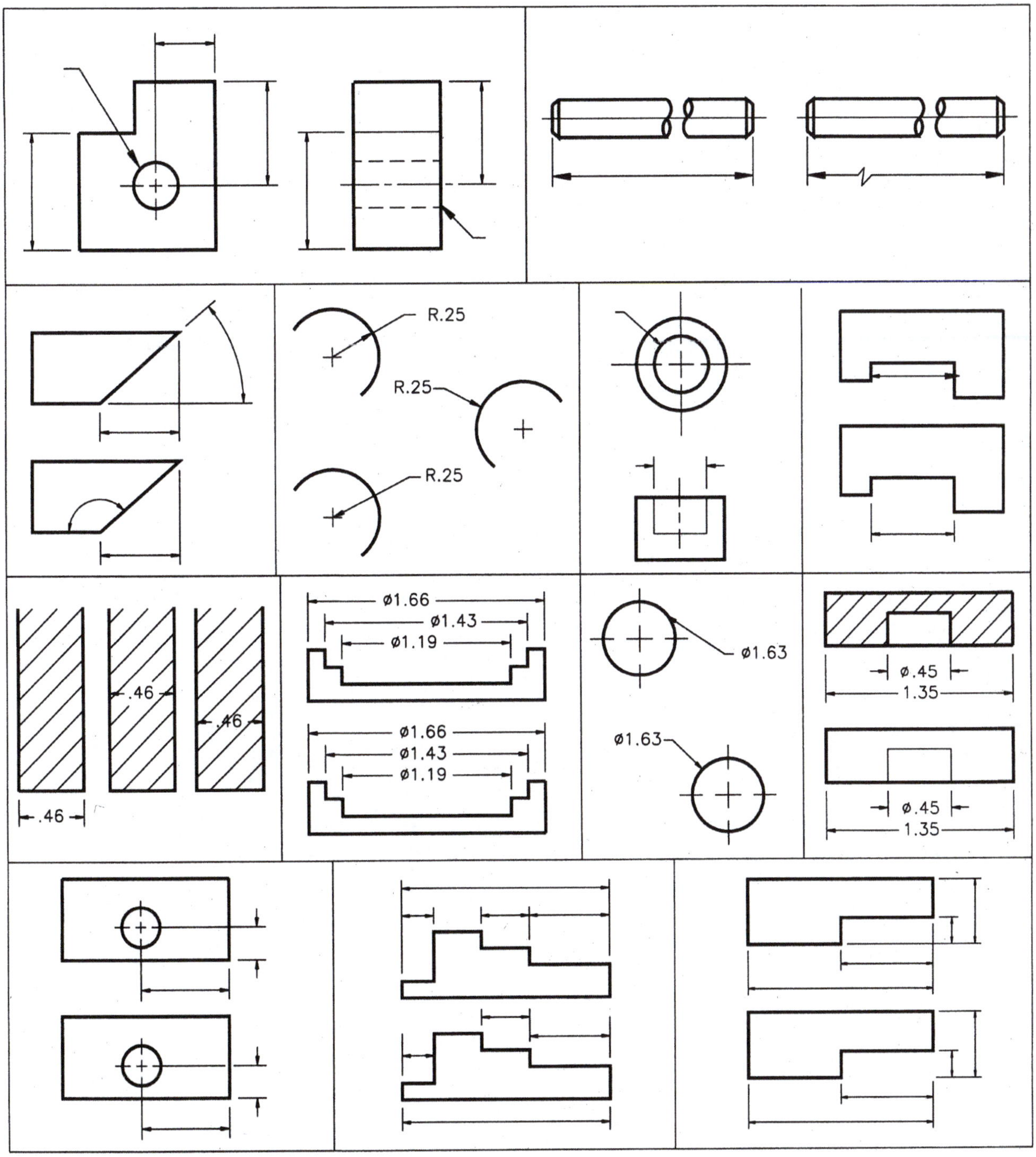

COURSE________ STUDENT____________________ DATE________ PROBLEM 10-1

Problem 10-2 Dimensioning

Measure and dimension, unidirectionally, only the features noted in the captions at full size. Use accepted practices, decimal inches to two places, and decimal angles to two places. Measure to the centers of lines.

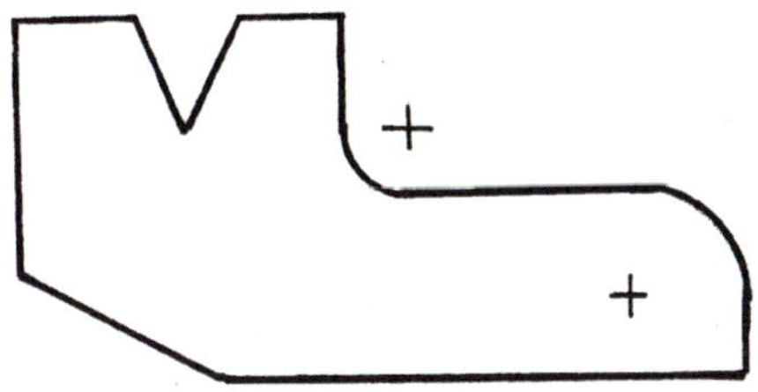

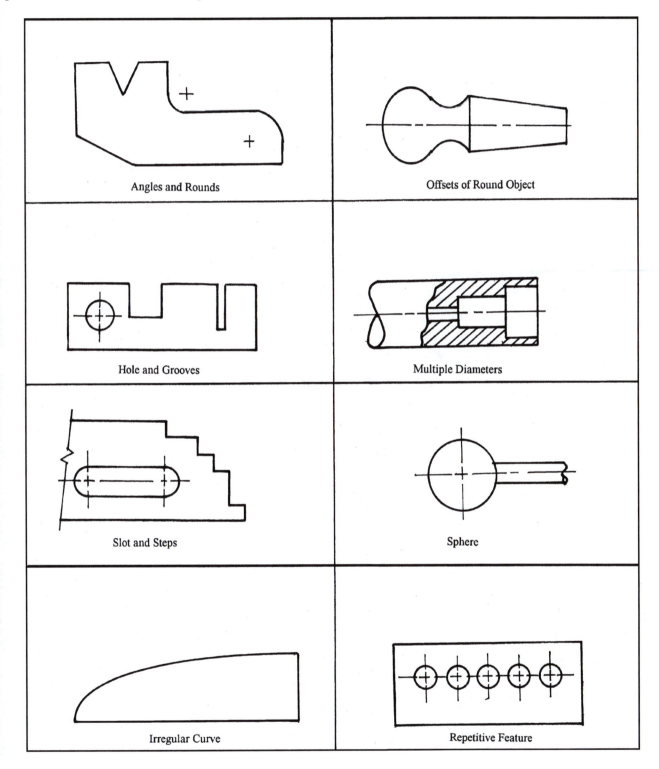

Angles and Rounds

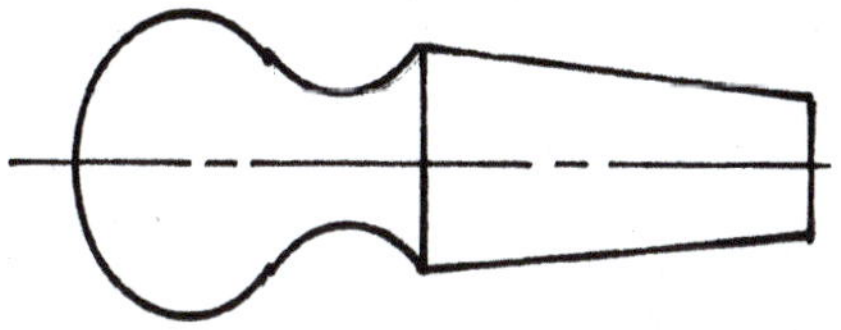

Offsets of Round Object

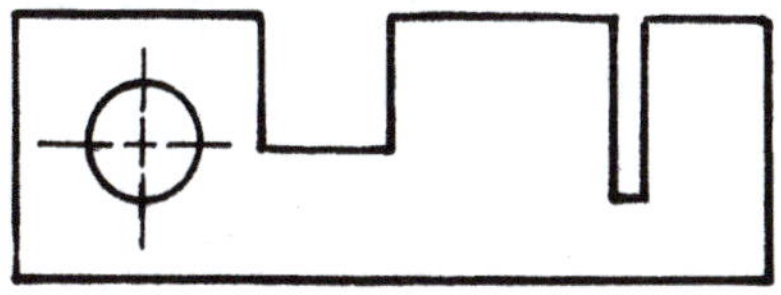

Hole and Grooves

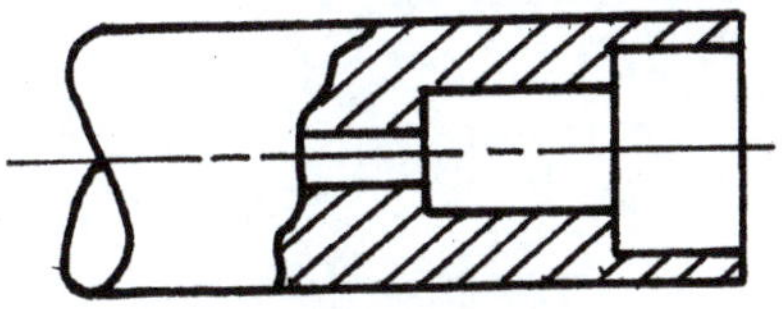

Multiple Diameters

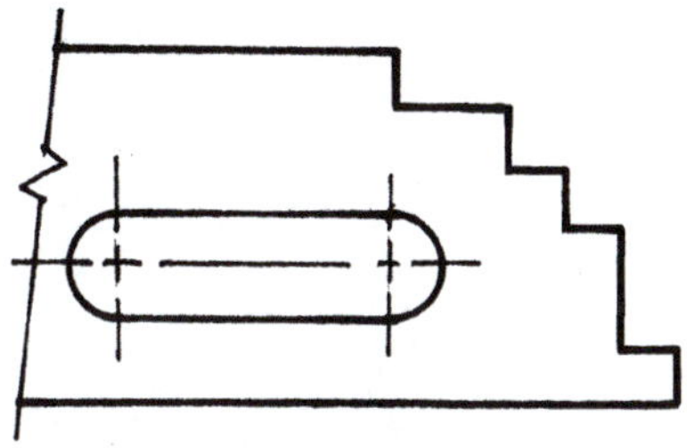

Slot and Steps

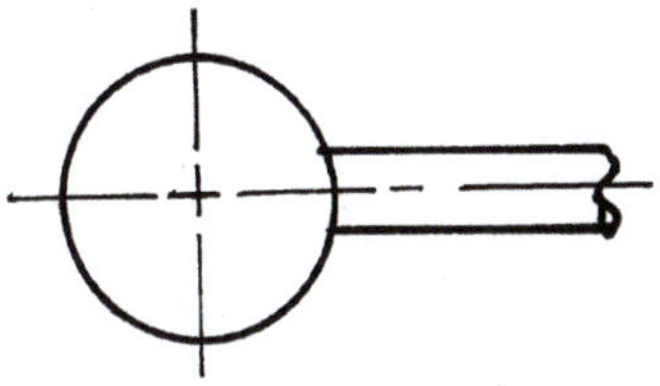

Sphere

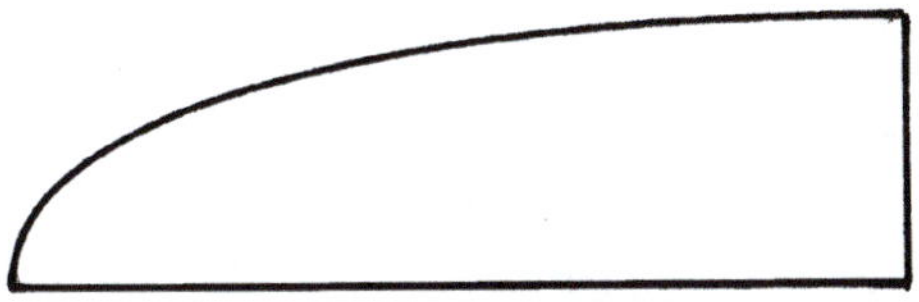

Irregular Curve

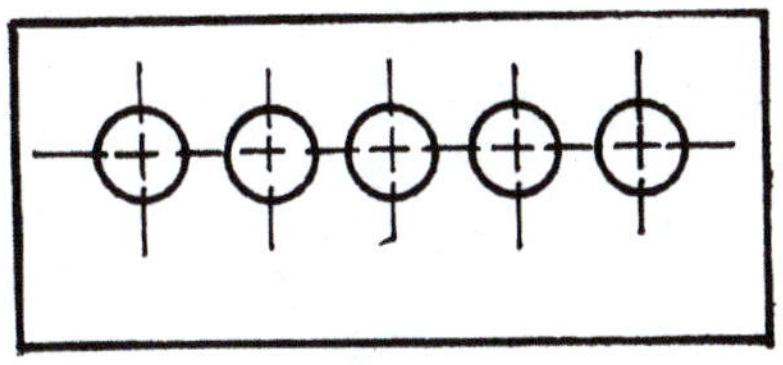

Repetitive Feature

COURSE________ STUDENT________________________ DATE________ PROBLEM 10-2

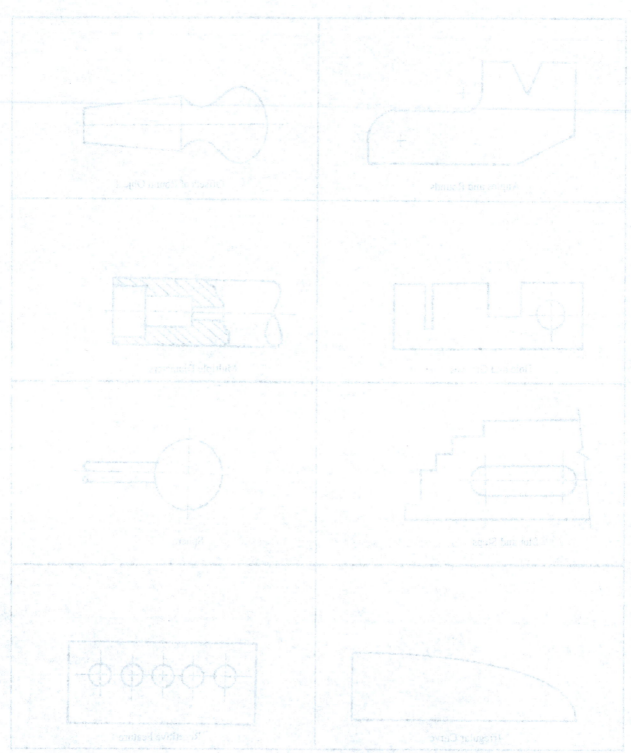

Problem 10-3 Dimensioning

Use the grid to determine all dimensions in decimal inches to two places. Redraw the views shown in the spaces indicated. Assume the grid has ½" spacing. Leave 1" actual spaces between views. Scale 1" = 3".

CHECKLIST FOR DIMENSIONING

Size dimensions
Location dimensions
Placement
Spacing
Extension lines
Dimension lines
Leaders
Arrowheads
Lettering
Figures

A

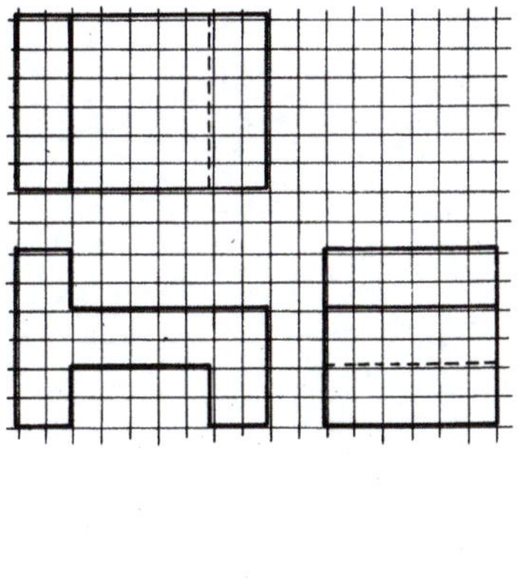

B

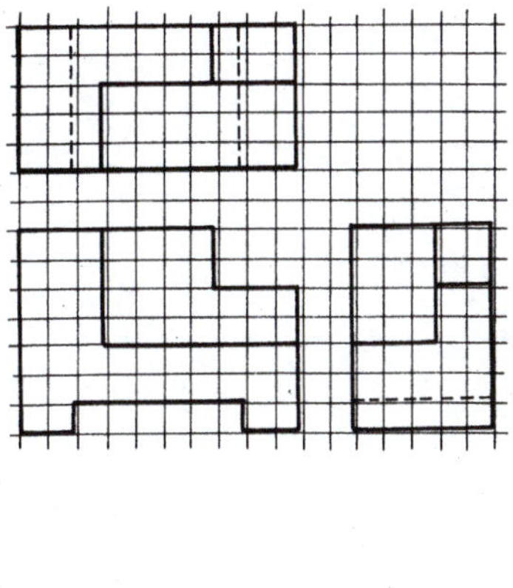

COURSE ________	STUDENT ______________________	DATE ________	PROBLEM 10-3

Problem 10-4 Dimensioning

Use the grid to determine all dimensions in decimal inches to two places. Redraw the views shown in the spaces indicated. Assume the grid has ½" spacing. Leave 1" actual spaces between views. Scale 1" = 2".

CHECKLIST FOR DIMENSIONING

Size dimensions
Location dimensions
Placement
Spacing
Extension lines
Dimension lines
Leaders
Arrowheads
Lettering
Figures

A

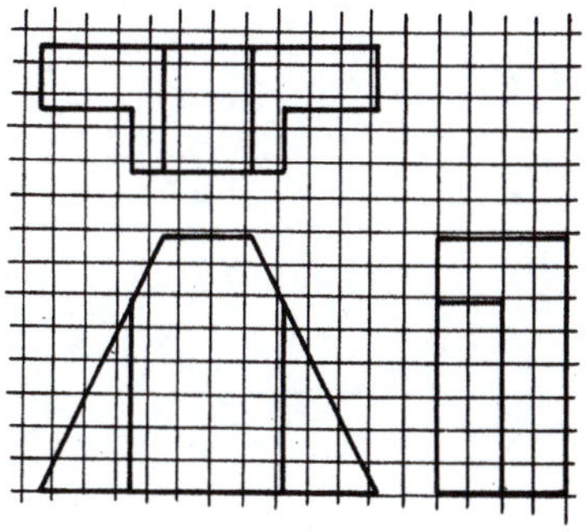

B

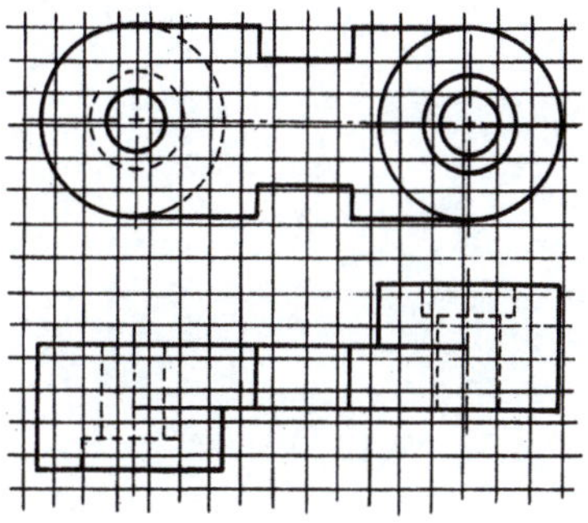

COURSE ________	STUDENT ________________	DATE ________	PROBLEM 10-4

Problem 10-5 Dimensioning

Use the grid to determine all dimensions in decimal inches to two places. Redraw the views shown in the spaces indicated. Assume the grid has ½" spacing. Leave 1" actual spaces between views. Scale 1" = 2".

CHECKLIST FOR DIMENSIONING

Size dimensions
Location dimensions
Placement
Spacing
Extension lines
Dimension lines
Leaders
Arrowheads
Lettering
Figures

A

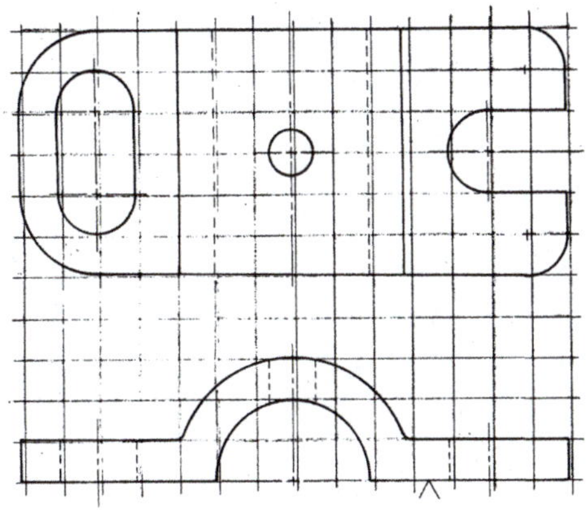

B

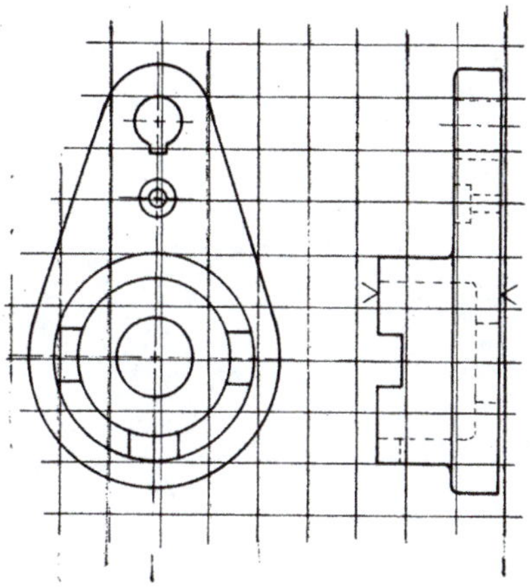

COURSE________	STUDENT____________________	DATE________	PROBLEM 10-5

Problem 10-6 Dimensioning

Use the grid to determine all dimensions in decimal inches to two places. Redraw the views shown in the spaces indicated. Assume the grid has ½" spacing. Leave 1" actual spaces between views. Scale 1" = 2".

CHECKLIST FOR DIMENSIONING

Size dimensions

Location dimensions

Placement

Spacing

Extension lines

Dimension lines

Leaders

Arrowheads

Lettering

Figures

A

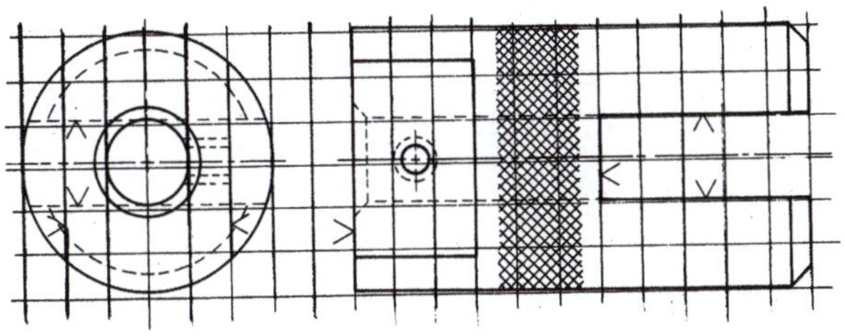

B

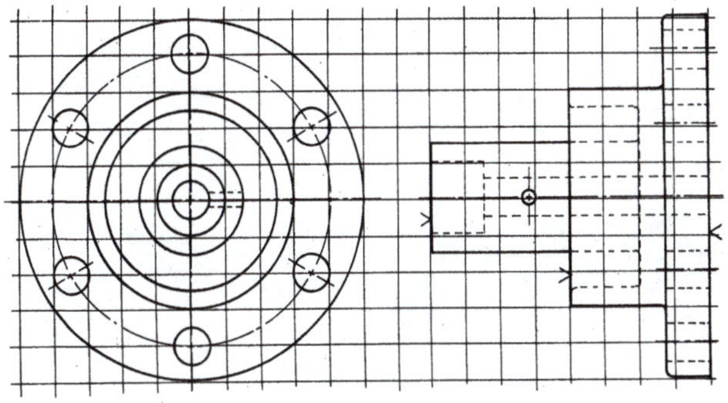

COURSE ________	STUDENT ______________________	DATE ________	PROBLEM 10-6

Problem 10-7 Dimensioning

CHECKLIST FOR DIMENSIONING

- Size dimensions
- Location dimensions
- Placement
- Spacing
- Extension lines
- Dimension lines
- Leaders
- Arrowheads
- Lettering
- Figures

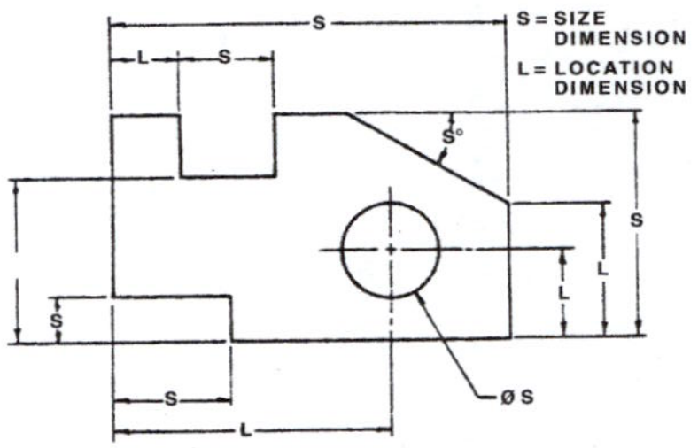

SIZE AND LOCATION DIMENSIONS

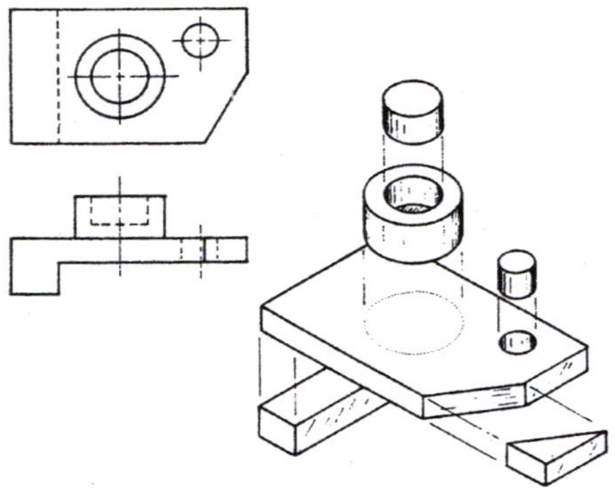

GEOMETRIC BREAKDOWN OF AN OBJECT

A Dimension the object shown using S = size and L = location dimensions. Use standard dimensioning practices.

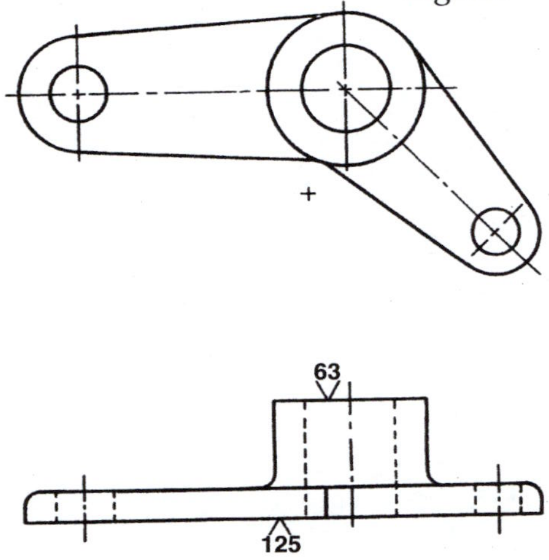

B Dimension the object shown using S = size and L = location dimensions. Use standard dimensioning practices.

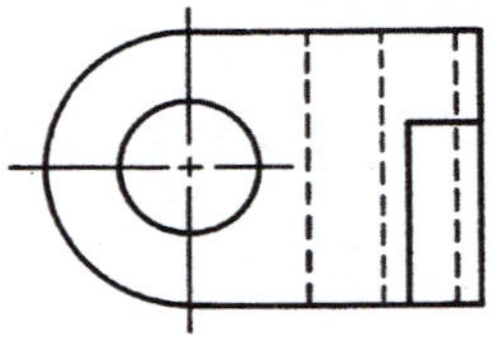

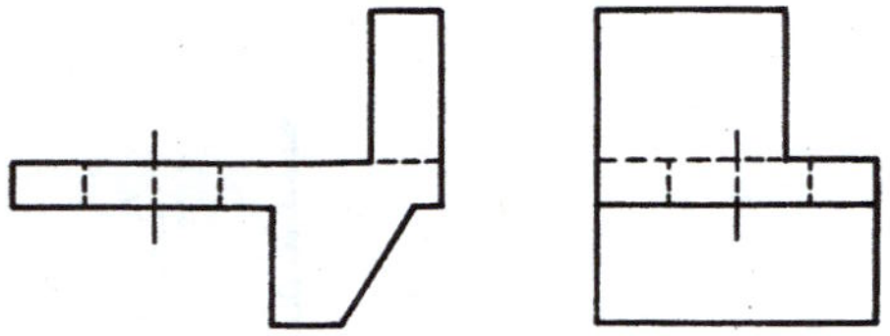

COURSE______ STUDENT______ DATE______ PROBLEM 10-7

Problem 10-8 Dimensioning

Place the following NOTES AND CALLOUTS in the appropriate locations on the drawing below:

1. ALL DIMENSIONS IN INCHES
2. ALL UNMARKED RADII, R.06
3. Φ.625/.620 THRU
4. Φ.44-20 UNF-2 INLINE WITH Φ.625/.620 HOLE
5. Φ.38 THRU
6. Φ.75 ↓ .25
7. R.62
8. Φ1.25

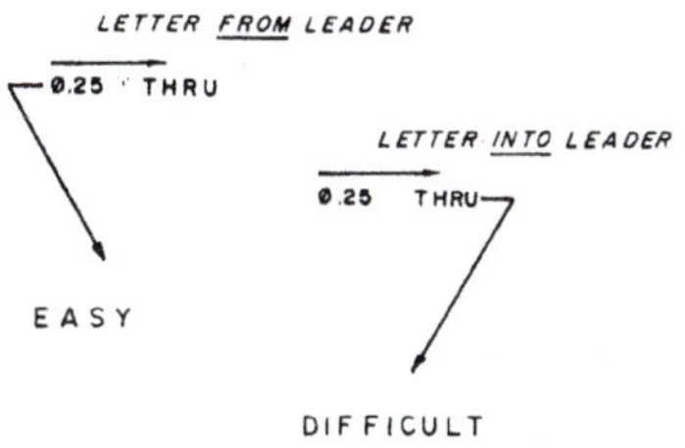

LETTERING AND LEADERS

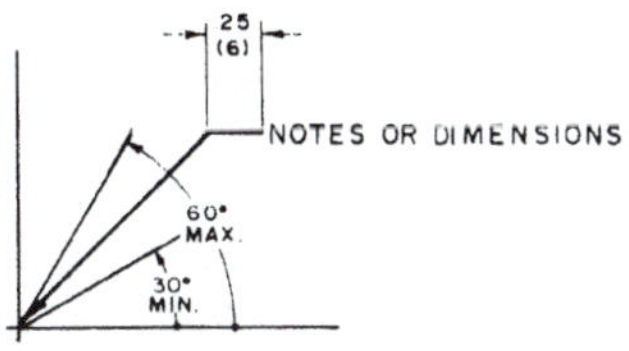

ANGLE OF A LEADERLINE

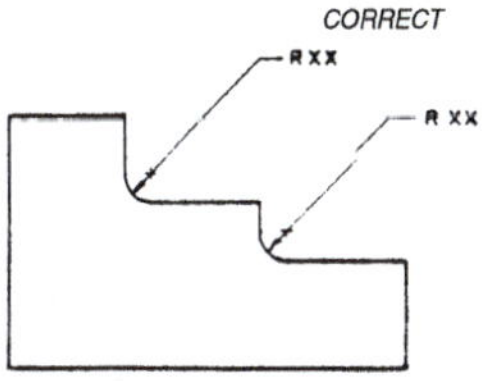

ONE DIMENSION PER LEADER LINE IS PREFERRED

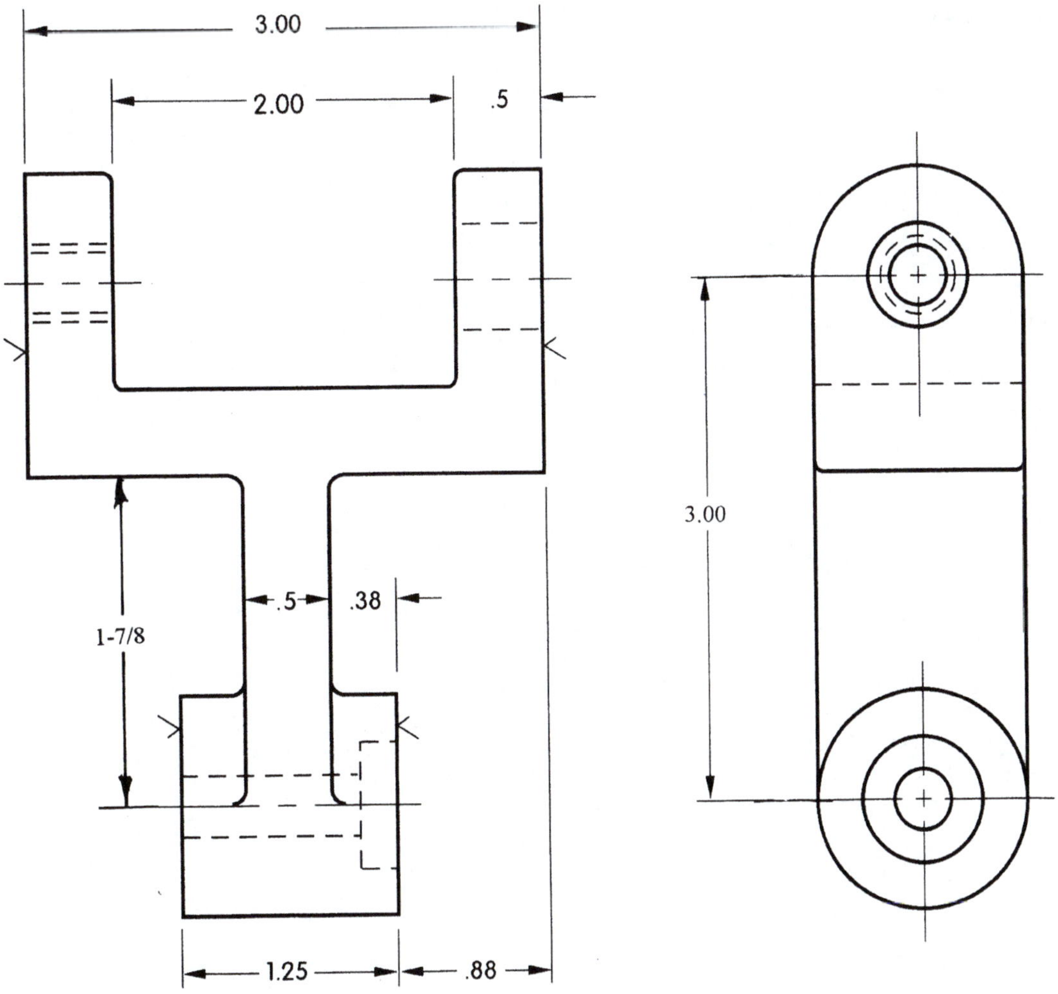

COURSE ______	STUDENT ______	DATE ______	PROBLEM 10-8

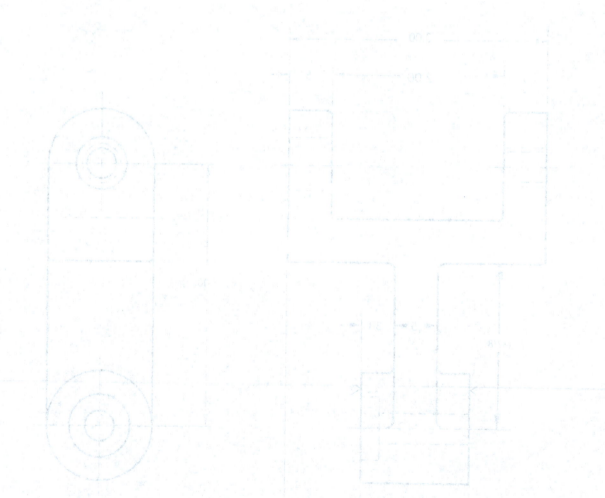

Problem 10-9 Dimensioning

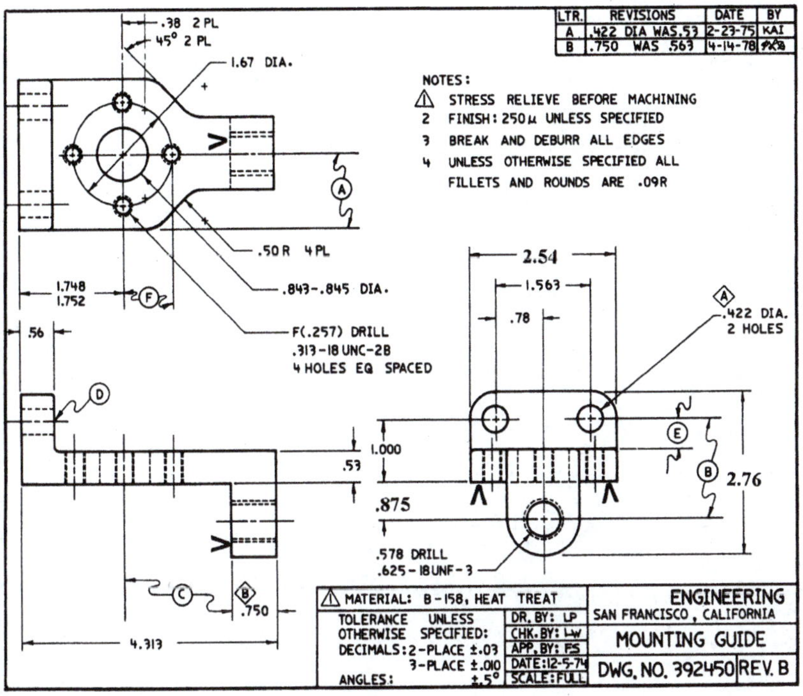

Study the drawing of the MOUNTING GUIDE and answer the following questions:

1. How much would a MOUNTING GUIDE weigh if it were made of
   a. steel at 0.28 lb/cu in.?
   b. aluminum at 0.10 lb/cu in.?
   c. brass at 0.30 lb/cu in.?

   Show your calculations.
2. What is the overall height of the guide?
3. What is dimension A?
4. What tolerance is expected for the two-place dimensions?
5. What tolerance is expected for the three-place dimensions?
6. What is the maximum dimension of F?
7. What tolerance do angles have?
8. What does the V mark mean? What is the current practice for these marks?

ANSWERS

1. _____
   a. _____
   b. _____
   c. _____
2. _____
3. _____
4. _____
5. _____
6. _____
7. _____
8. _____

COURSE________ STUDENT________________ DATE________ PROBLEM 10-9

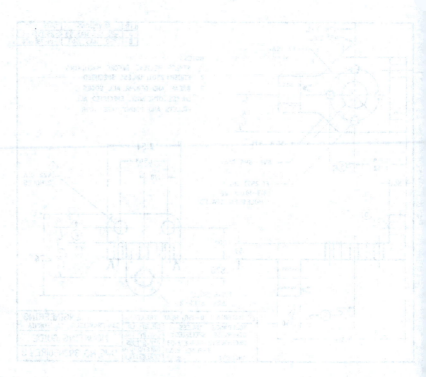

Problem 10-10 Dimensioning

A. Dimension the object shown using standard practices. Assume full scale in inches. Use accuracy to two places except for the large hole. Measure to centers of lines. The large hole is to have limit dimensions for an RC4 fit, HOLE BASIS. Show your calculations.

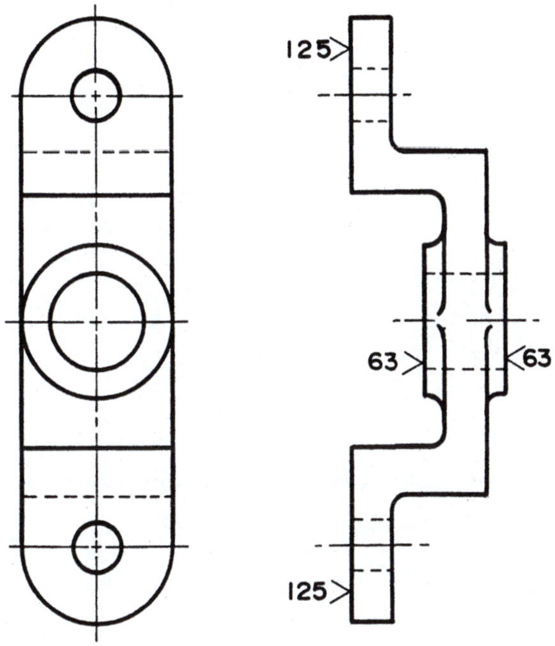

B. Assume that a ball bearing with an outside diameter of .5624 ±.0001" is to be installed in the large hole. Use the SHAFT BASIS method to determine limit dimensions for an FN2 fit. Show your calculations.

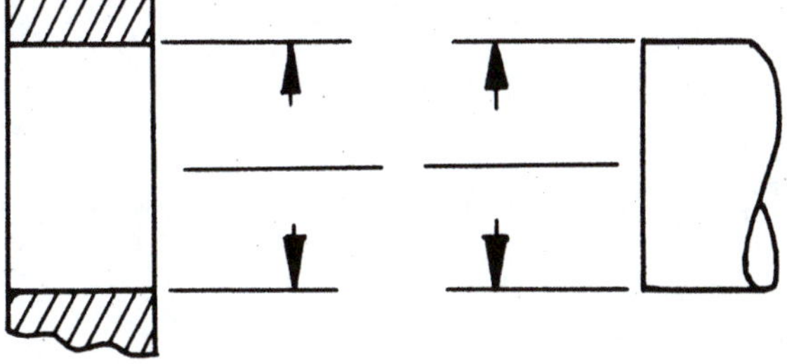

COURSE ________	STUDENT ____________________	DATE ________	PROBLEM 10-10

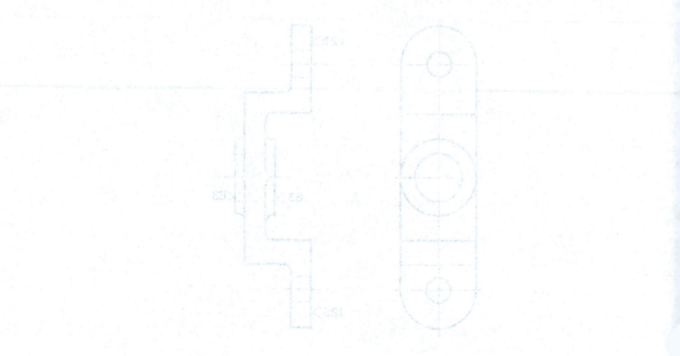

Problem 10-11 Dimensioning

LIMIT DIMENSIONING AND FITS FOR MATING COMPONENTS

EXAMPLE SHAFT BASIS

COMMENT: One typical application for the SHAFT BASIS for establishing limit dimensions occurs in bearing installations.

GIVEN: An off-the-shelf ball bearing as shown.

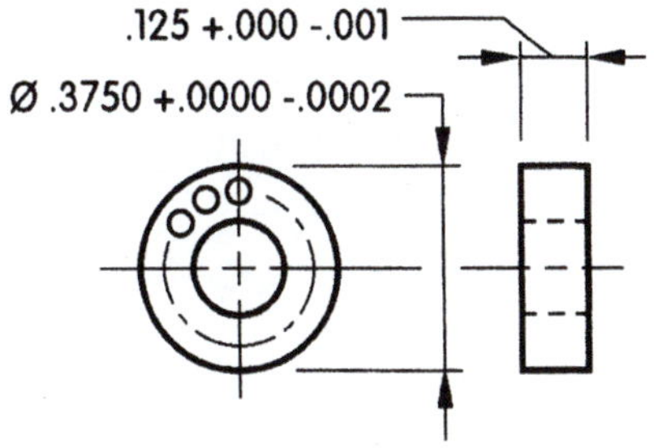

Next check ALLOWANCE and CLEARANCE using equations.

ALLOWANCE = .3736 − .3750 = −.0014 or 1.4*interference
CLEARANCE = .3744 − .3748 = −.0004 or .4* interference

* = thousandths of an inch

TO FIND: Limit dimensions for the two holes in the linkage shown for an FN2 interference fit for the bearings, SHAFT BASIS. Note that interference values, not clearance values, are listed in the FN2 table.

Use ALLOWANCE and CLEARANCE equations to determine the limit dimensions.

ALLOWANCE:
tightest fit (most interference) = smallest hole − largest shaft
−.0014 = smallest hole − .3750
smallest hole = .3736

CLEARANCE:
loosest fit (least interference) = largest hole − smallest shaft
−.0004 = largest hole −.3748
Largest hole = .3744

Minus signs indicate interference.

Note: The dimension the machinist comes to first is on top.

PROBLEM

Assume that the bearings have an outside diameter of .6249 ±.0001, determine limit dimensions for the linkage holes, using the SHAFT BASIS, and an FN2 fit. Show your calculations.

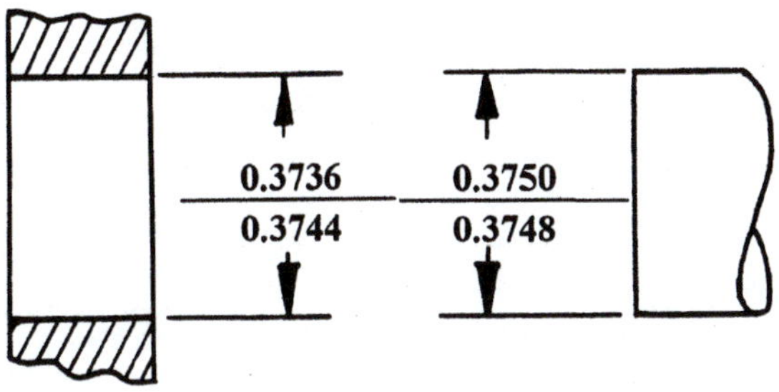

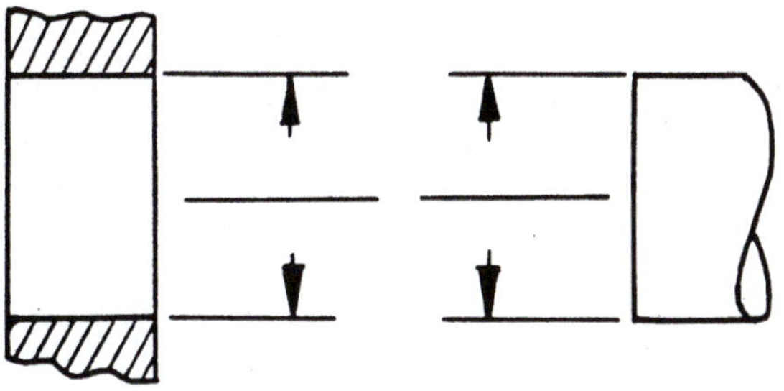

COURSE________ STUDENT________________________ DATE________ PROBLEM 10-11

Problem 10-12 Tolerancing

A. Dimension the object shown using standard practices and callouts. Decimal millimeters. Accuracy, one place. Measure to centers of lines. Scale 1 mm = 2 mm.

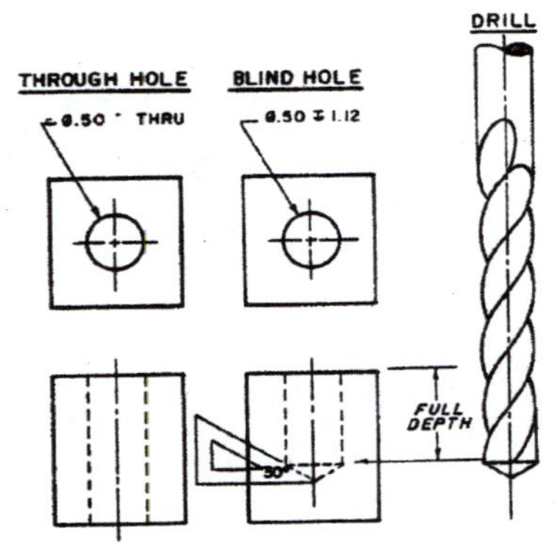

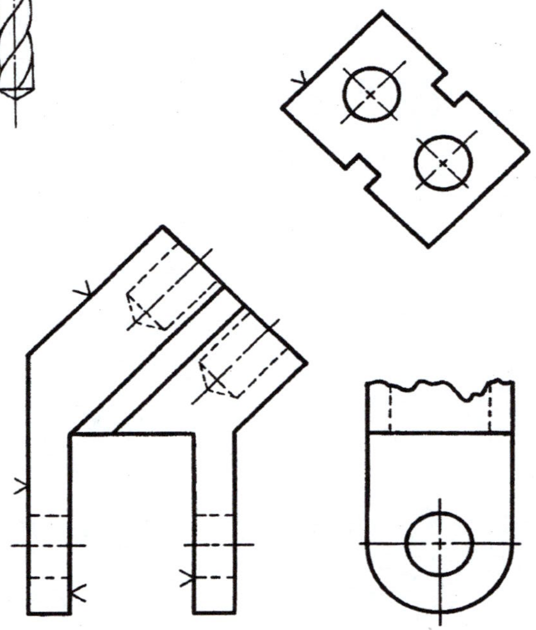

CHECKLIST FOR DIMENSIONING

Size dimensions
Location dimensions
Placement
Spacing
Extension lines
Dimension lines
Leaders
Arrowheads
Lettering
Figures

B. Dimension the object shown using standard practices. Decimal inches. Accuracy to two places. Measure to centers of lines. Scale 1" = 4"

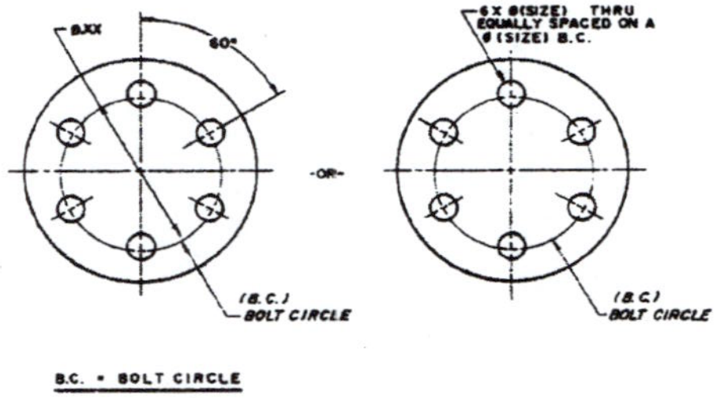

Dimensioning holes on a bolt center diameter

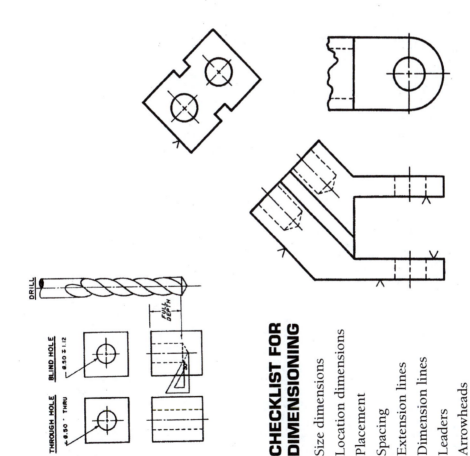

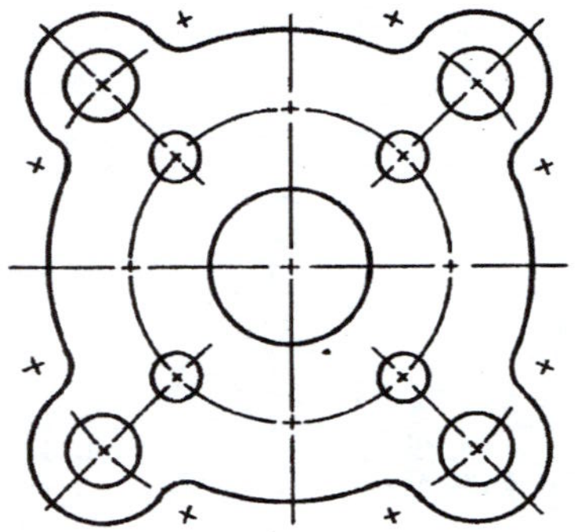

COURSE ________ STUDENT ________________ DATE ________ PROBLEM 10-12

Problem 10-13 Tolerancing

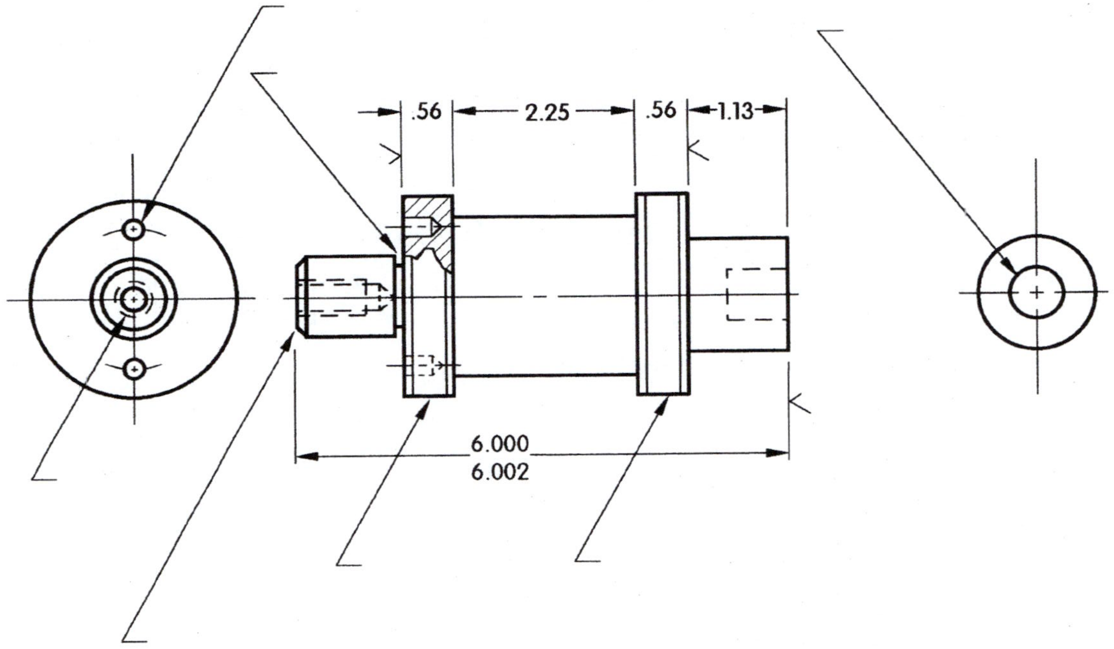

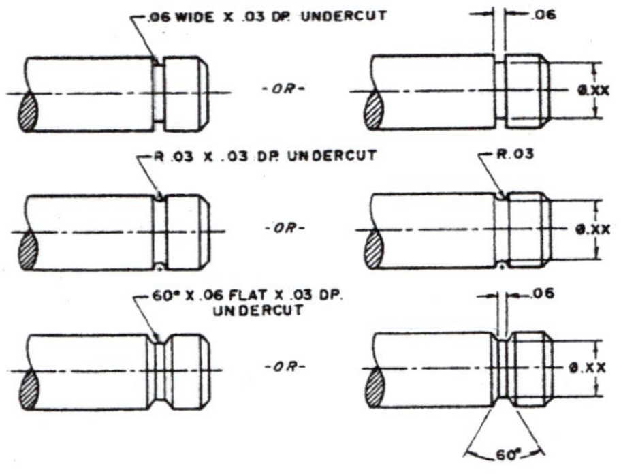

Dimensioning necks and undercuts

Place the following NOTES AND CALLOUTS in the appropriate locations in the following drawing:

1. ALL DIMENSIONS IN INCHES
2. KNURL 64 DP
3. Φ.38-24 UNF-2
4. .06 × 45° CHAMFER
5. UNDERCUT .09 × .06 DEEP
6. 2 × .252/.250 ↓ .38
7. Φ.62 ↓ .75

Note: Some dimensions of the object have been omitted for this problem.

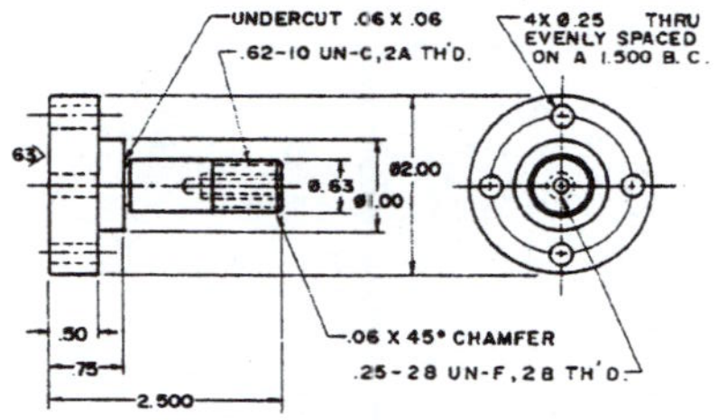

Notes are placed horizontally

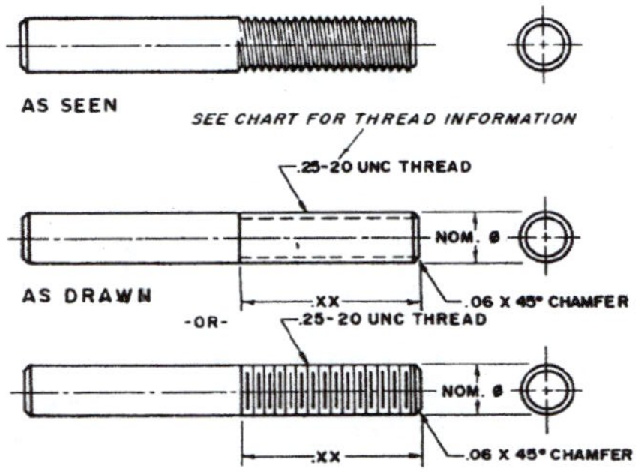

Dimensioning threads

COURSE ________	STUDENT ________	DATE ________	PROBLEM 10-13

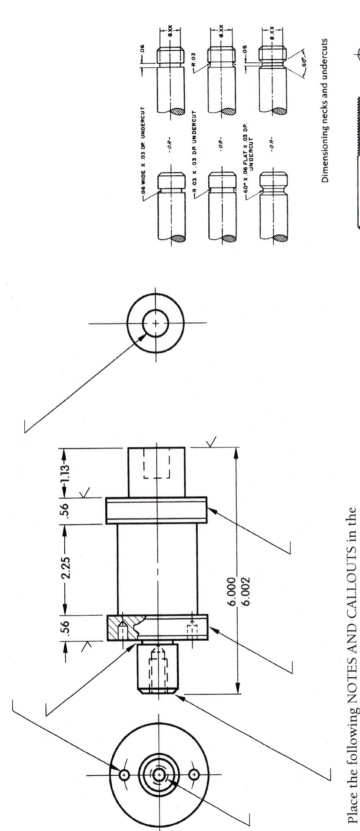

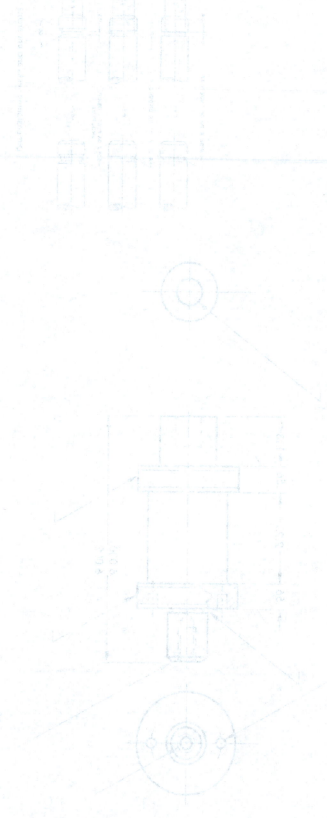

Problem 10-14 Tolerancing

LIMIT DIMENSIONING AND FITS FOR MATING COMPONENTS

EXAMPLE:

GIVEN: American National Standard Running and Sliding Fits
ANSI B4.1-1967 (R1994) (courtesy ASME)
Excerpt from RC3 shown below.

TO FIND: Limit dimensions for Φ½" shaft in Φ½" hole.

Nominal Size Range Inches	Class RC3		
	Clearance	Standard Tolerance Limits	
		Hole H7	Shaft f6
	Values shown are in thousandths of an inch		
0.40–0.71	0.6 1.7	+0.7 0	-0.6 -1.0

PROBLEM

Determine limit dimensions for Φ ⅝" hole/shaft, RC3. Show your calculations.

HOLE BASIS: (Note: For the HOLE BASIS, one limit dimension of the hole is equal to the decimal equivalent of the nominal size—e.g., Φ½" = 0.5000.)

0.5000	0.05000	0.5000	0.5000
+0.0007	0.0000	−0.0006	−0.0010
0.5007	0.5000	0.4994	0.4990

Note: The dimension the machinist comes to first is on top.

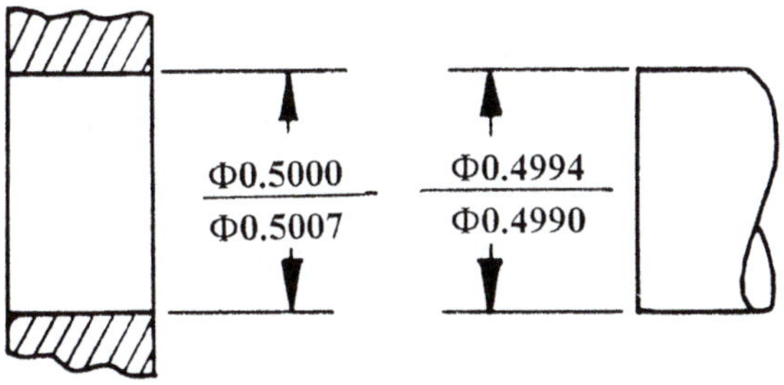

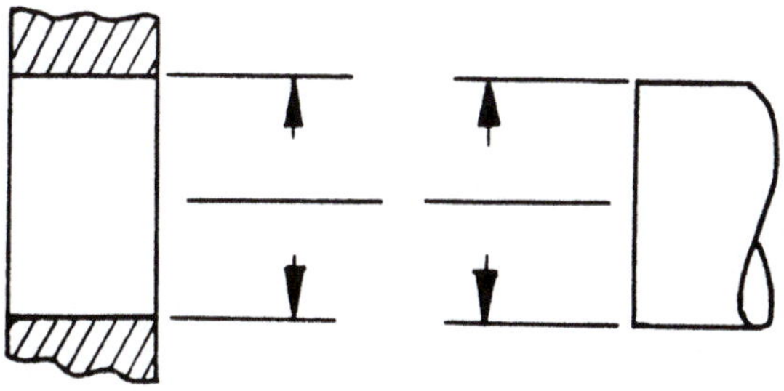

Next check ALLOWANCE and CLEARANCE using equations.

ALLOWANCE = tightest fit = smallest hole − largest shaft
= 0.5000 − 0.4994 = 0.0006 or .6

CLEARANCE = loosest fit = largest hole − smallest shaft
= 0.5007 − 0.4990 = 0.0017 or 1.7

COURSE ________	STUDENT ______________________________	DATE ________	PROBLEM 10-14

roblem 10-15 Tolerancing

LIMIT DIMENSIONING AND FITS FOR MATING COMPONENTS

EXAMPLE SHIFT BASIS

GIVEN: A shaft Φ¾" nominal, .7500 ± .0002.

TO FIND: Limit dimensions for class RC2 fit.

Use ALLOWANCE and CLEARANCE equations to determine the limit dimensions.

PROBLEM

Determine limit dimensions for Φ ⅝" .6250 ± .0002, SHAFT BASIS, RC2 fit. Show your calculations.

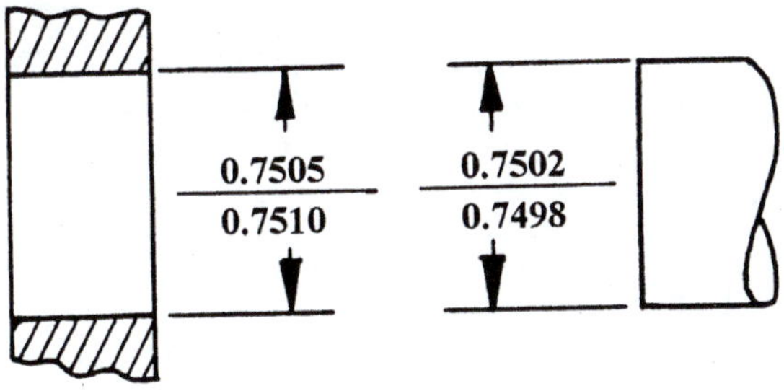

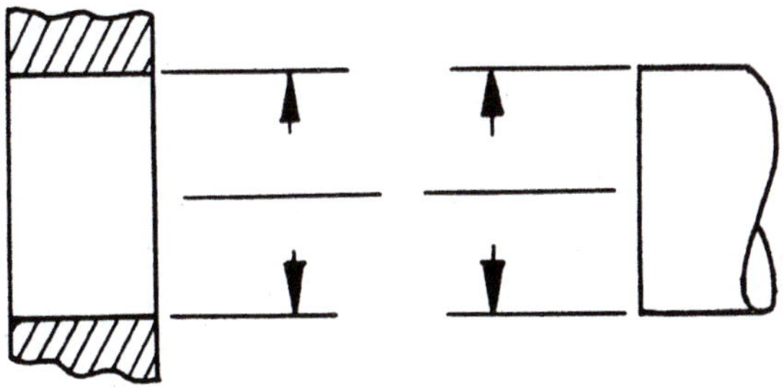

Next check ALLOWANCE and CLEARANCE using equations.

ALLOWANCE = .7505 − .7502 = .0003 or .3*

CLEARANCE = .7510 − .7498 = .0012 or 1.2*

* = thousandths of an inch

COURSE______	STUDENT______________	DATE______	PROBLEM 10-15

CHAPTER 11

Fasteners and Springs

Problem 11-1 Spring and Fasteners

Name the following items. Letter your answers in the guide lines provided below.

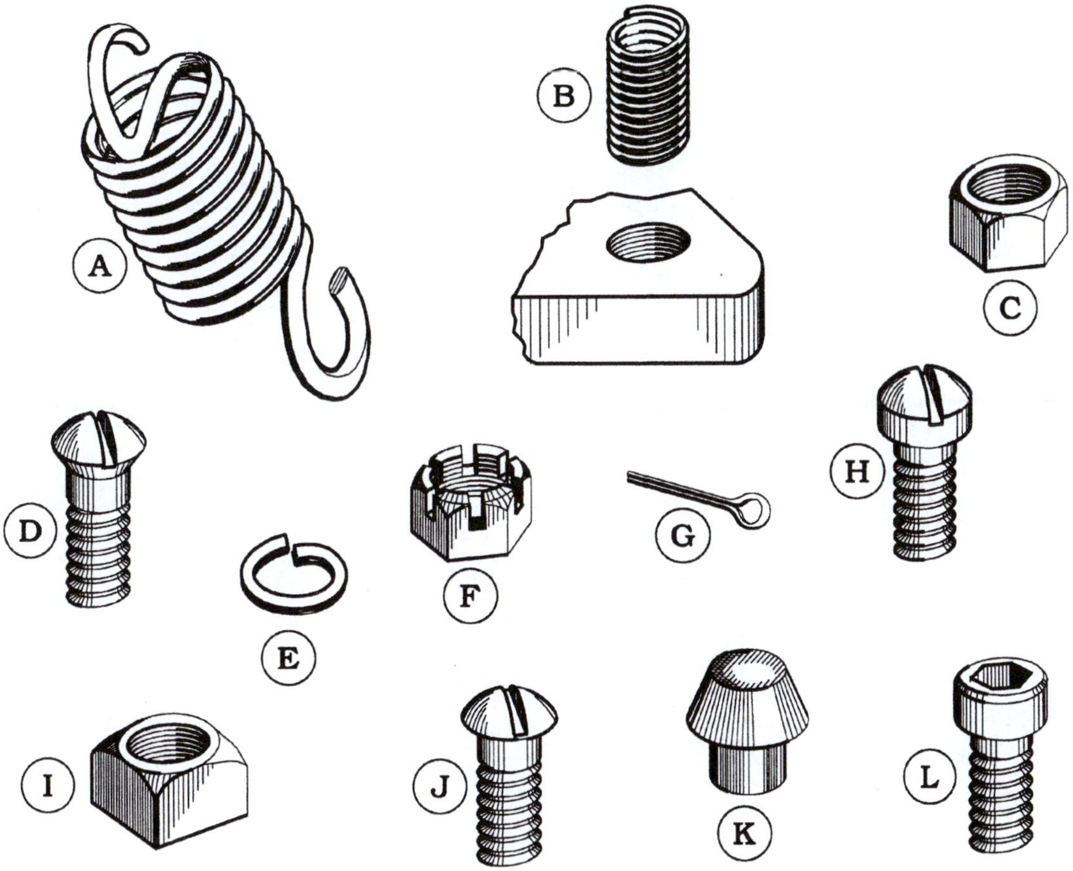

A. ______________________________
B. ______________________________
C. ______________________________
D. ______________________________
E. ______________________________
F. ______________________________
G. ______________________________
H. ______________________________
I. ______________________________
J. ______________________________
K. ______________________________
L. ______________________________

COURSE ________ STUDENT ______________________ DATE ________ PROBLEM 11-1

Problem 11-2 Fasteners

On the following centerlines, draw two views (end view and side view) of the fasteners:

A

Flat head machine screw #10–24 UNC–2A × 1.25 long

B

Round head cap screw .31–24 UNF–2A × 2.00 long

C

Oval head cap screw .75–10 UNC–2A × 1.75 long

D

Socket head cap screw (Allen head) .50–28 UNF–2A × 3.00 long

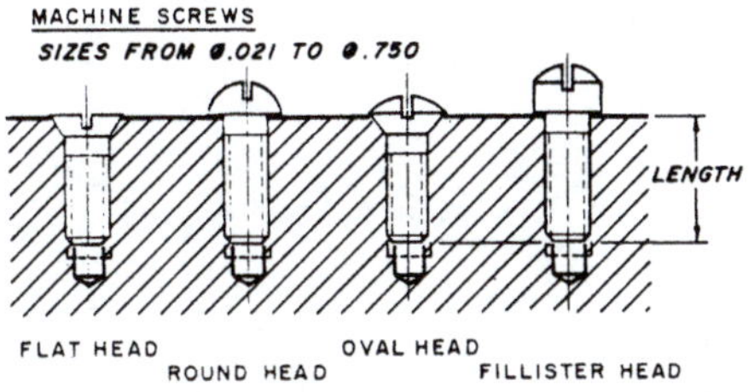

Machine screws

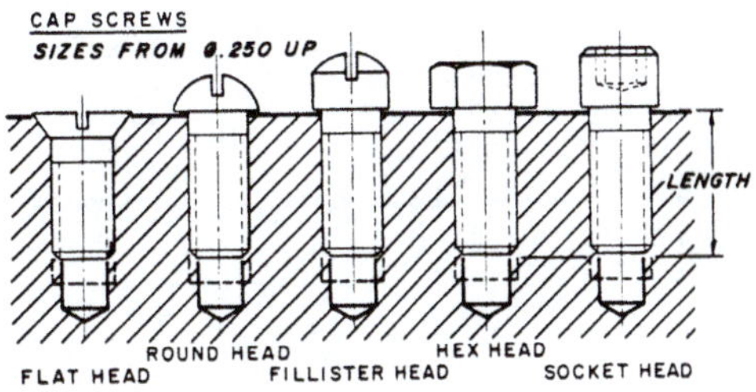

Cap screws

COURSE ________	STUDENT ______________________	DATE ________	PROBLEM 11-2

Problem 11-3 Complete the Views

Complete the three views. Dimension them and use standard callouts for the threads and chamfers.

A. Thread at hole A: ϕ1.00 coarse thread, average fit, 1.00" deep.

B. Thread at B: ϕ1.5 fine, from left end to shoulder, average fit. Add a thread relief next to the shoulder.

C. Thread at C: ϕ2.00 coarse thread, average fit. Chamfer at right end to depth of thread.

D. Thread at hold D: ϕ.50 fine thread, average fit, 1.25" deep, full thread. Add a counterbore to accommodate a ϕ.50 fillister head cap screw, 1.0" long (head to be flush with right-end surface).

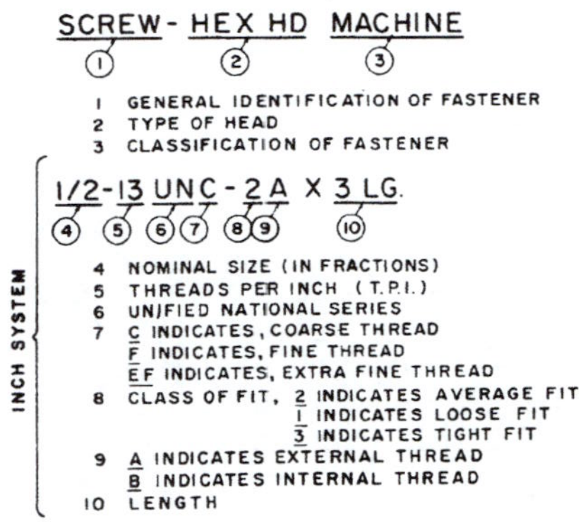

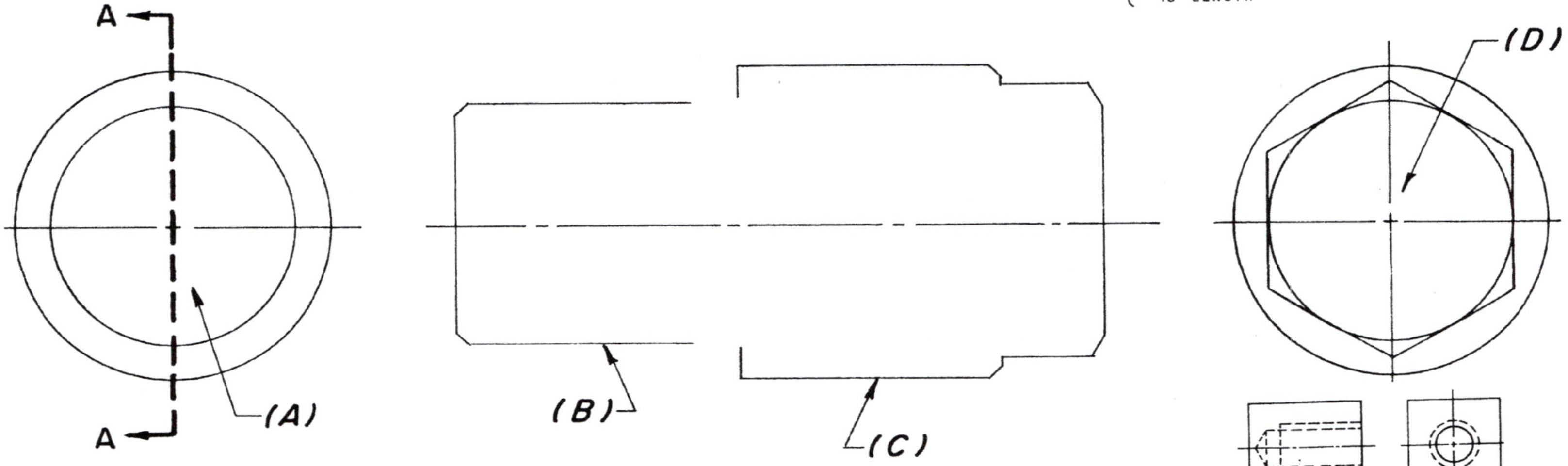

SECTION A-A

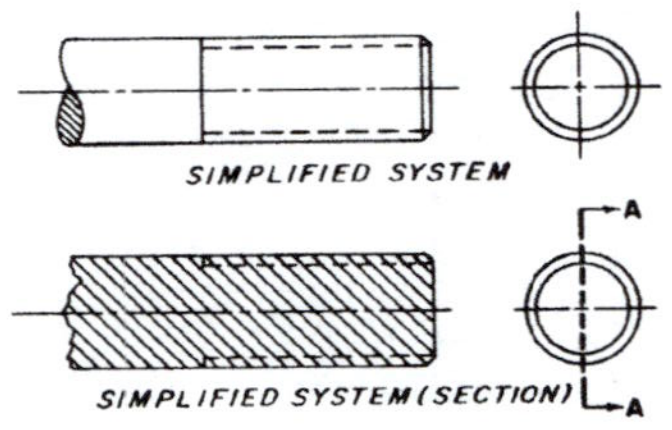

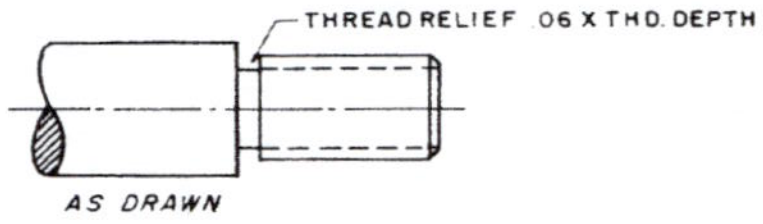

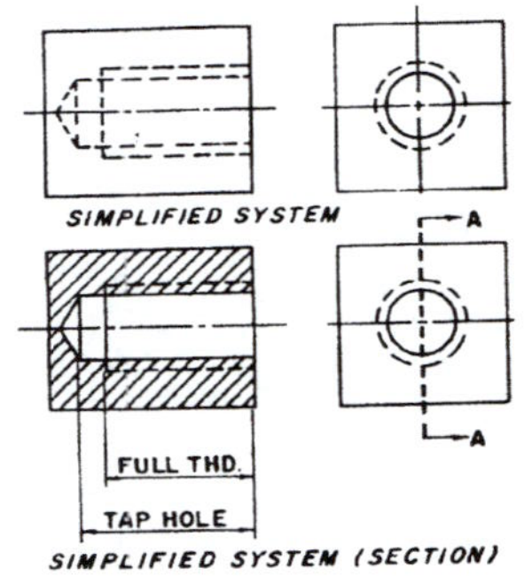

roblem 11-4 Machine Screws

Give the type of head of each of the following machine screws. Also give the length and thread notes of each screw. Refer to the tables to find additional information.

A

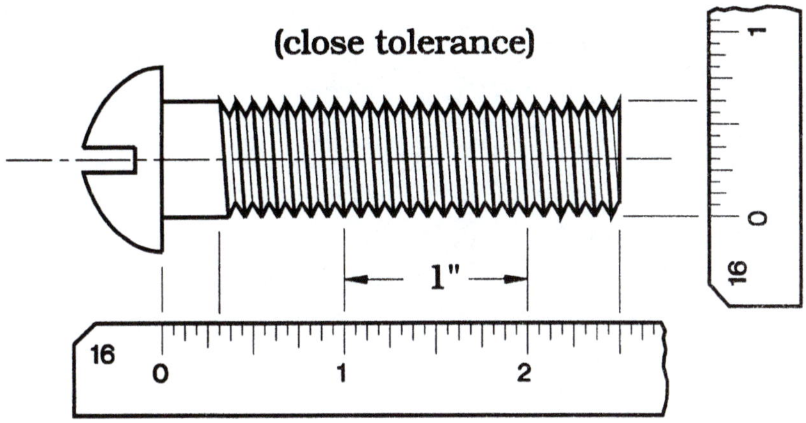

B

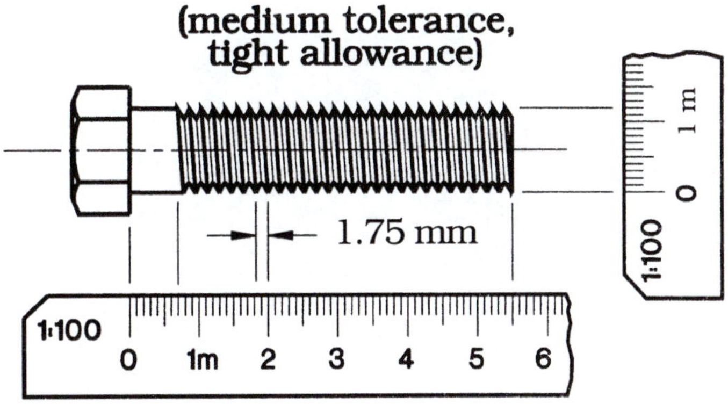

C

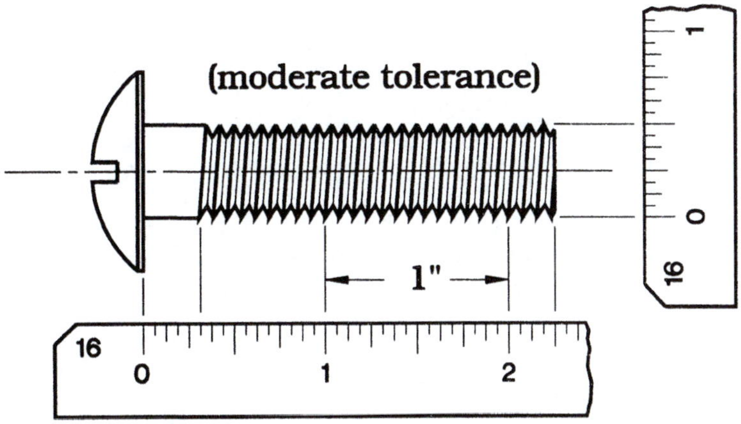

COURSE ______ STUDENT ______ DATE ______ PROBLEM 11-4

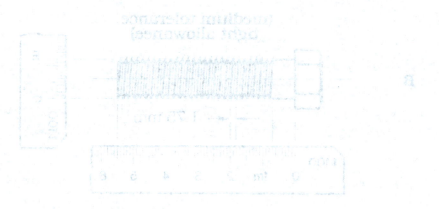

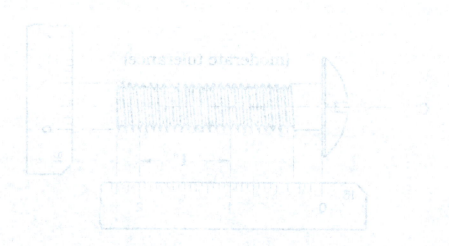

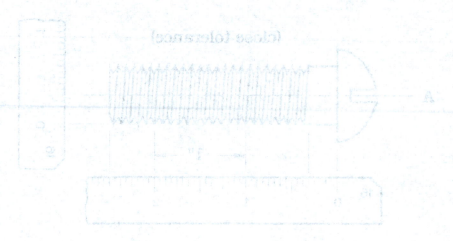

Problem 11-5 Machine Screws

Give the type of head of each of the following machine screws. Also give the length and thread notes of each screw. Refer to the tables to find additional information.

A

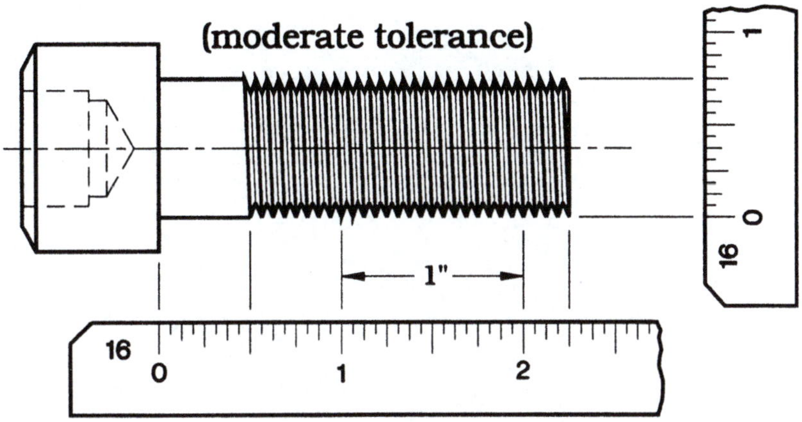

B

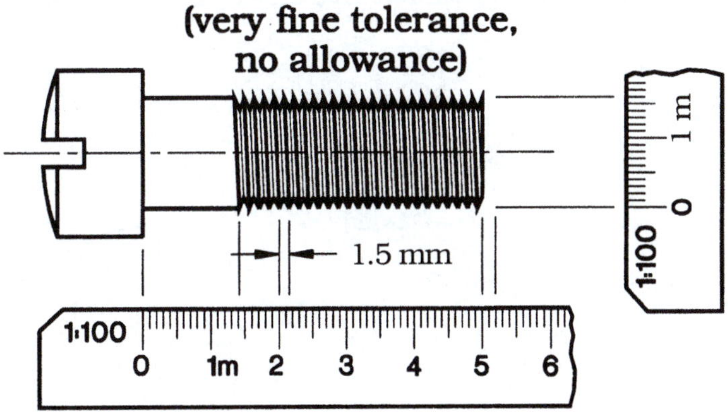

C

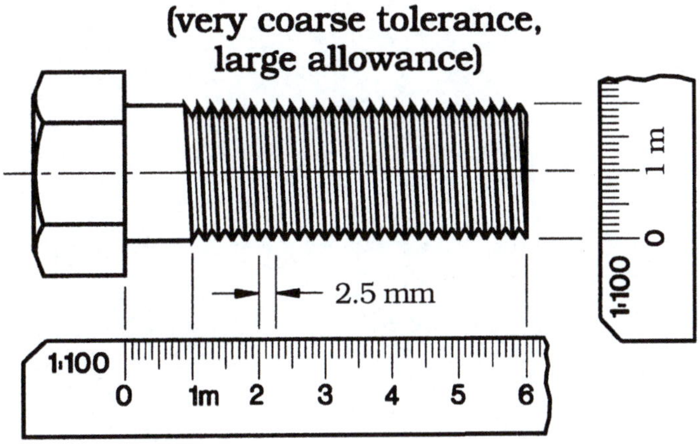

COURSE ______ STUDENT ______ DATE ______ PROBLEM 11-5

Problem 11-6 Thread Notes

Give the thread notes for each of the following drawings. Refer to the tables to find additional information.

A

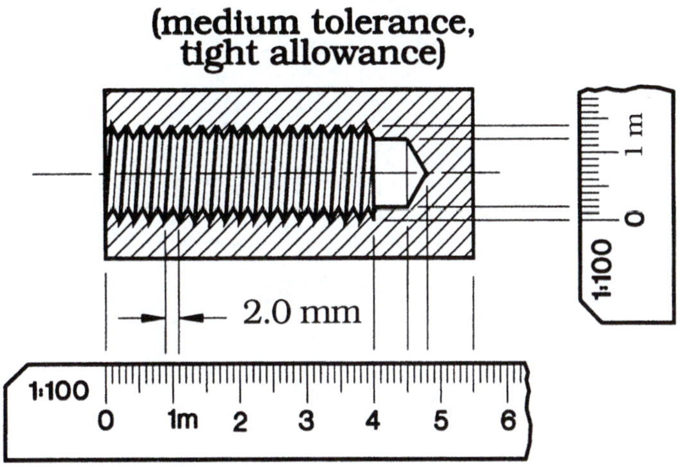

____

B

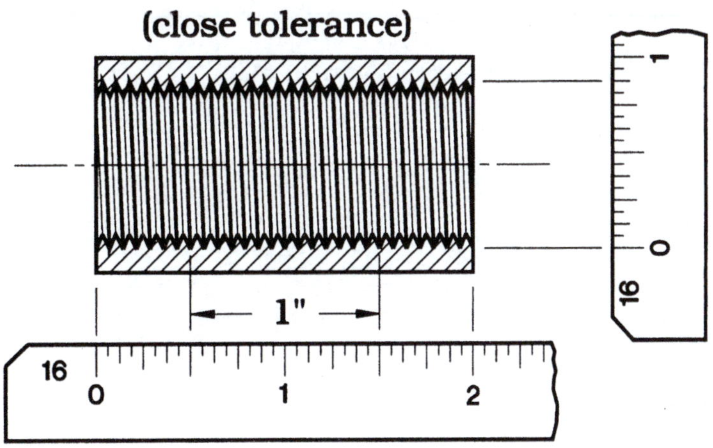

____

C

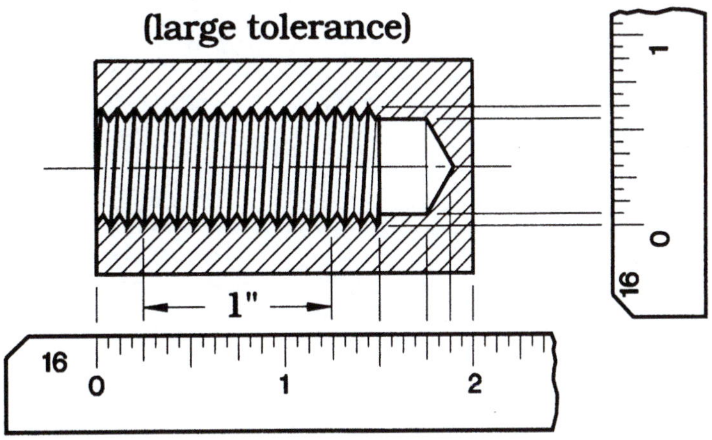

____

COURSE ________	STUDENT ________________	DATE ________	PROBLEM 11-6

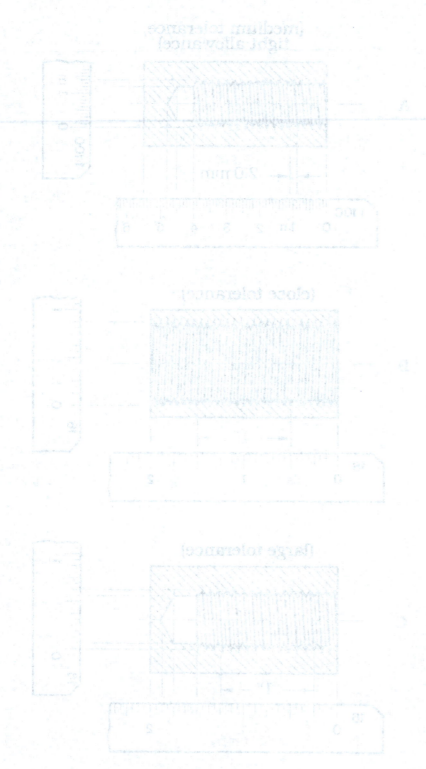

Problem 11-7 Springs

Follow instructions below.

A. Label the terminology of the spring below.

What type of spring is this?

____

____

____

____

1 ______________________
2 ______________________
3 ______________________
4 ______________________
5 ______________________
6 ______________________

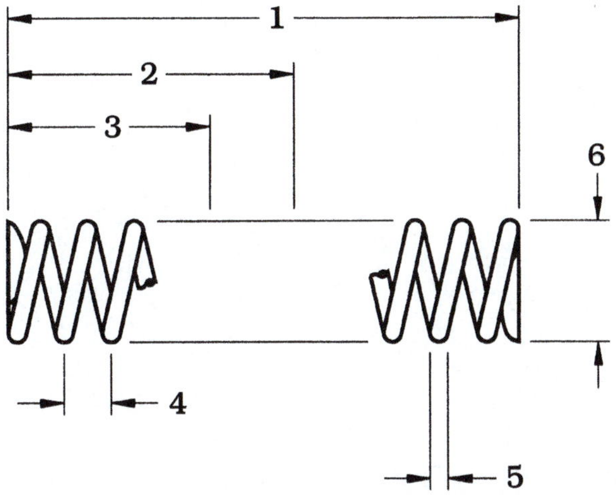

B. At half scale, sketch the front view and section view of a 3" long hex bolt. The bolt is .75" in diameter and has 16 threads per inch. The bolt has a left-hand thread, and the thread designation is Unified National Fine. The drill depth = 3.74", the drill diameter = 0.6329", and the thread (or tap) depth = 3.435". Make a simplified representation.

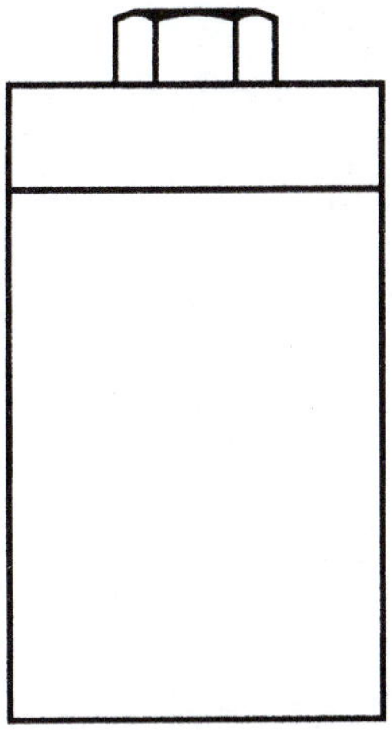

Front View

Section View

C. What will be the designation for this bolt? (Example: .50-13 UNC-2A)

____

____

What will be the designation for the internal screw threads? (Include both drill depth and thread depth in your answer.)

____

____

____

____

____

____

____

COURSE ________	STUDENT ____________________	DATE ________	PROBLEM 11-7

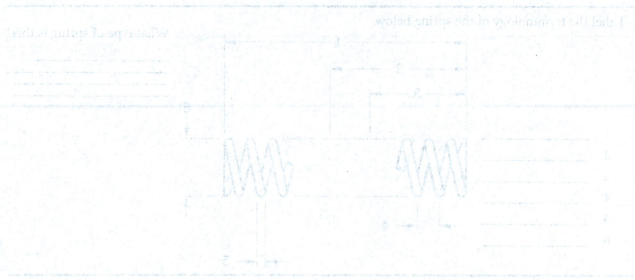

Problem 11-8 Springs

Draw the specified spring, which is to be used with an automobile valve mechanism, twice the size with the following specifications:

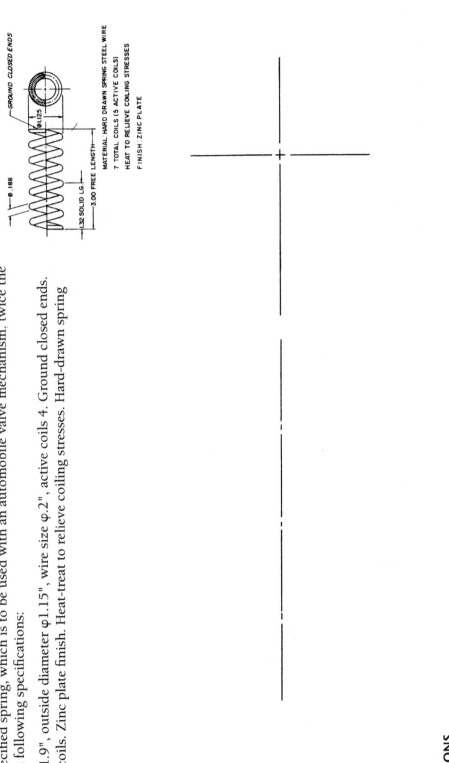

Free length 1.9", outside diameter φ1.15", wire size φ.2", active coils 4. Ground closed ends. Right-hand coils. Zinc plate finish. Heat-treat to relieve coiling stresses. Hard-drawn spring wire.

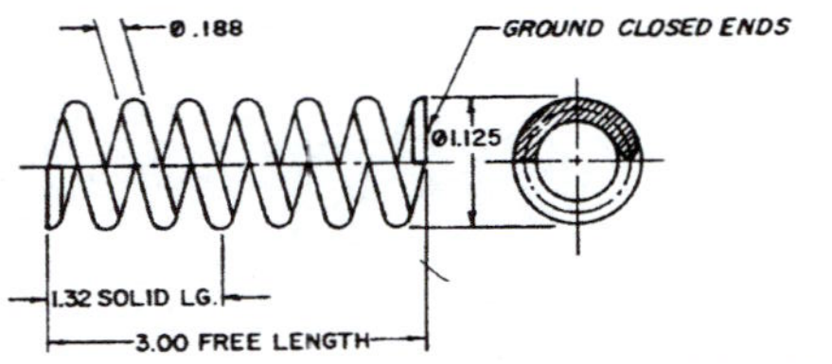

SPECIFICATIONS

____
____
____
____
____
____

COURSE ________ STUDENT ______________________ DATE ________ PROBLEM 11-8

Problem 11-9 Springs

Draw the specified spring, which is to be used with a child's pogo stick, half size with the following specifications:

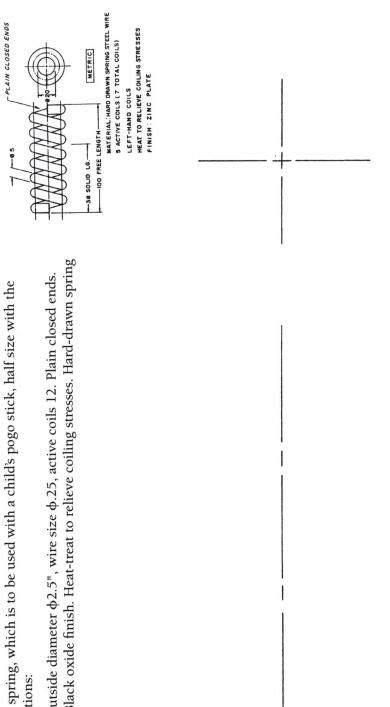

Free length 8.5", outside diameter ϕ2.5", wire size ϕ.25, active coils 12. Plain closed ends. Right-hand coils. Black oxide finish. Heat-treat to relieve coiling stresses. Hard-drawn spring wire.

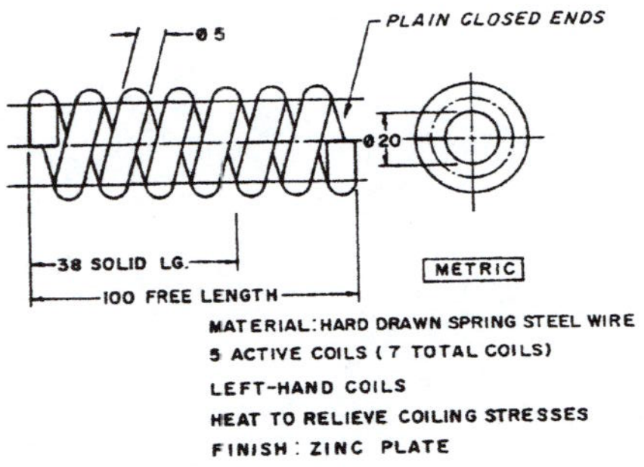

SPECIFICATIONS

COURSE______ STUDENT______ DATE______ PROBLEM 11-9

Problem 11-10 Springs

A. Draw a spring that would be used to return a pump handle to its original state, which is up. Follow the specifications given:

1. Free length = 7 1/2"
2. OD = 3 1/4"
3. Wire size = 1/4"
4. Active coils = 10
5. Plain closed end
6. Right-hand coils
7. Zinc plated
8. Heat to relieve cooling stresses

COURSE ________	STUDENT ______________________	DATE ________	PROBLEM 11-10

3. As an engineering technician, you are asked to design or select the springs to be used in the construction of a swimming pool gate. The gate should have two hooks connected to it that allow the gate too close automatically so children cannot enter through an open door. Design and or select an appropriate spring for this gate. You may want to consider visiting a local hardware store for some ideas. In the following space, show your work.

COURSE________ STUDENT________________________________ DATE________ PROBLEM 11-10

CHAPTER 12

Sections, Revolutions, and Conventional Breaks

Problem 12-1 Section Identification

Identify each type of section and describe the dominant characteristic of each.

A

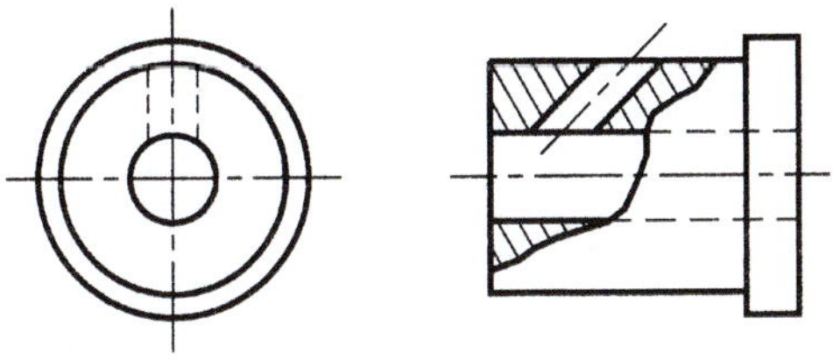

B

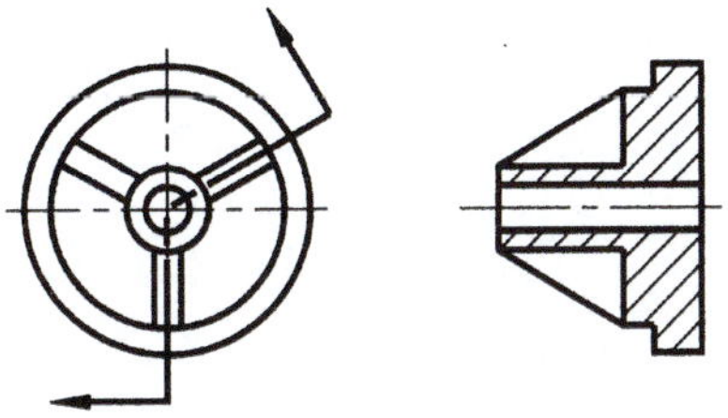

C

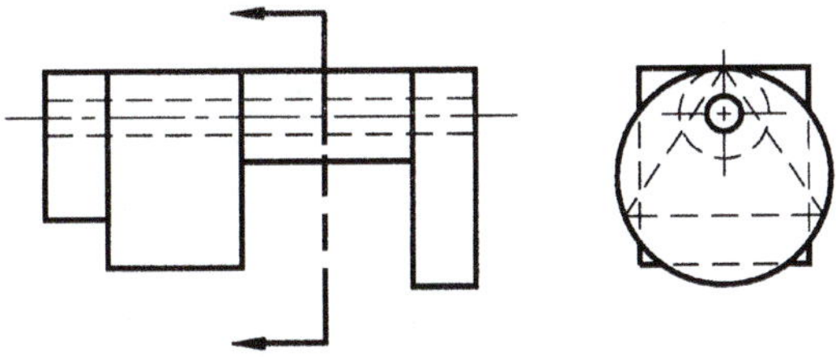

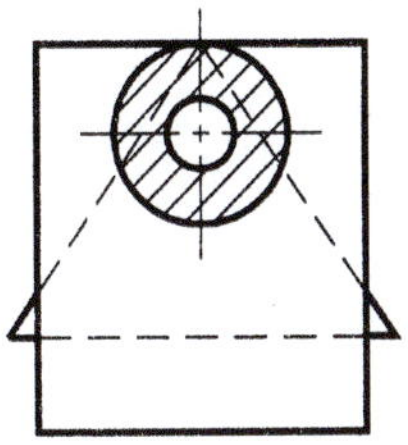

D

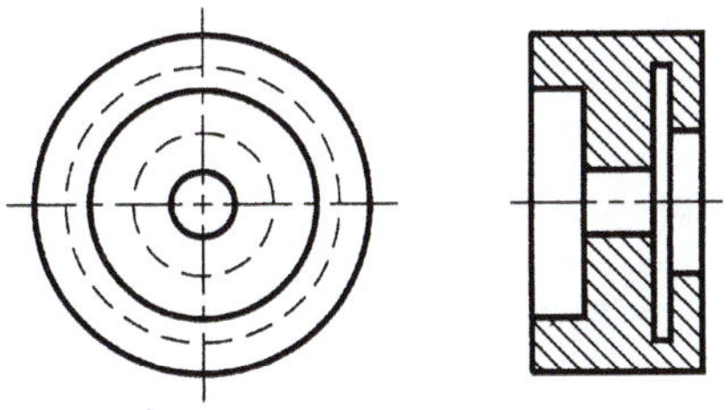

E

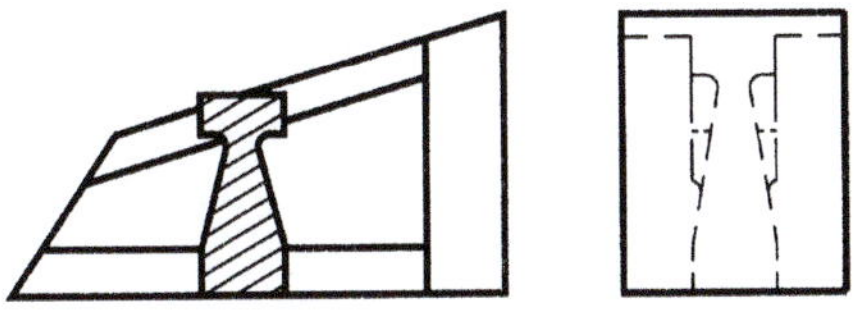

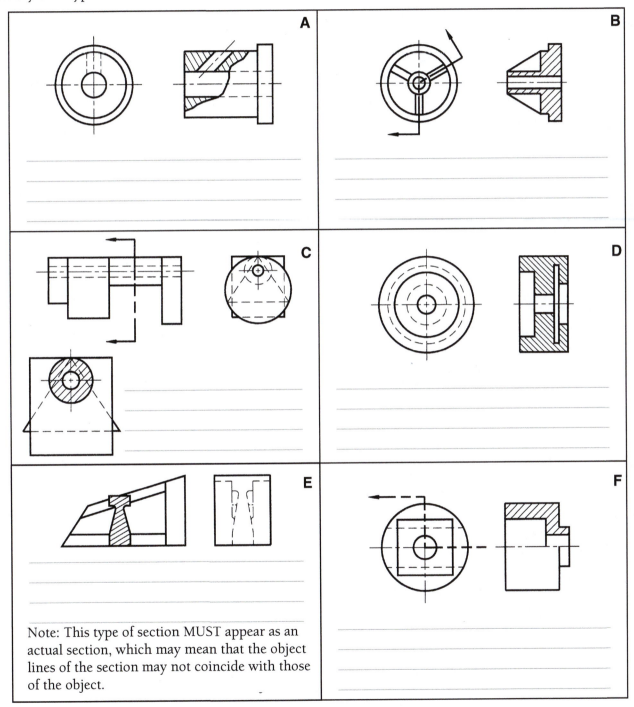

Note: This type of section MUST appear as an actual section, which may mean that the object lines of the section may not coincide with those of the object.

F

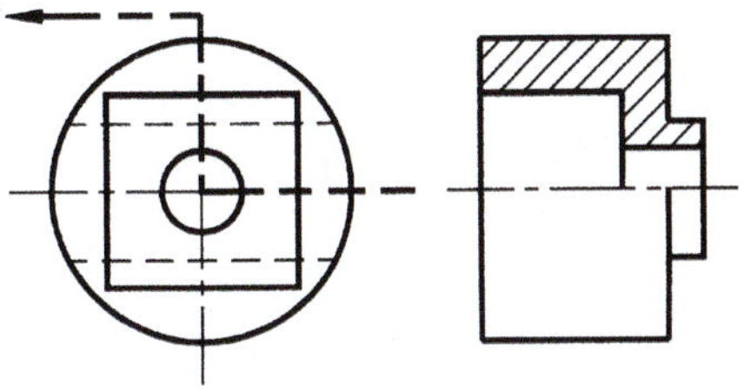

COURSE________ STUDENT______________________________ DATE________ PROBLEM 12-1

Problem 12-2 Section Identification

Follow the instructions below.

Circle the correct representation of these various sections and views.

A

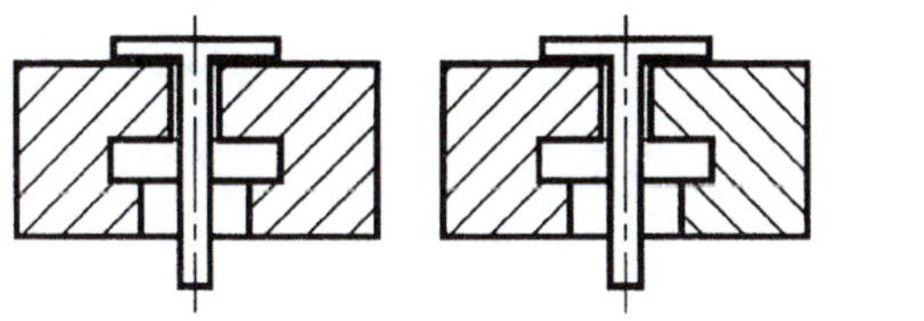

B

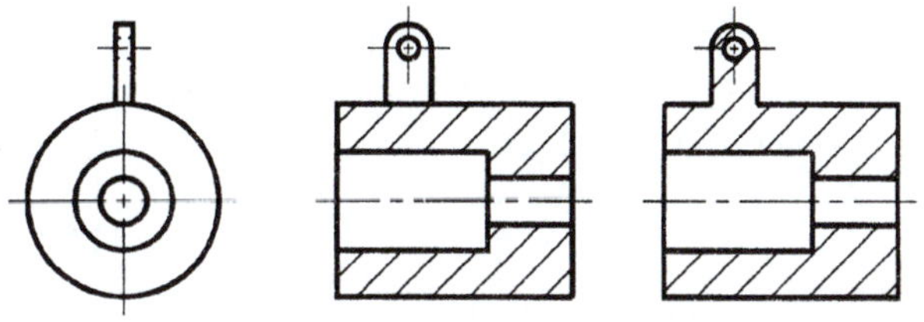

C

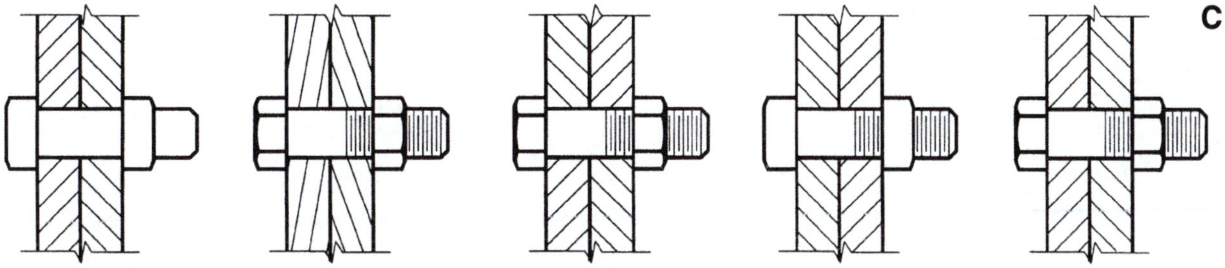

D

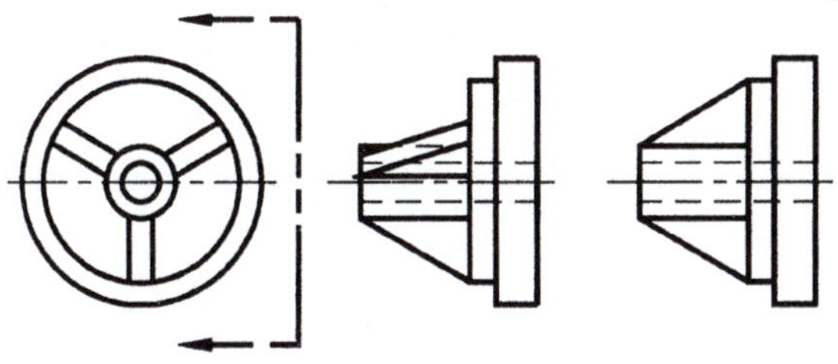

E

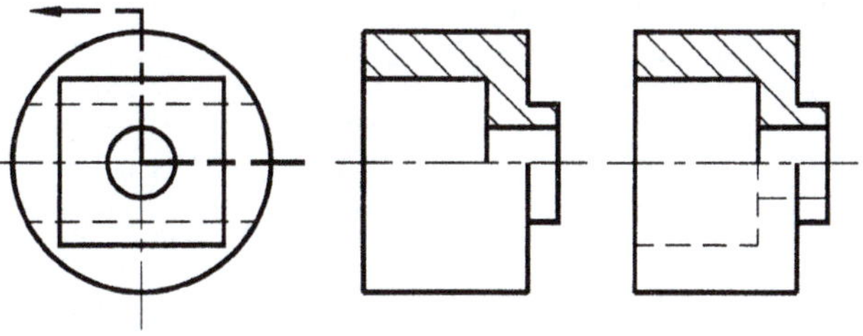

Circle the correct representation of section lines. For the other examples, explain why they are incorrect.

F

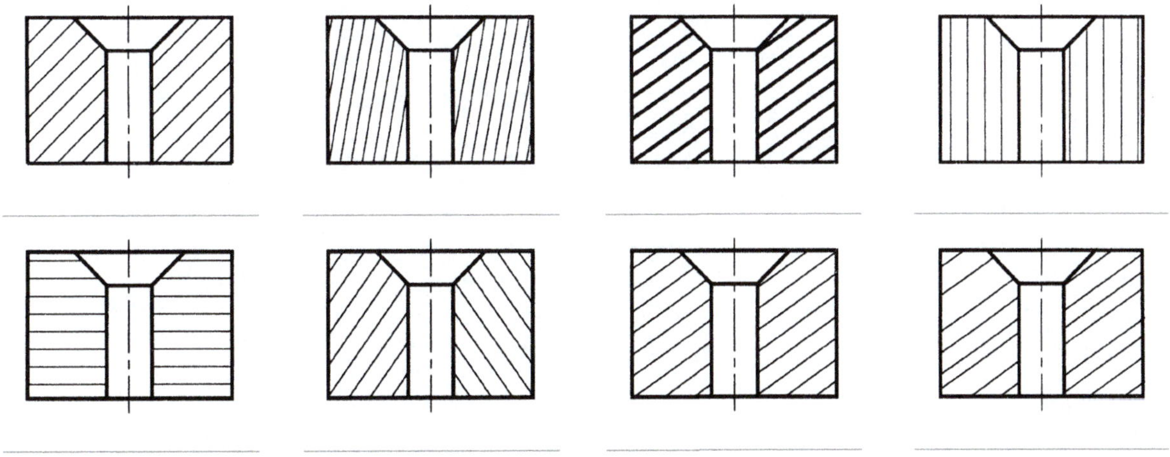

COURSE________ STUDENT____________________ DATE________ PROBLEM 12-2

Problem 12-3 Full Sections

Complete the right-side view, making it into a full section. Use correct line thickness.

A

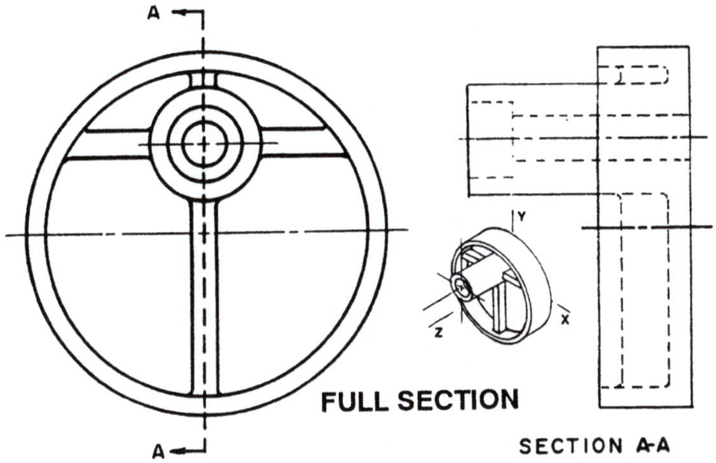

B

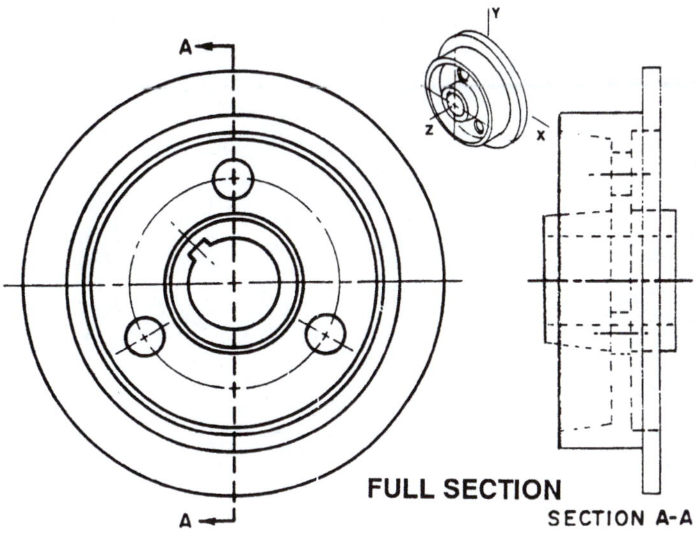

C

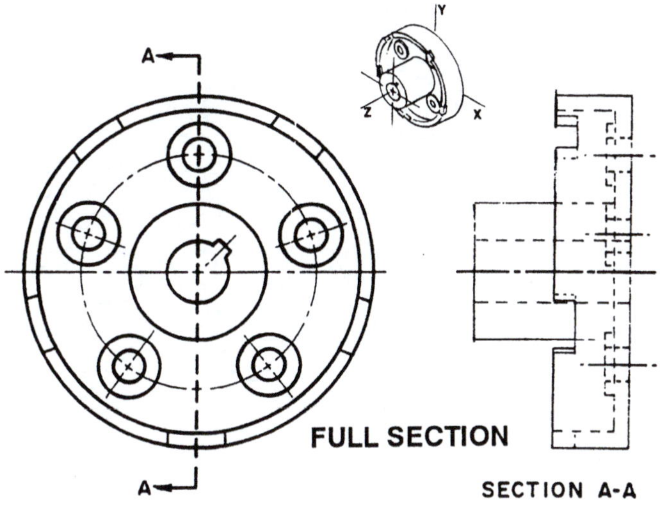

COURSE ________	STUDENT ______________________	DATE ________	PROBLEM 12-3

Problem 12-4 Half Sections

A. Complete the right-side view, making it into a half section. Use correct line thickness.

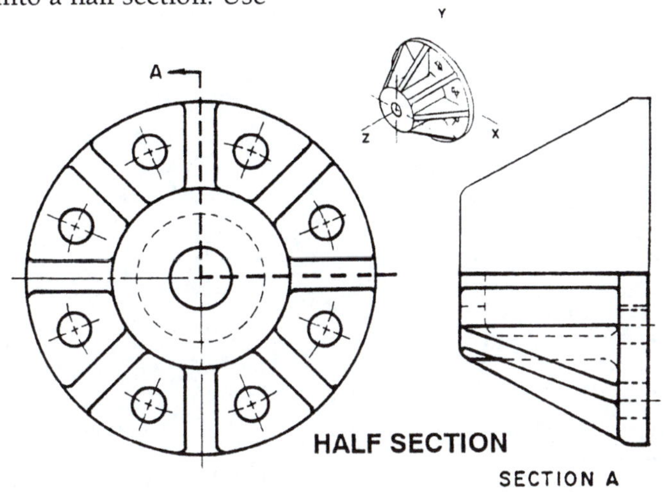

B. Complete the right-side view, making it into a half section. Use correct line thickness.

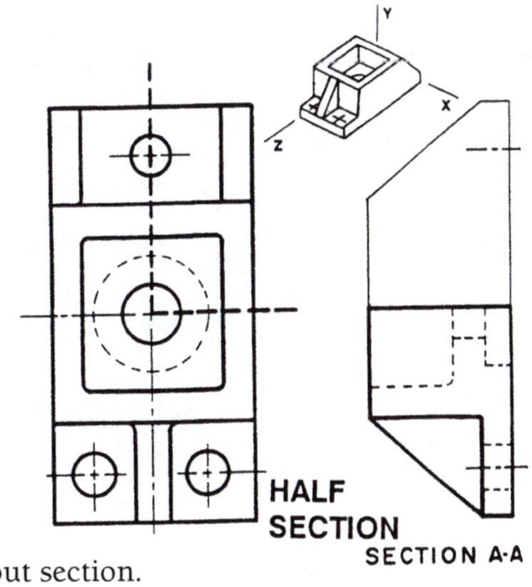

C. Make the front (top half) view into a broken out section. Use correct line thickness.

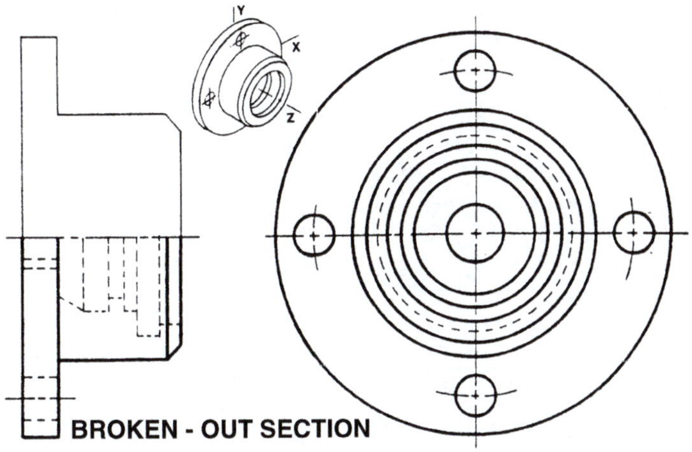

COURSE ______	STUDENT ______	DATE ______	PROBLEM 12-4

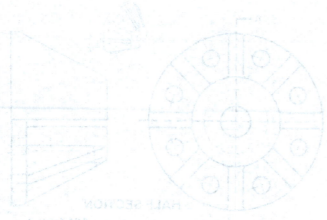

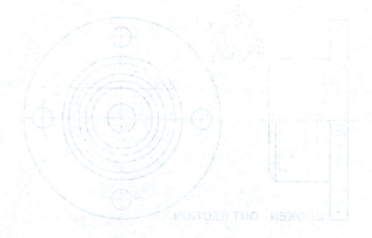

Problem 12-5 Sections

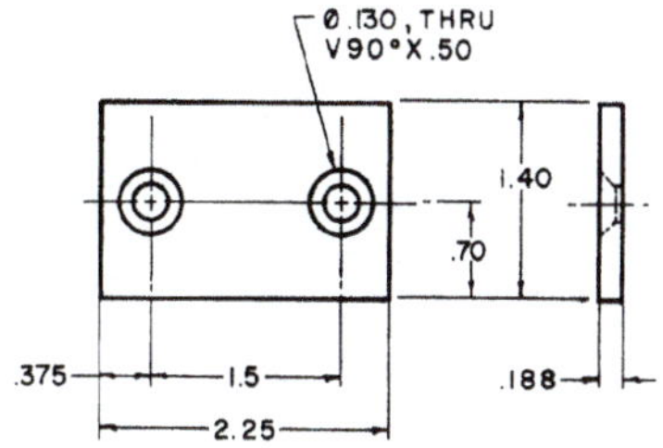

Part 6

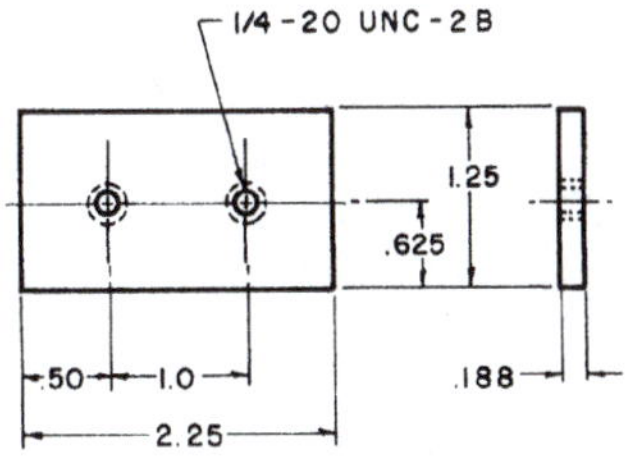

Part 7

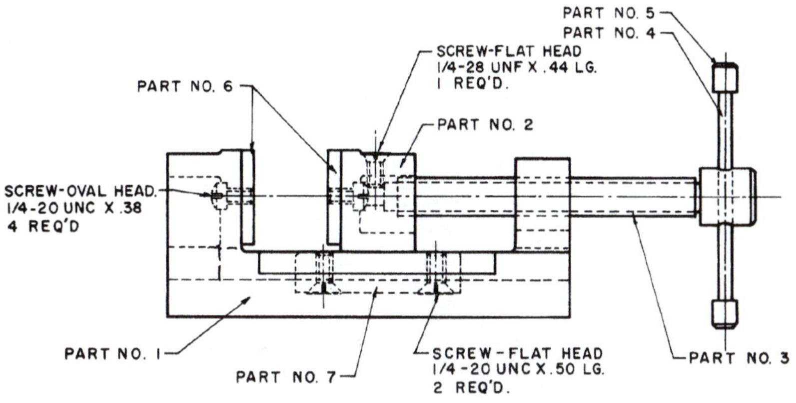

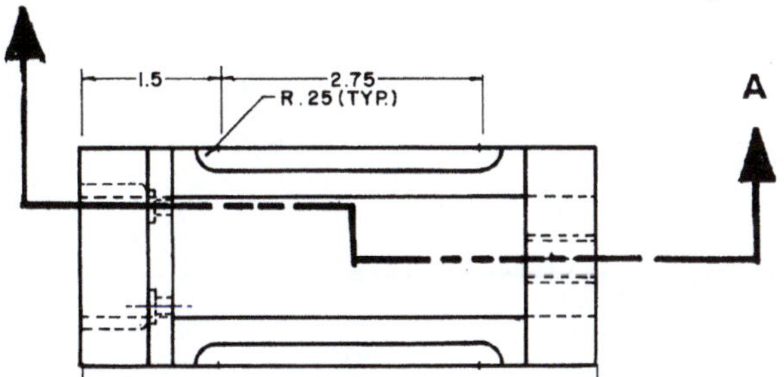

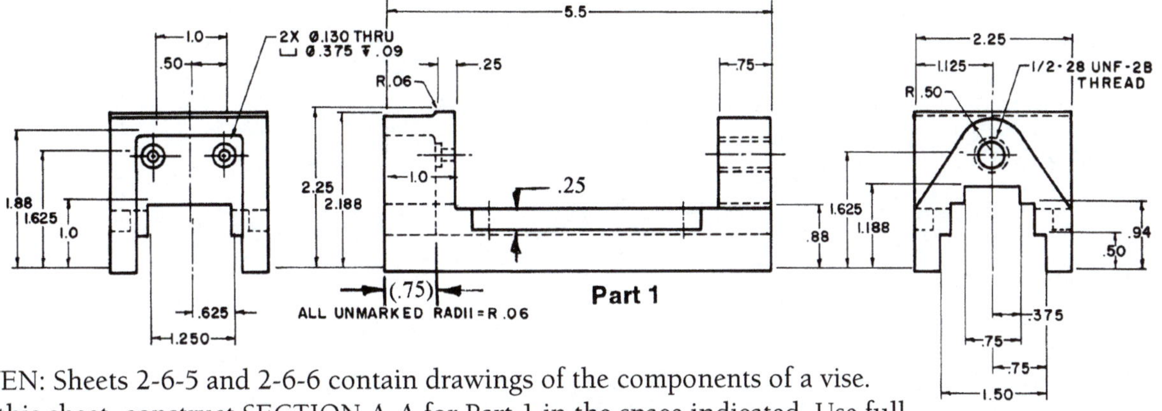

Part 1

GIVEN: Sheets 2-6-5 and 2-6-6 contain drawings of the components of a vise. On this sheet, construct SECTION A-A for Part 1 in the space indicated. Use full scale and omit dimensions.

SECTION A-A

COURSE ________	STUDENT ______________________	DATE ________	PROBLEM 12-5

Problem 12-6 Sections

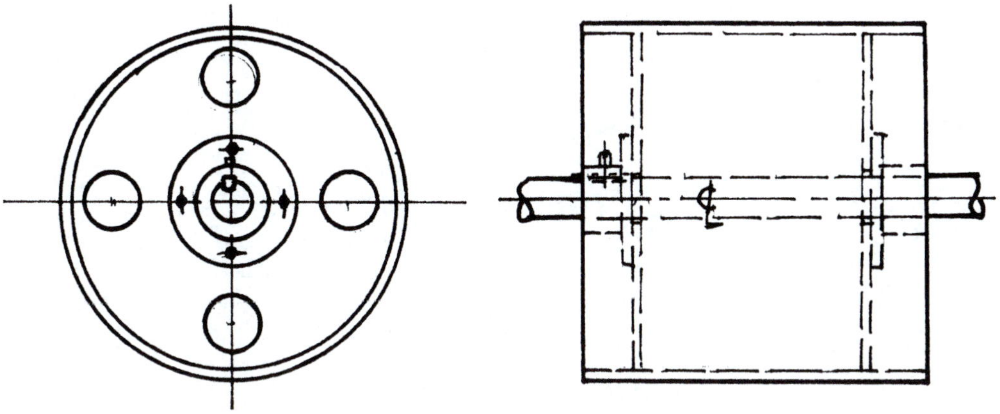

BELT DRIVE

A square key held in place by a set screw transmits the turning moment from the shaft to the hub of the welded steel pulley shown in the upper two halves. The hub drives the pulley through four bolts and the pulley drives a belt of a belt conveyor.

The example broken out section shown in the lower view is a similar shaft-key-hub-pulley assembly.

Prepare a broken out section 3 × size of the upper pulley in the space indicated. Use the two-view drawing for measurements and information. Note the access holes for assembly of the four bolts.

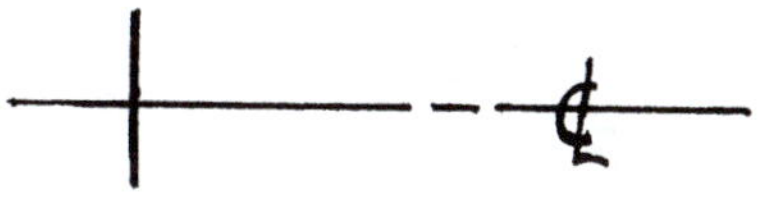

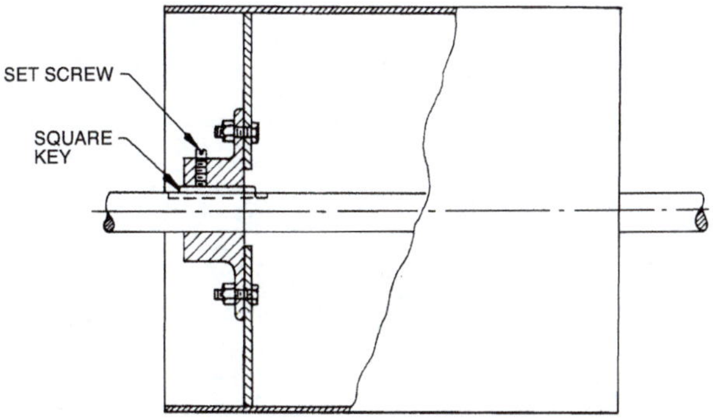

COURSE ________ STUDENT ______________________________ DATE ________ PROBLEM 12-6

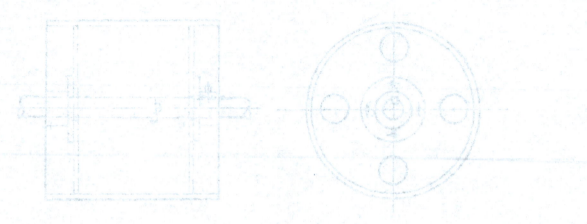

Problem 12-7 Freehand Sketch Section

The following freehand sketch specifies edge radius requirements at three stations (cross sections) of a blowtorch nozzle. At each station along the nozzle, the cross-sectional area must be the same for best flow conditions. The cross-sectional area at station 1 is

$$\frac{\Pi(.8)^2}{4} = .503\ sq.\ in.$$

Show the removed cross sections for each station (similar to the example drawing). Show inside dimensions for each cross section. Use full size. Show your calculations.

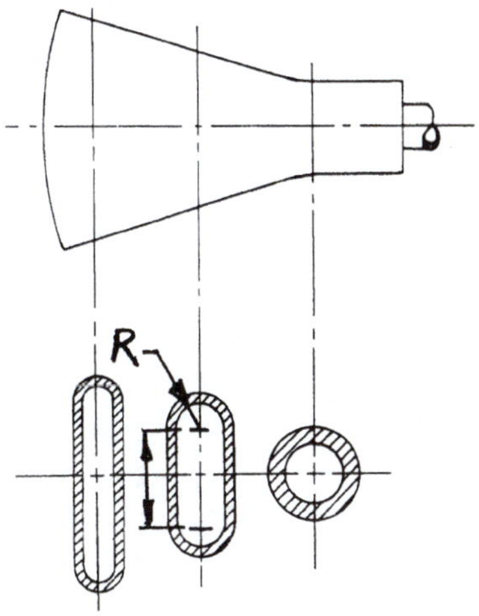

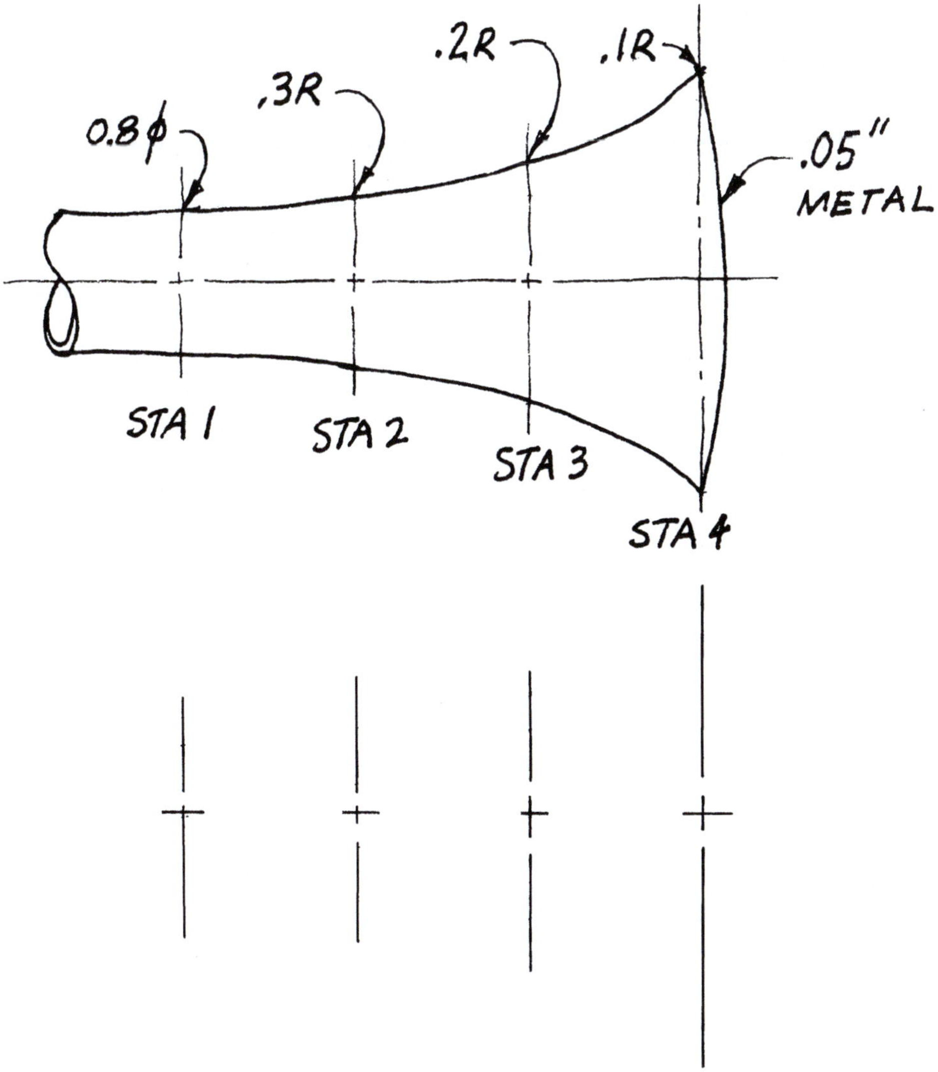

COURSE ________	STUDENT ________________	DATE ________	PROBLEM 12-7

Problem 12-8 Section

A

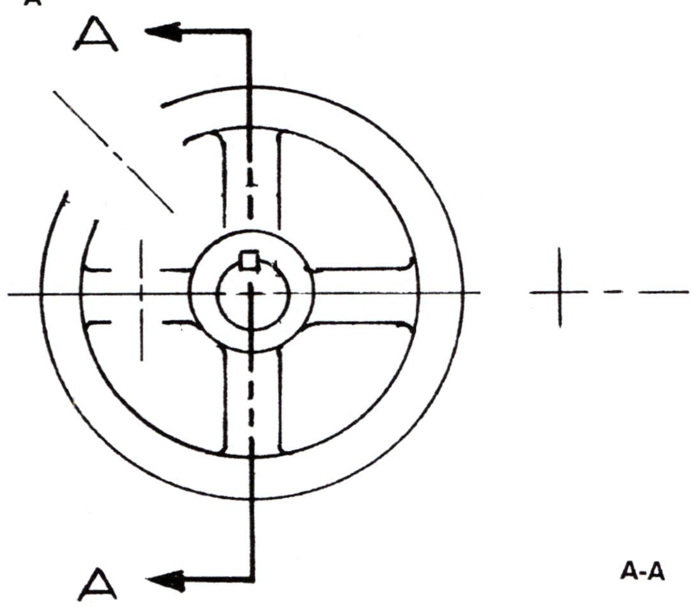

HANDWHEEL SPECIFICATIONS

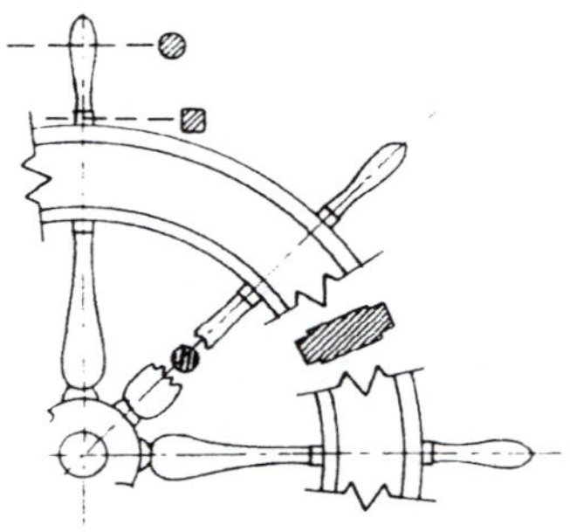

Examples

Rim: Circular cross section
Spokes: Major dia. 3 × minor dia.
Hub length = Hub dia.
Hub bore = Shaft dia.
Key = 1/8 of shaft dia.; square key

Scale the handwheel shown and do the following in the spaces indicated:

a. Construct section A-A.

b. Do a revolved section in the handwheel rim.

c. Do a revolved section in a spoke.

B

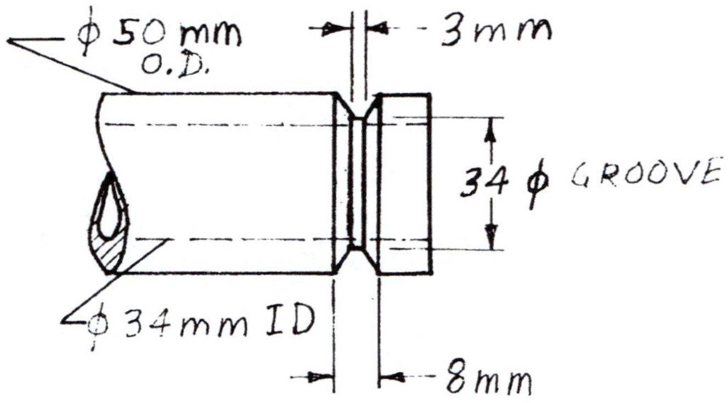

Use the information on the drawing of the grooved tube and prepare a 4 × enlarged removed-section of the groove.

COURSE ______	STUDENT ______	DATE ______	PROBLEM 12-8

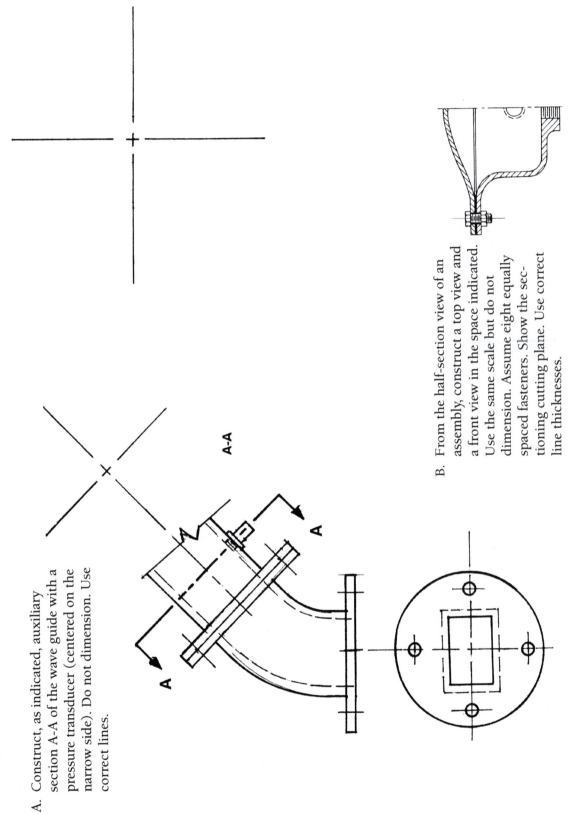

Problem 12-9 Section

A. Construct, as indicated, auxiliary section A-A of the wave guide with a pressure transducer (centered on the narrow side). Do not dimension. Use correct lines.

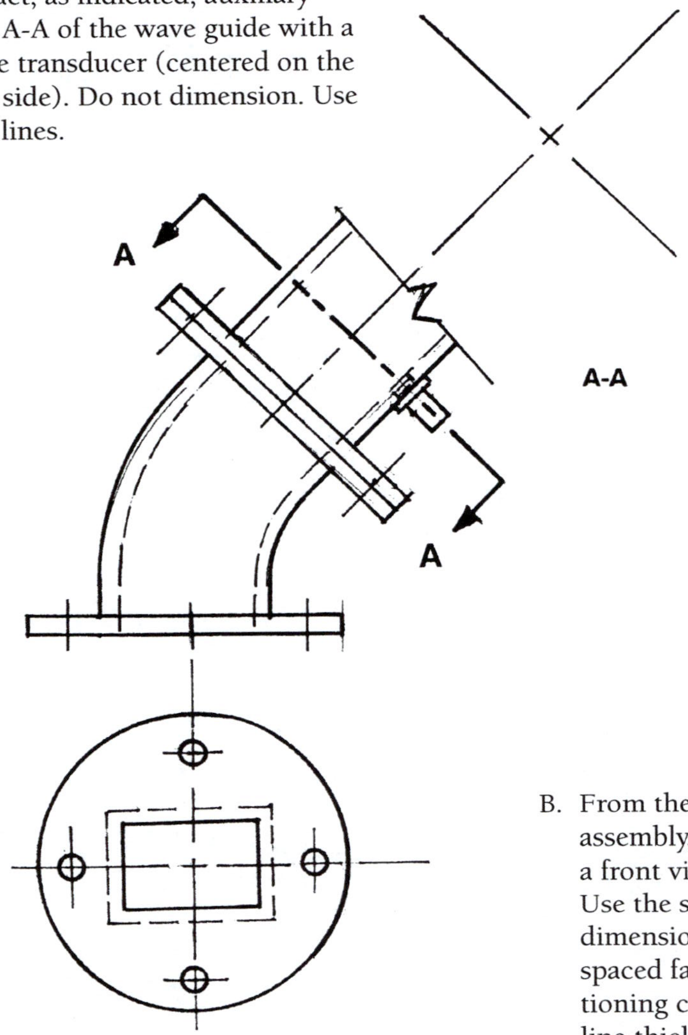

B. From the half-section view of an assembly, construct a top view and a front view in the space indicated. Use the same scale but do not dimension. Assume eight equally spaced fasteners. Show the sectioning cutting plane. Use correct line thicknesses.

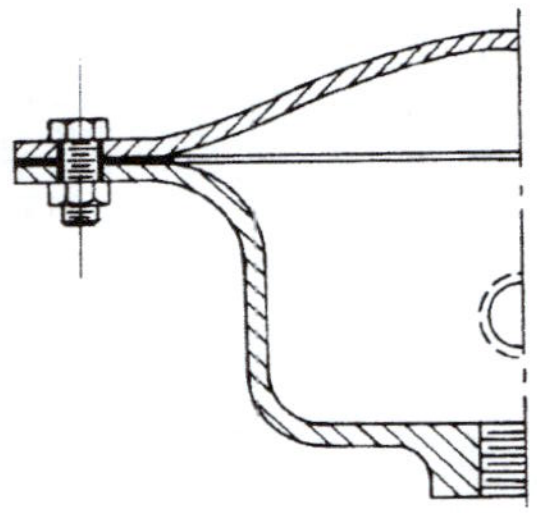

COURSE________ STUDENT____________________ DATE________ PROBLEM 12-9

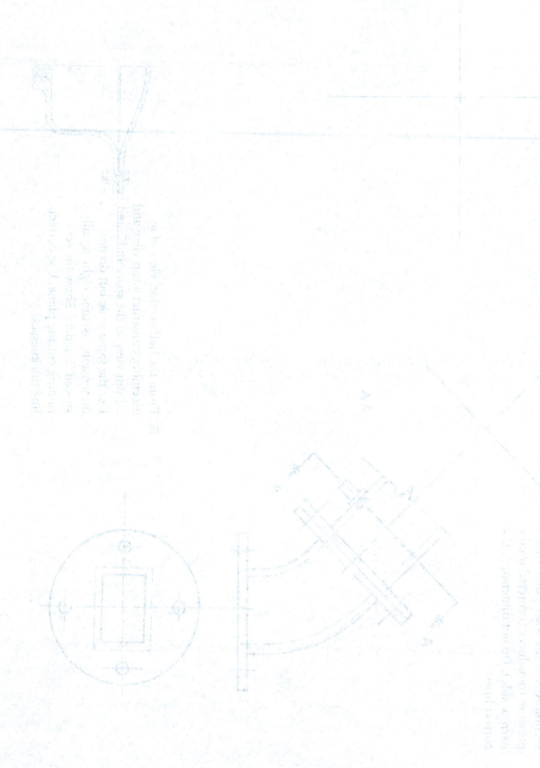

Geometric Dimensioning and Tolerancing

Problem 13-1 Geometric Tolerancing

Fill in the missing geometric tolerancing information in the following exercises.

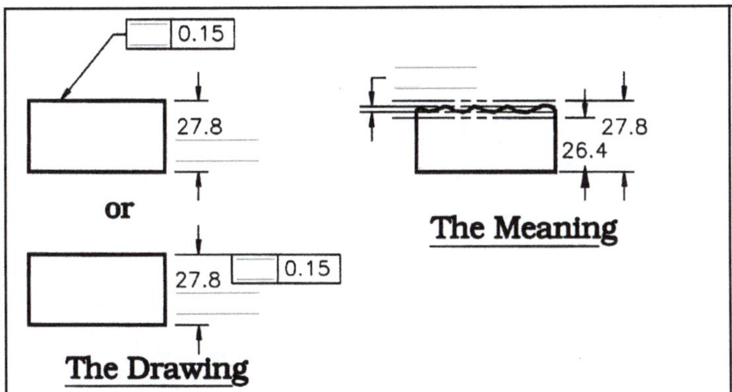

What representation is this? ______

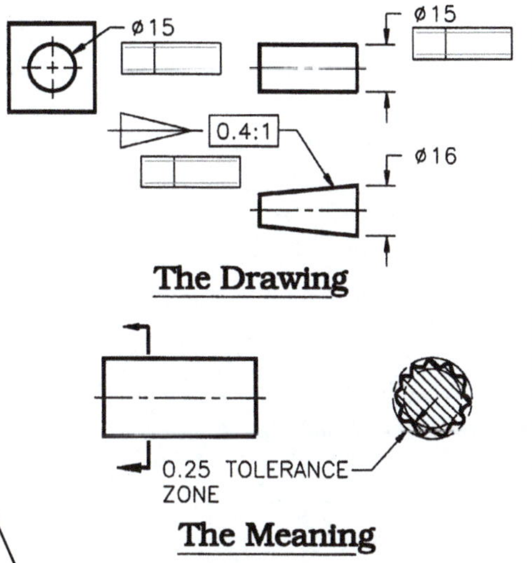

What representation is this? ______

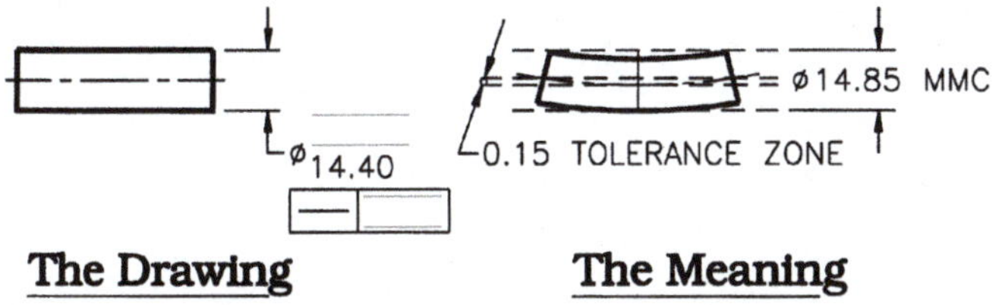

What representation is this? ______

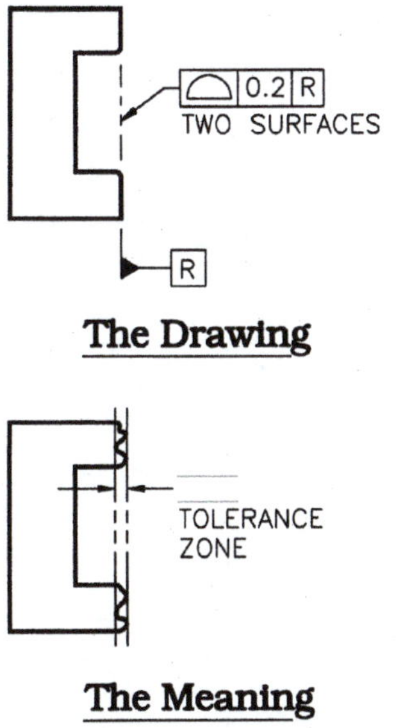

What representation is this? ______

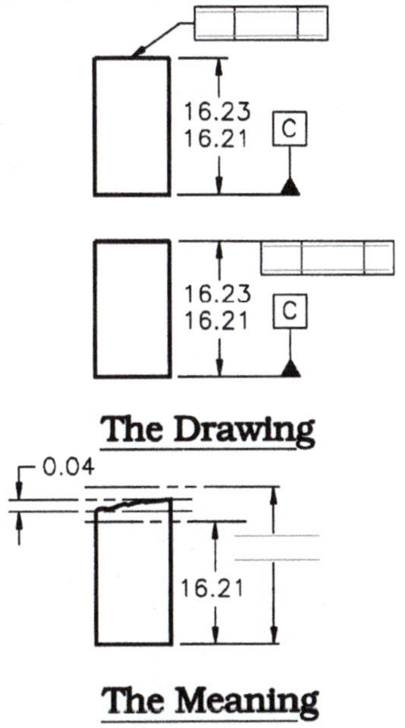

What representation is this? ______

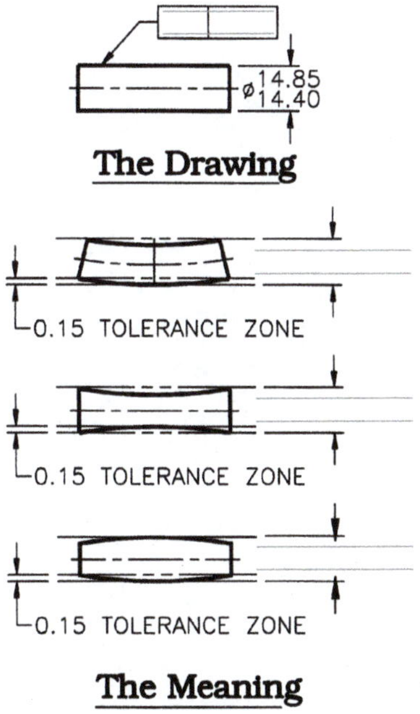

What representation is this? ______

COURSE ______ STUDENT ______ DATE ______ PROBLEM 13-1

Problem 13-2 MMC and LMC

Fill in the following MMC and LMC.

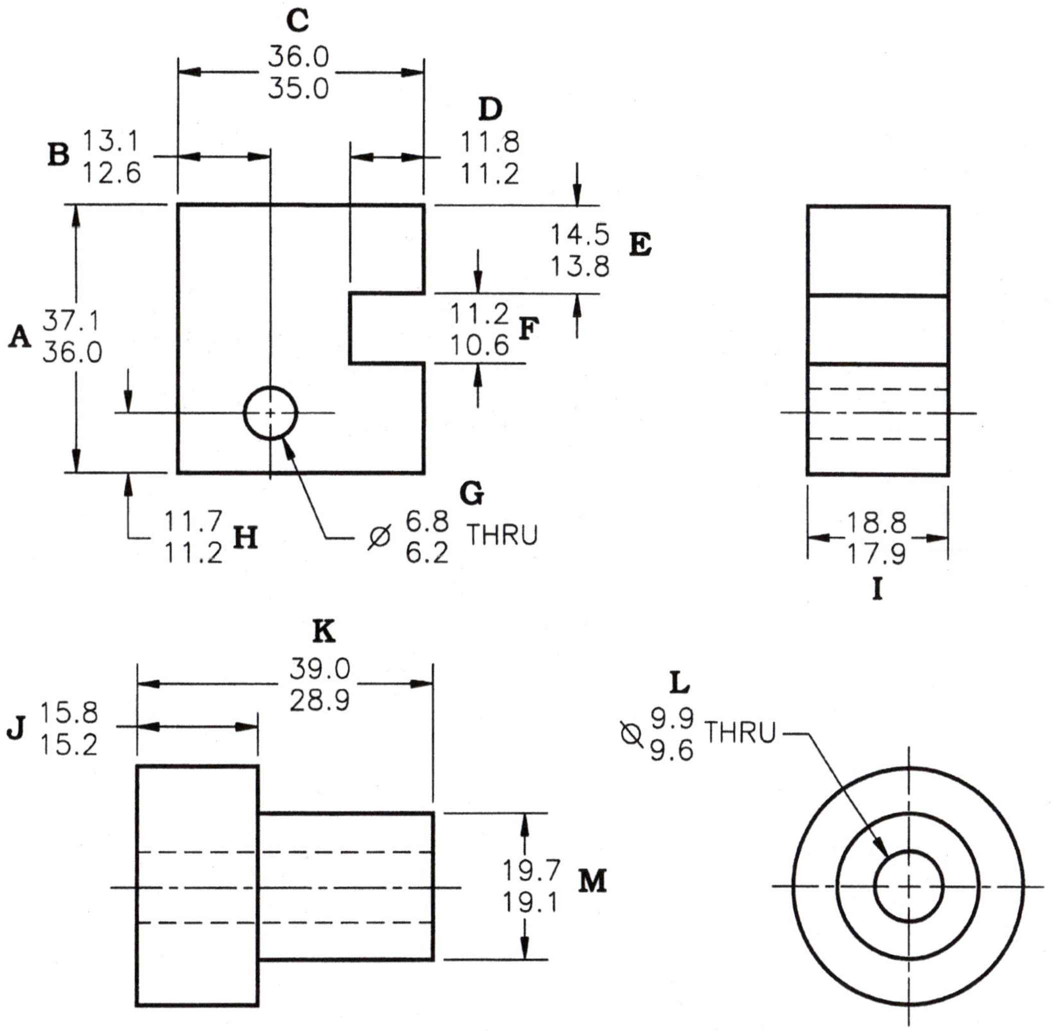

Letter	MMC	LMC
A		
B		
C		
D		
E		
F		
G		

Letter	MMC	LMC
H		
I		
J		
K		
L		
M		

COURSE________ STUDENT______________________________ DATE________ PROBLEM 13-2

Problem 13-3 Geometric Characteristic

Refer to the drawing and answer the following questions:

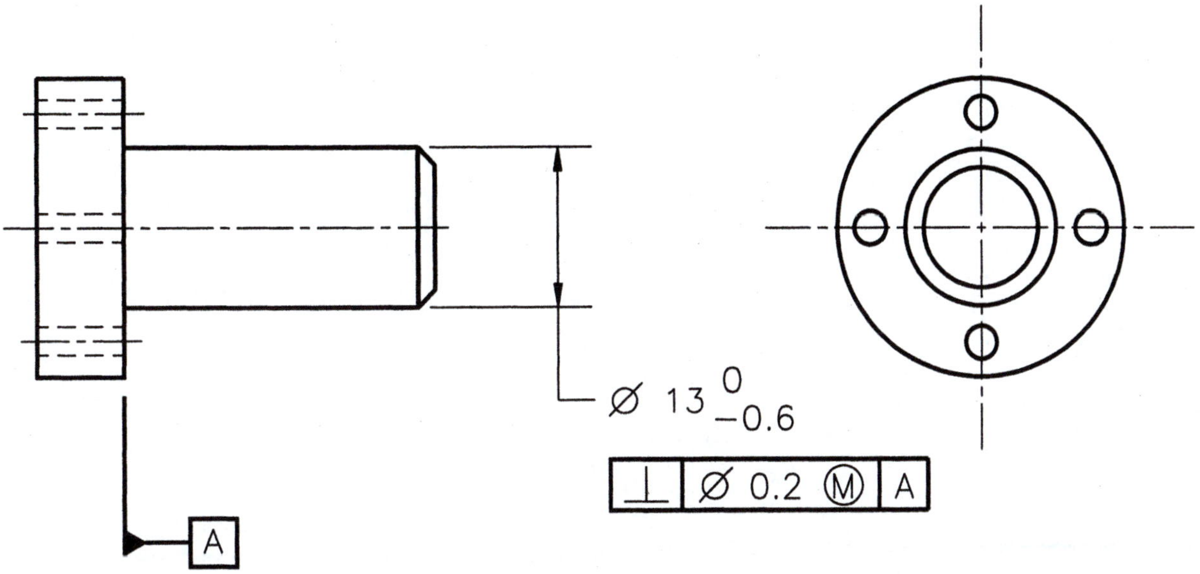

1. What is the geometric characteristic? ______________________

2. What type of geometric tolerance is applied? ______________________

3. What does the symbol Ⓜ represent? ______________________

4. What is the MMC of the feature? ______________________

5. What is the geometric tolerance at MMC feature? ______________________

6. What is the virtual condition? ______________________

7. What would the geometric tolerance be at LMC? ______________________

COURSE ________	STUDENT ______________________	DATE ________	PROBLEM 13-3

Problem 13-4 Geometric Tolerancing

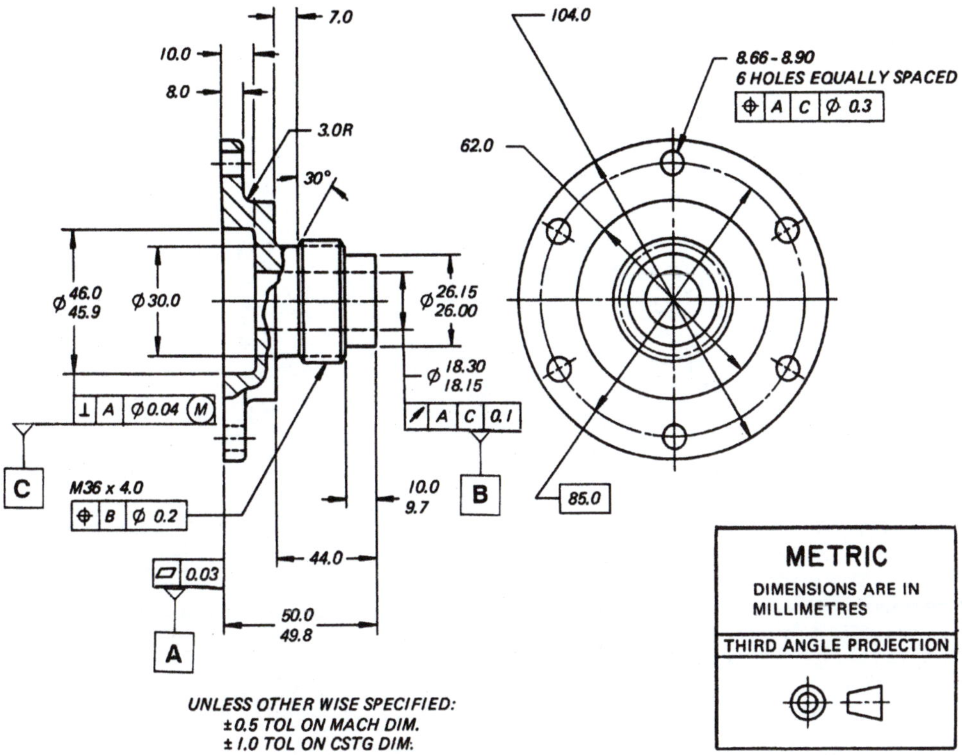

Refer to the above drawing and answer the following questions:

1. What general tolerances are specified for the part? ______________________________
2. What feature of the part is datum A? datum B? datum C? ______________________________
3. What relationship must exist between the six equally spaced holes and datums A and C? ______________________________
4. What specification is given for surface A? ______________________________
5. Interpret the feature control symbol for the 18.30–18.15 Dia. hole. ______________________________
6. What relationship must exist between the features identified as datum A and datum C? ______________________________
7. What does the rectangular frame around the 85.0 dimension designate? ______________________________
8. Interpret the dimension M36 × 4.0. ______________________________
9. What relationship must exist between the threaded feature and datum B? ______________________________
10. What does the symbol (M) designate? ______________________________

COURSE __________ STUDENT ______________________________ DATE __________ PROBLEM 13-4

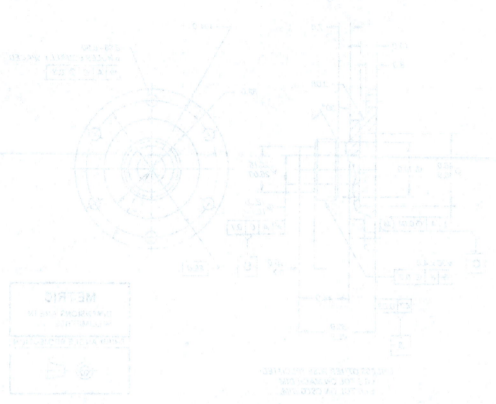

Problem 13-5 Geometric Tolerancing

Complete the dimensioning of this INDEX STOP to comply with the following design requirements. Use proper symbols and lettering.

1. The small cylinder is cylindrical within .05 and is concentric with the large cylinder within .10.
2. The large cylinder is cylindrical within .05.
3. The left end is perpendicular to the large cylinder within .10 and parallel to the other end of the large cylinder within .15.
4. The flat surface on the large cylinder is parallel to the centerline of the large cylinder within .05 and flat within .02.

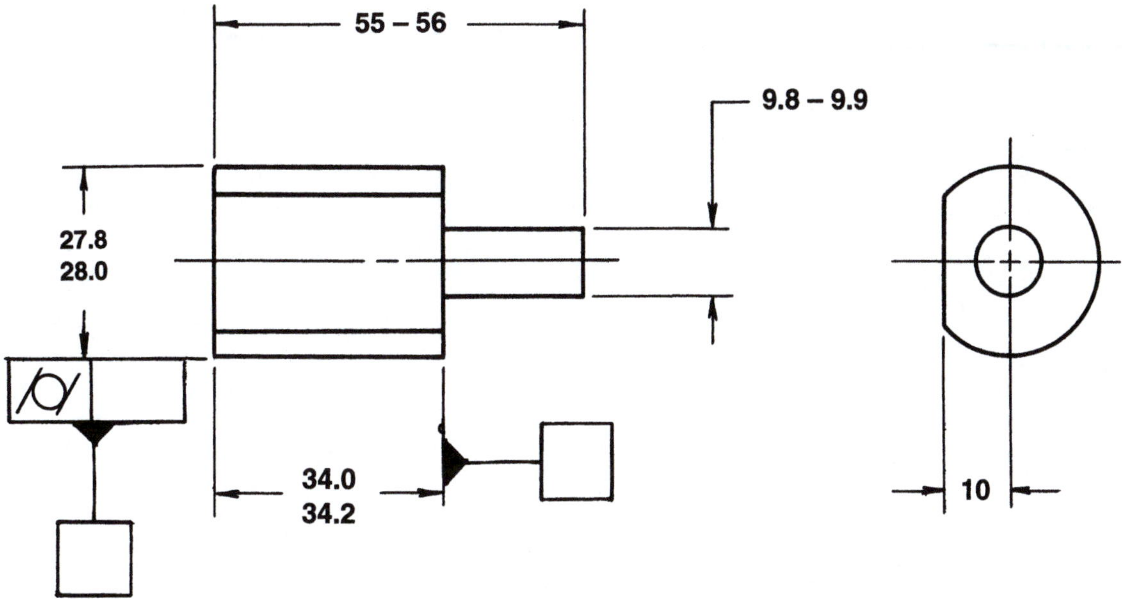

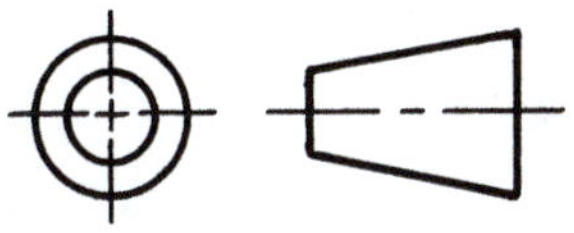

METRIC

Third-Angle Projection

COURSE	STUDENT	DATE	PROBLEM 13-5

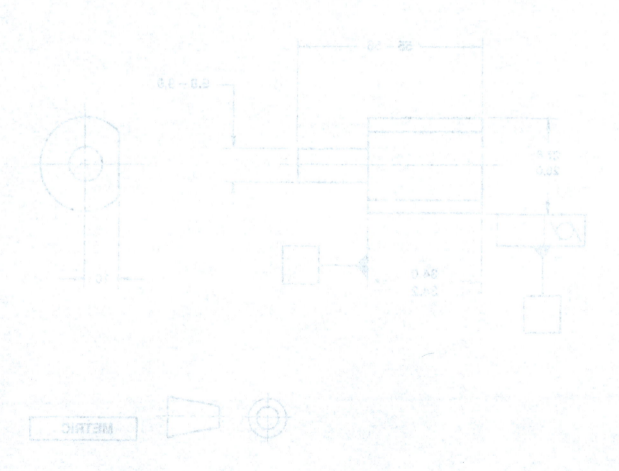

Problem 13-6 Geometric Dimensioning

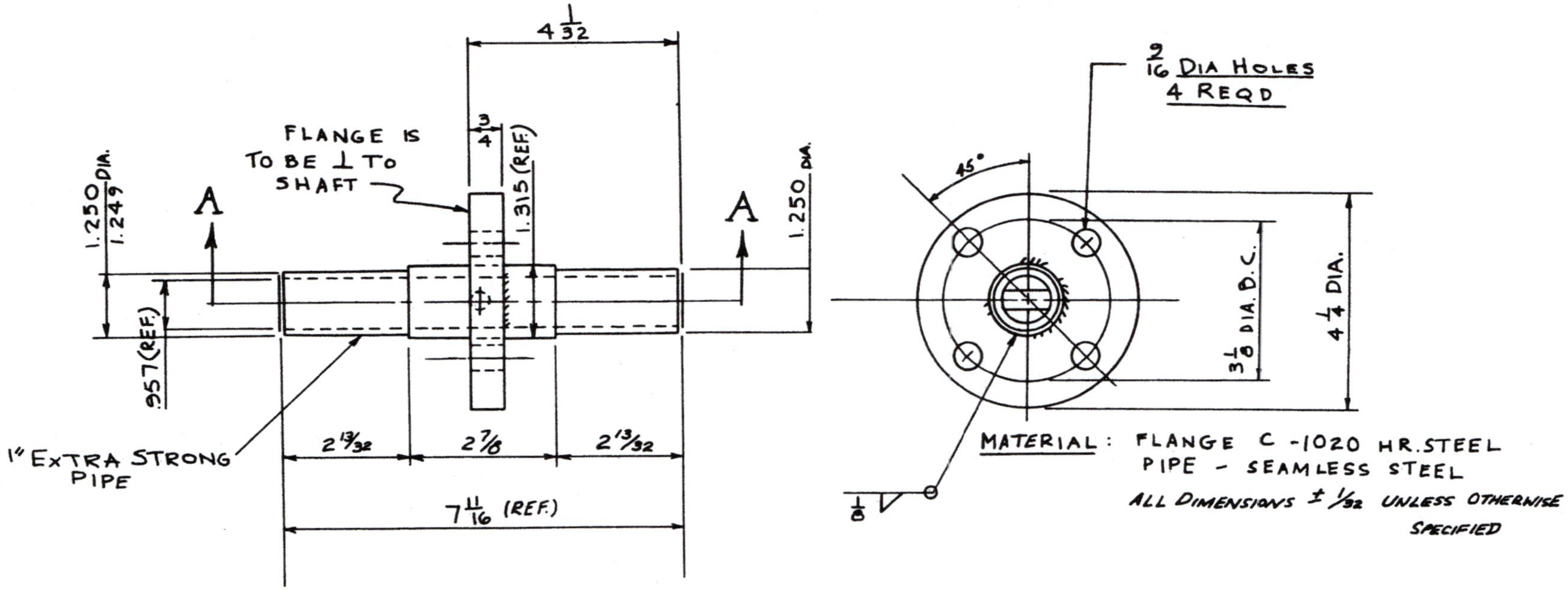

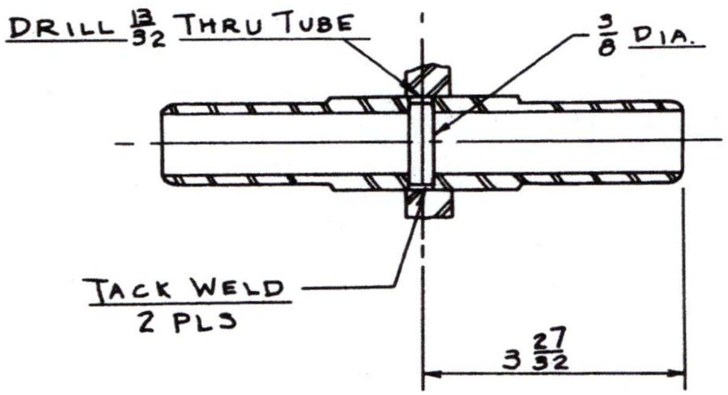

The two views and section drawing of the control-blade-shaft were used in a machine shop that was in the same building as the designer who prepared the drawings. Any questions that the shop had were resolved readily because the designer frequently visited the shop. For example, the designer informed the machinist that the left side of the 4-1/4" diameter flange had to be perpendicular to the centerline of the tubular shaft and that the four holes in the flange were to accurately match a blade (not shown) with the same four-hole pattern.

In the future, the control-blade-shaft is to be made by a shop in another city, and it must assemble with the blades made by the designer's company, so more precise dimensioning is required. Therefore, redo the drawing using the 1.250" diameter on the left side as datum A, and use geometric dimensioning and tolerancing to ensure that the flange will meet the requirements.

COURSE________ STUDENT____________________ DATE________ PROBLEM 13-6

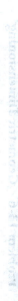

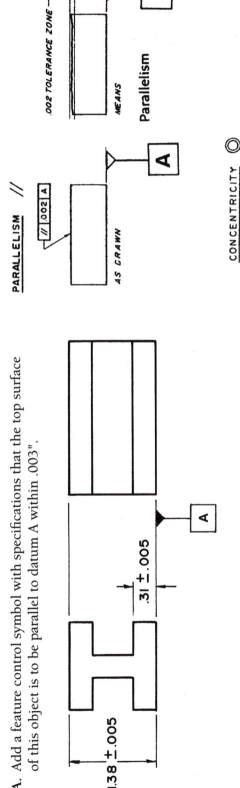

Problem 13-7 Geometric Dimensioning

A. Add a feature control symbol with specifications that the top surface of this object is to be parallel to datum A within .003".

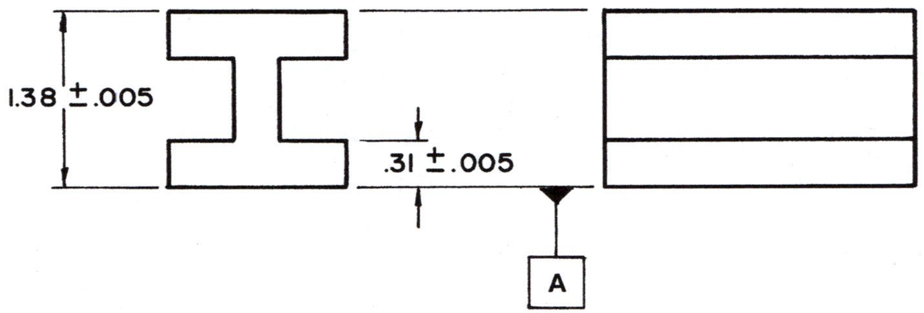

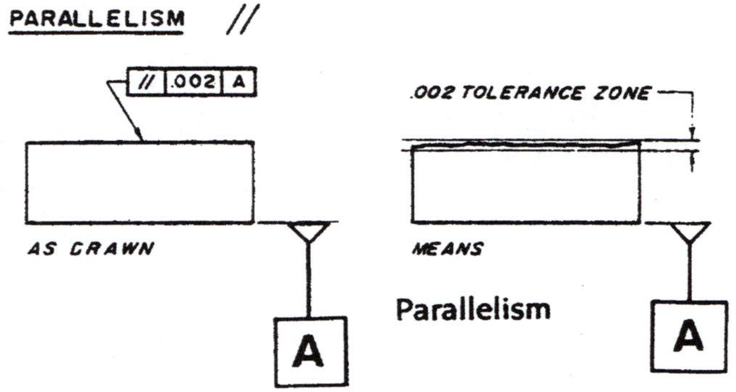

B. Add a feature control symbol with specifications that the smaller diameter of this object is to be concentric to datum A within .003".

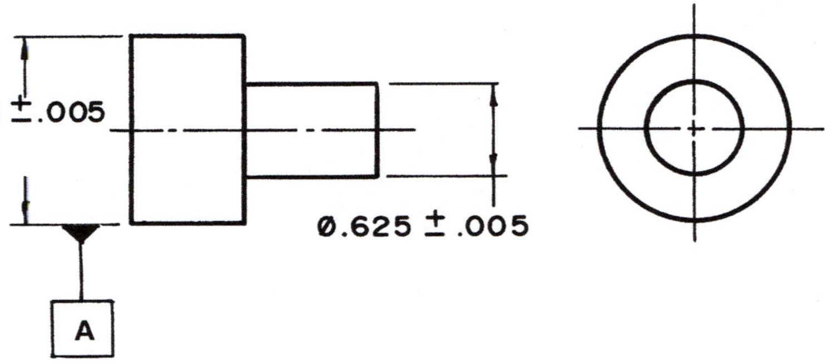

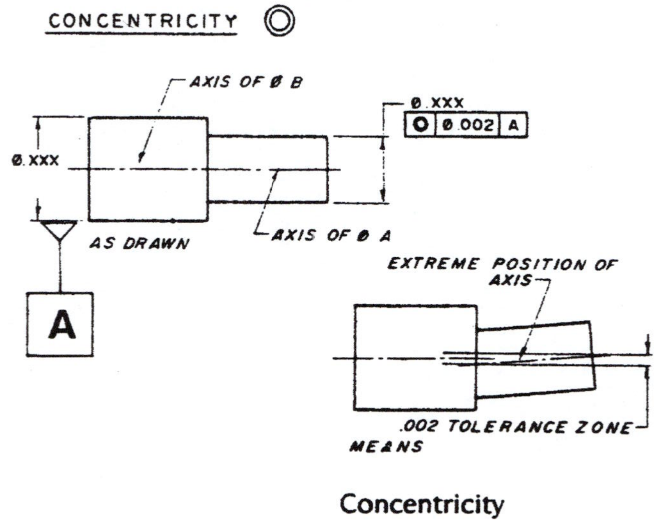

COURSE ______	STUDENT ______	DATE ______	PROBLEM 13-7

Problem 13-8 Adding Dimensions

Use DATUM TARGET symbols and arbitrarily locate, using BASIC DIMENSION symbols, three specific points on surface A, two specific points on surface B, and one specific point on surface C.

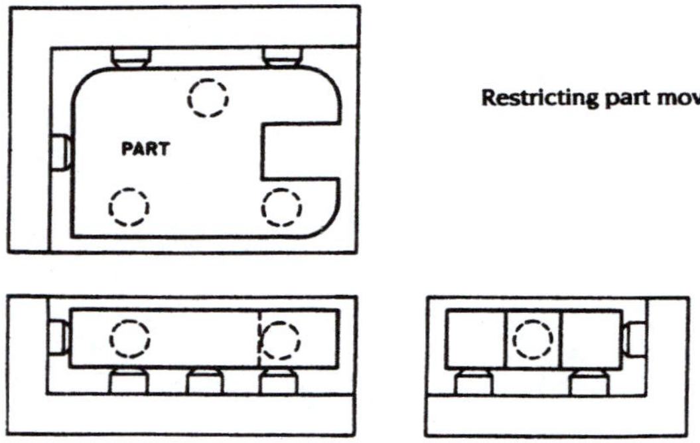

Restricting part movement

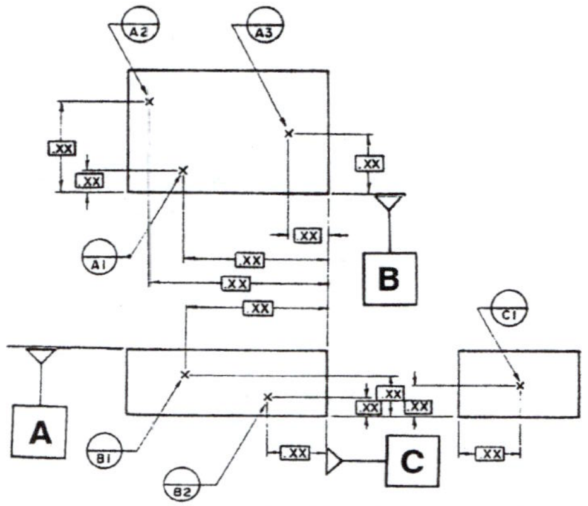

Dimensioning datum points

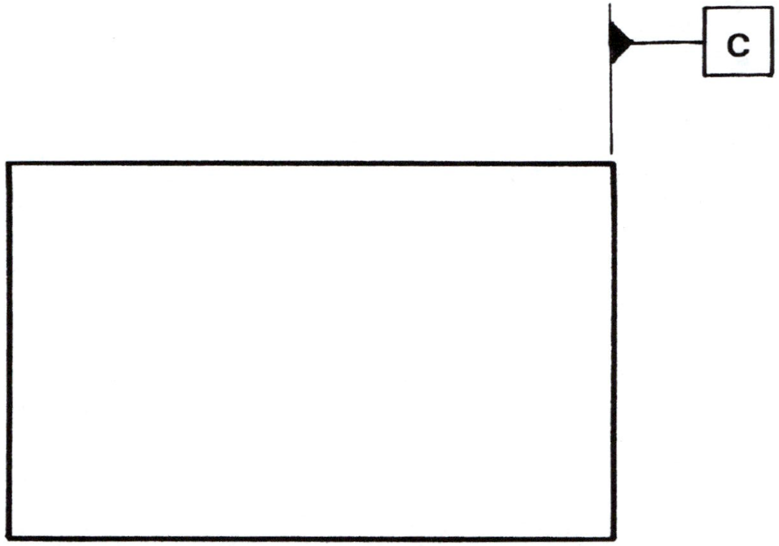

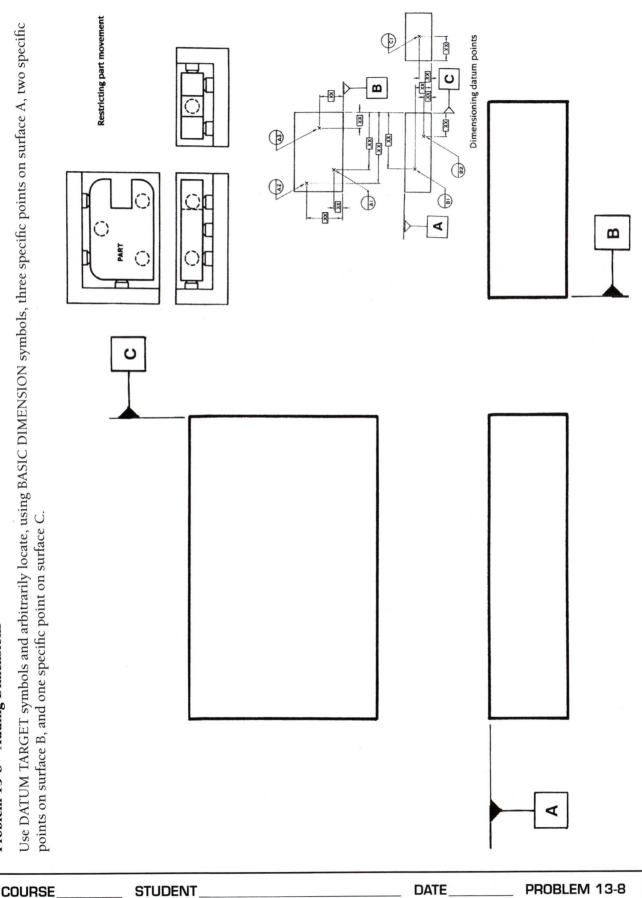

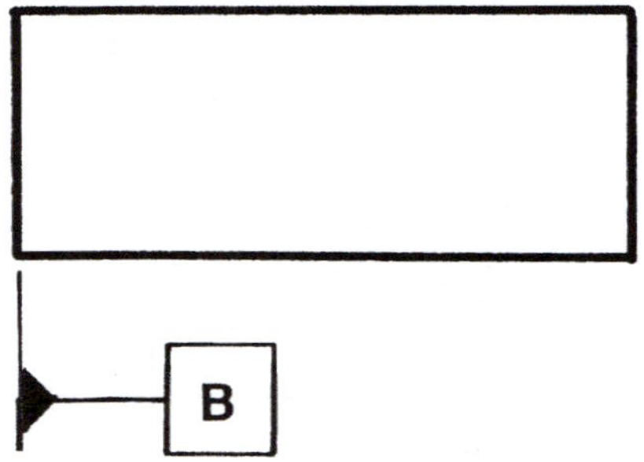

COURSE________ STUDENT____________________ DATE________ PROBLEM 13-8

Problem 13-9 Tolerancing

Add dimensions to the LASER LOCATOR drawing in accordance with the following:

1. Use unilateral tolerances for the height, which is to be machined to a range of .492 to .500".
2. Identify the bottom surface as datum A.
3. The top must be parallel to the bottom within .003".
4. The top must be flat within .0005.
5. The right surface must be perpendicular to the bottom within .002.
6. The back surface must be parallel to the front within .004 and perpendicular to the bottom within .001.

Scale the drawing for other dimensions. All dimensions in inches.

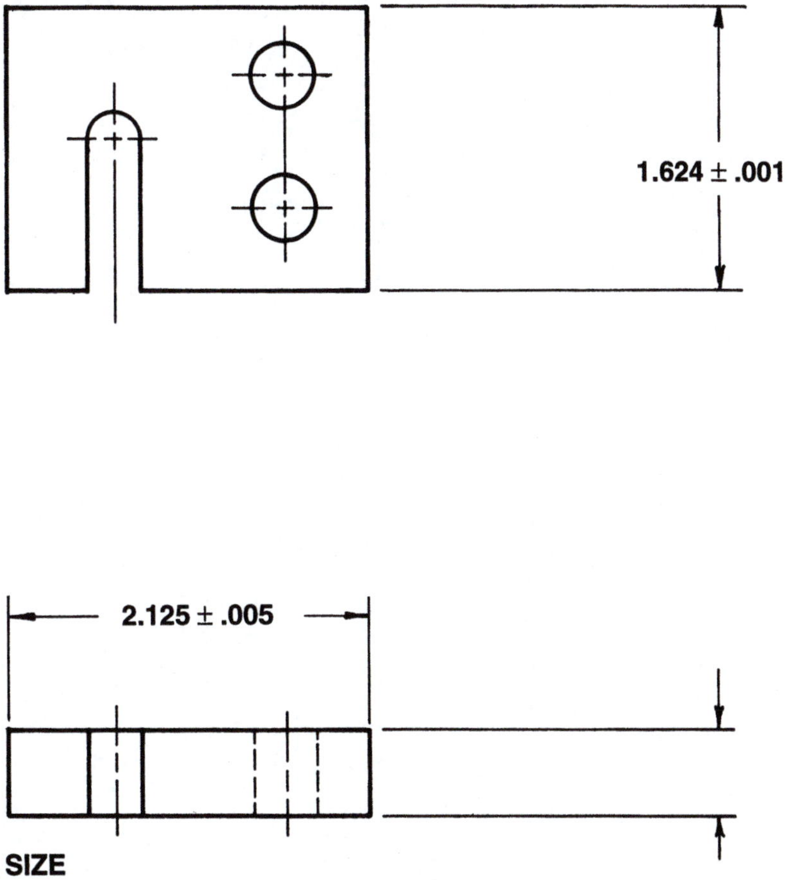

SCALE: FULL SIZE

COURSE________ STUDENT____________________ DATE________ PROBLEM 13-9

Problem 13-9 Tolerancing

Add dimensions to the LASER LOCATOR drawing in accordance with the following:

1. Use unilateral tolerancing for the height, which can be machined to a range of [illegible]
2. Identify the bottom surface as datum A.
3. The top must be parallel to the bottom within .002.
4. The top must be flat within .0005.
5. The right surface must be perpendicular to the bottom within .002.
6. The back surface must be parallel to the front within .004 and perpendicular to the bottom within .001.

Scale the drawing for other dimensions. All dimensions in inches.

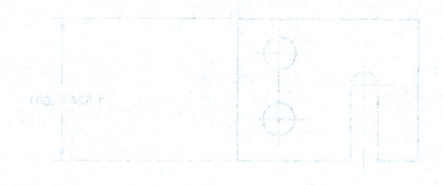

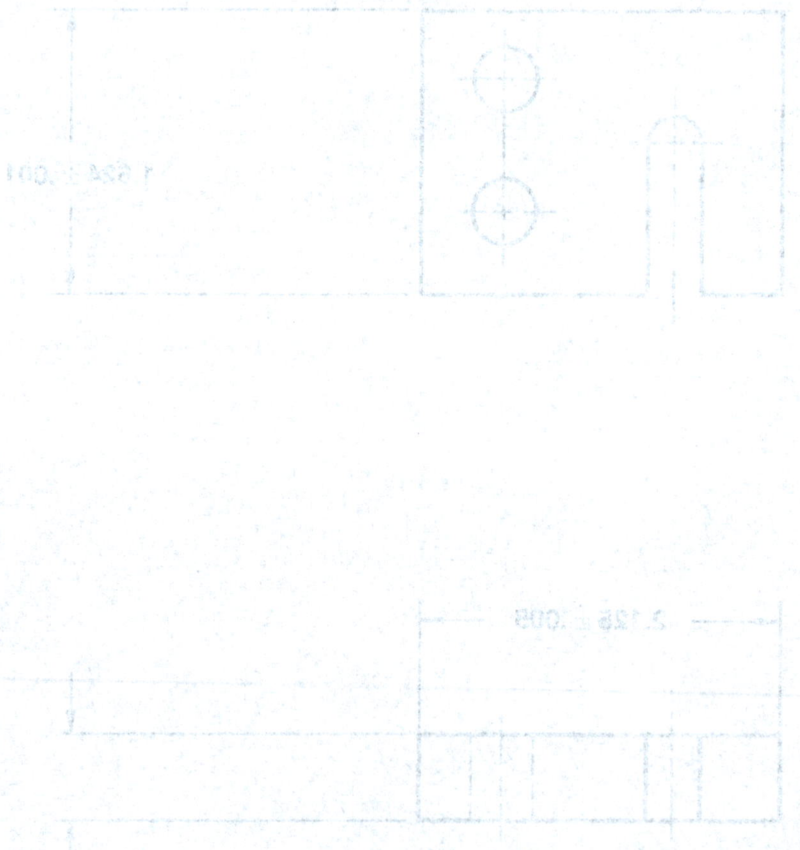

SCALE: FULL SIZE

COURSE ______ STUDENT ______ DATE ______ PROBLEM 13-9

Problem 13-10 Geometric Tolerancing

The more precise system of tolerancing called GEOMETRIC DIMENSIONING and POSITIONING TOLERANCING allows designers to ensure a component's function by establishing realistic tolerances. Some of the tolerances are tolerances of SIZE, tolerances of FORM, tolerances of ORIENTATION, and tolerances of LOCATION. This more precise system is recognized on technical drawings because the information is usually presented in three different BOXES. The three boxes shown are from ASME Y14.5-1994.

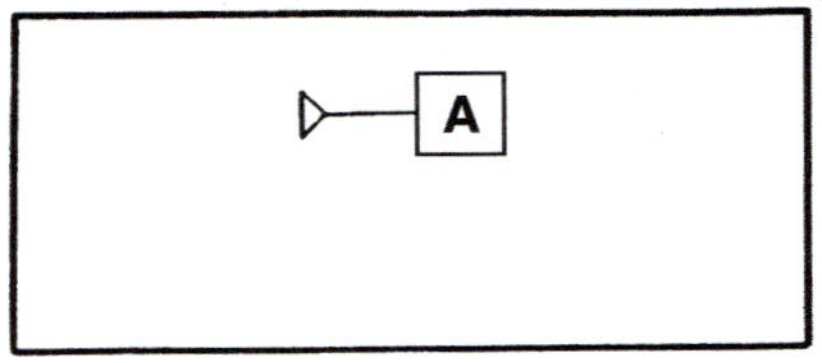

DATUM IDENTIFYING SYMBOL

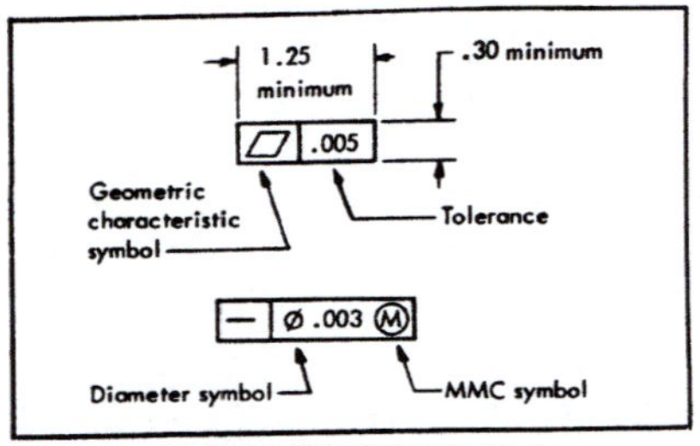

FEATURE CONTROL SYMBOLS

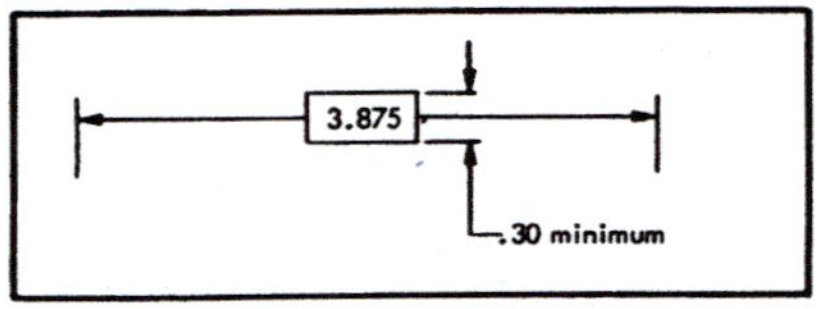

BASIC DIMENSION SYMBOL

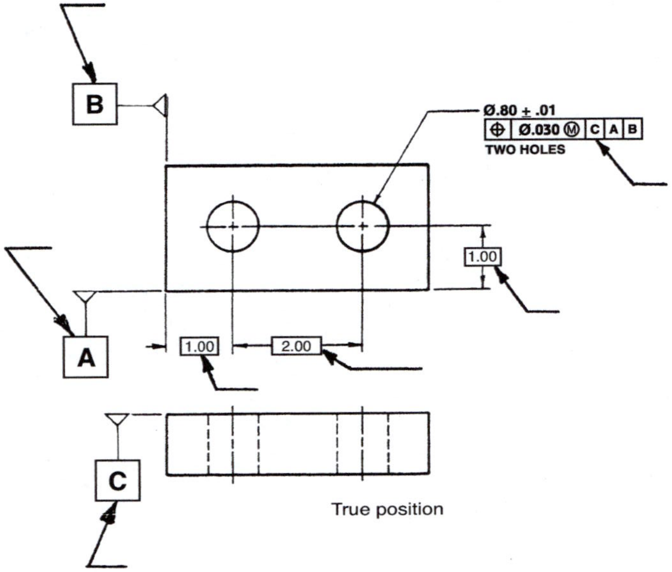

Do the following, related to the figure shown:

1. Print BASIC DIMENSION where basic dimensions occur.
2. Print DATUM where datums occur.
3. Print FEATURE CONTROL SYMBOL at the proper location.
4. Briefly explain the following:
   a. ø.80 ± .01 ______
   b. ⌖ ______
   c. ø.030 Ⓜ ______
5. What is the tolerance zone for the holes if the diameter of the holes is ø.81? Explain.

______

COURSE ______ STUDENT ______ DATE ______ PROBLEM 13-10

Problem 13-11 Geometric Tolerancing

GENERAL QUESTIONS: DATUMS

Fill in the following blanks.

The P, Q, and R planes shown are all ________________ to each other. In drawings, usually all dimensions are either ________________ or ________________ to the three planes.

Note: Actual datum planes are usually surfaces of machine tool tables, machine tool surfaces, or inspection equipment surfaces.

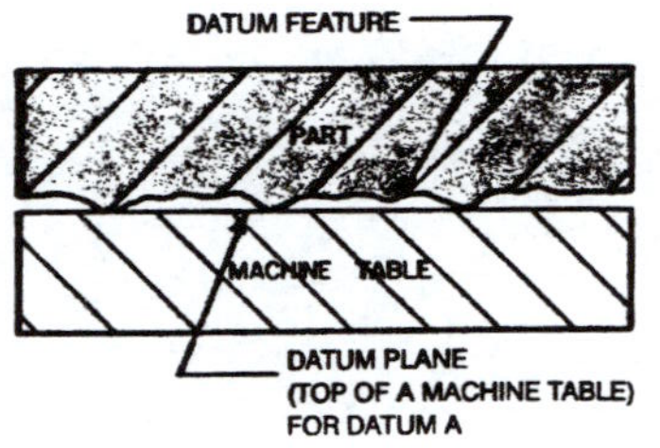

Establishing datums

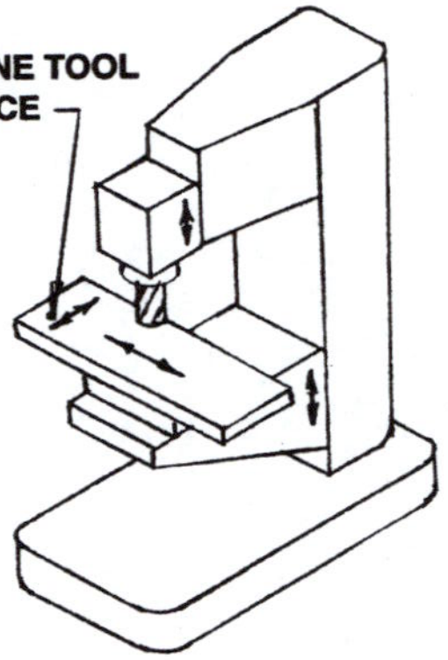

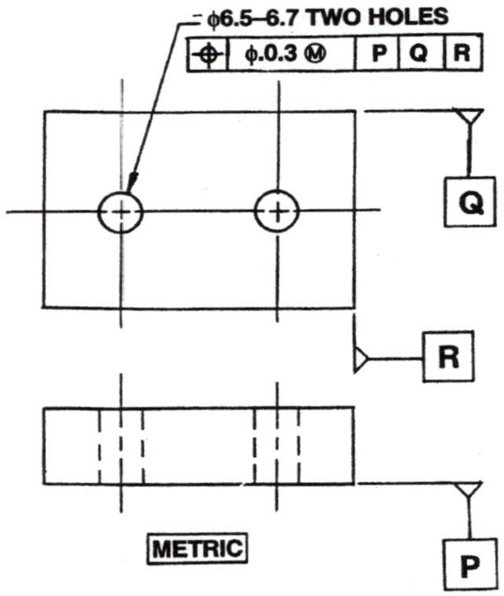

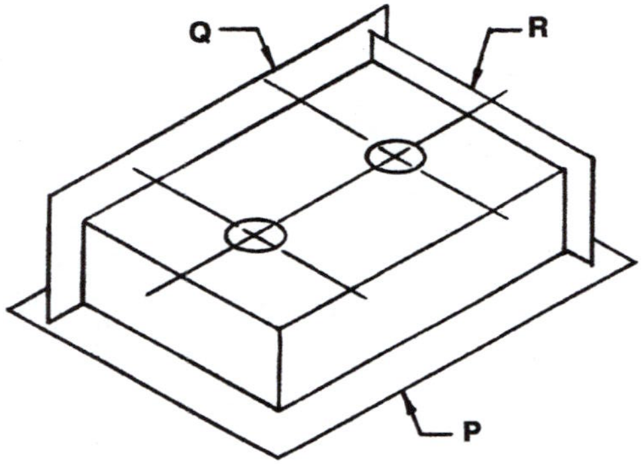

What would be the tolerance zone diameter for the hole at the left if the hole diameter was:

6.5 mm ________________?

6.7 mm ________________?

COURSE________ STUDENT________________ DATE________ PROBLEM 13-11

Problem 13-12 Geometric Tolerancing

A. Add a feature control symbol with specifications that the left-side surface of this object is to be perpendicular to datum A within .003".

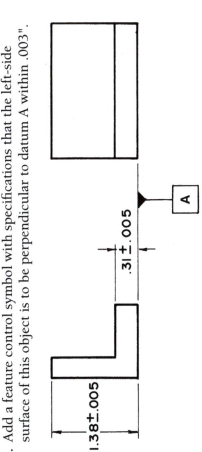

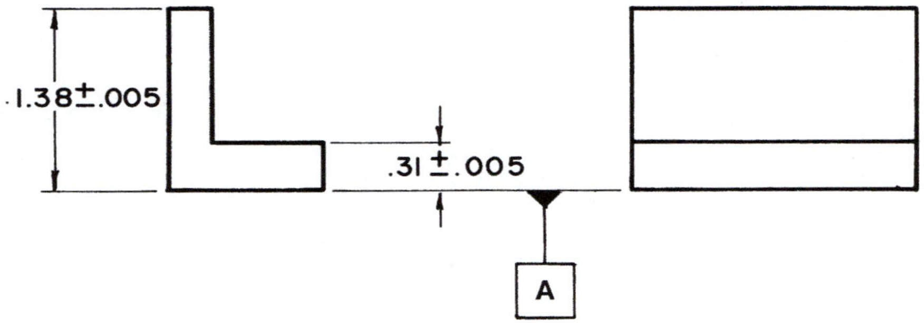

B. Add a feature control symbol with specifications that surface X is to have a total runout of .003" with respect to datum A.

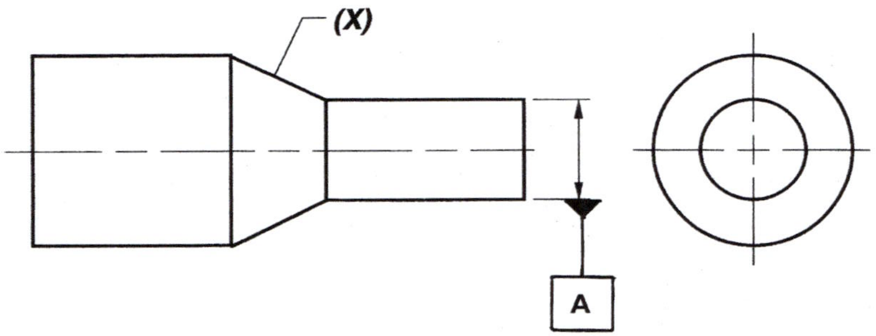

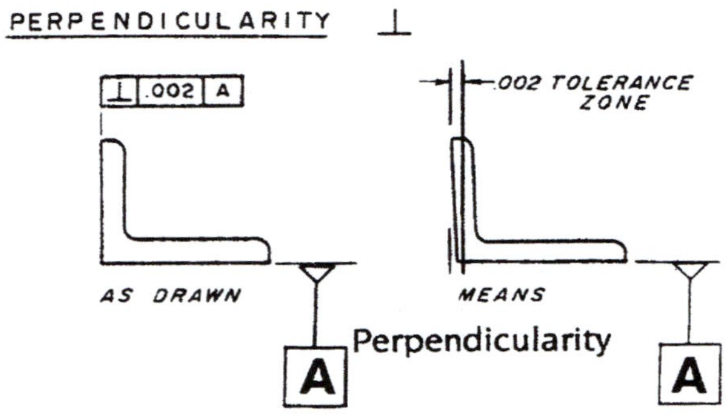

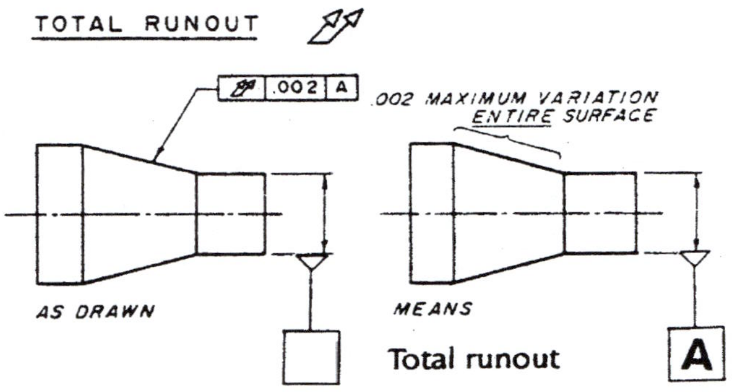

COURSE ______	STUDENT ______	DATE ______	PROBLEM 13-12

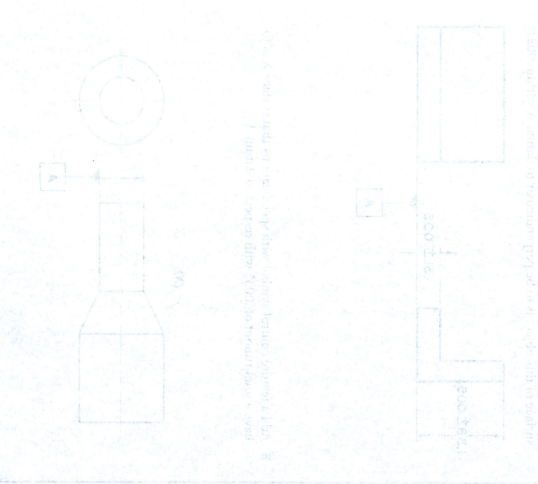

Problem 13-13 Geometric Tolerancing

Refer to the drawing and answer the following questions:

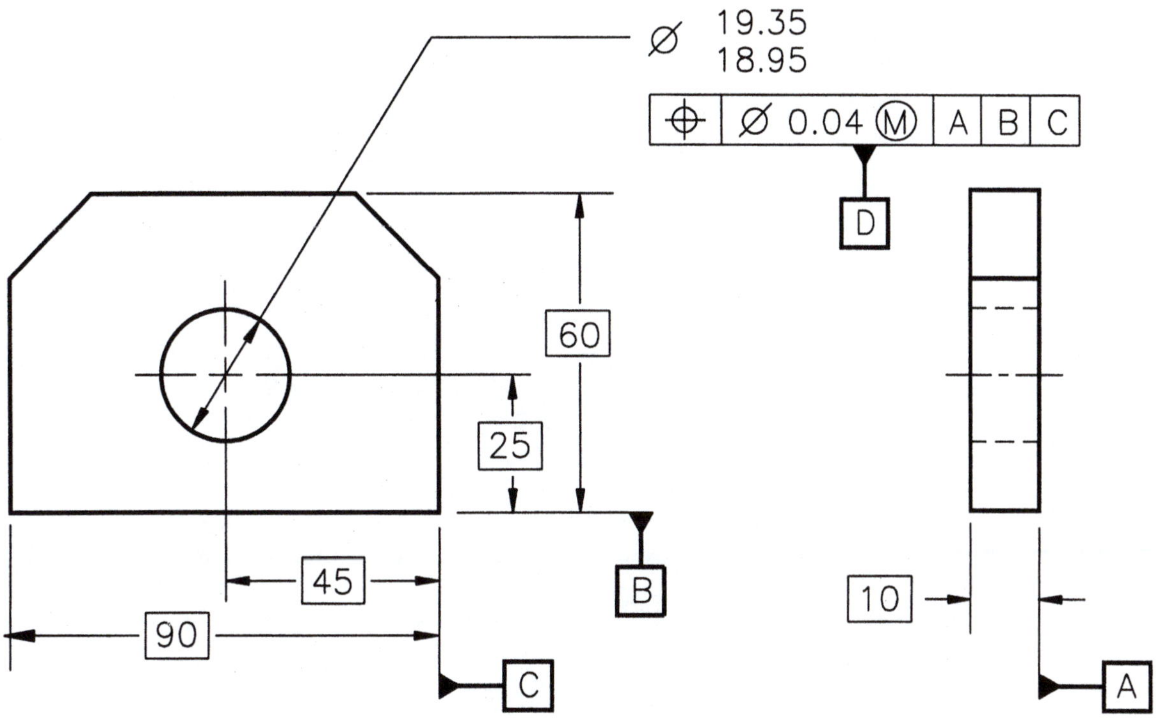

1. What is the geometric characteristic? ______________________

2. What type of geometric tolerance is applied? ______________________

3. What does the symbol Ⓜ represent? ______________________

4. What is the MMC of the feature? ______________________

5. What is the geometric tolerance at MMC Feature? ______________________

6. What is the virtual condition? ______________________

7. What would the geometric tolerance be at LMC? ______________________

COURSE ________	STUDENT ______________________	DATE ________	PROBLEM 13-13

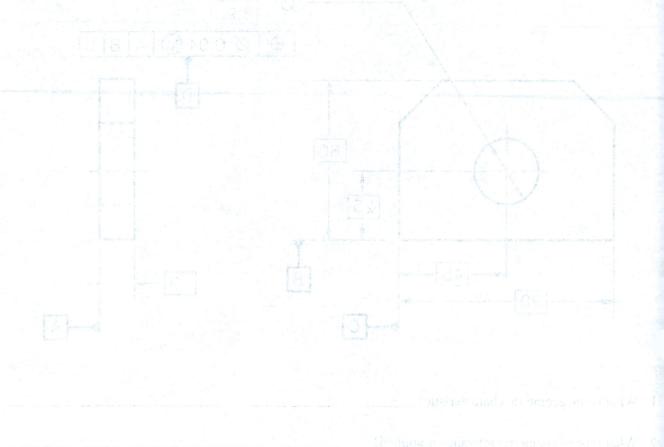

CHAPTER 14

Pictorial Drawings and Technical Illustrations

Problem 14-1 Using Dividers

Draw the following object in the given pictorial representations. Use dividers to transfer view projections. Use (and show) common approximate angles and scales.

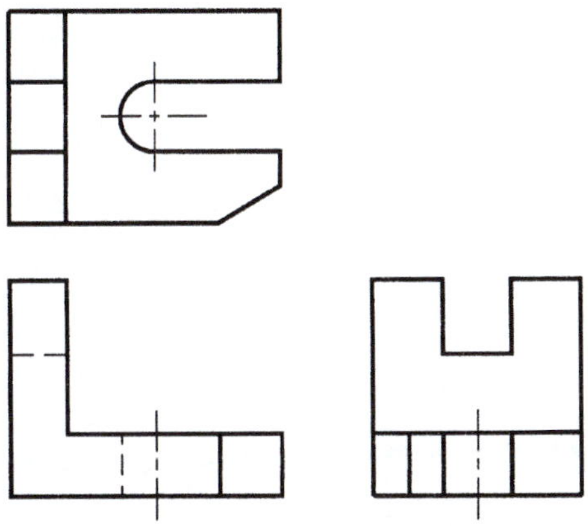

Trimetric Projection

Isometric Projection

Cavalier Oblique

Dimetric Projection

Cabinet Oblique

COURSE ________ STUDENT ______________________________ DATE ________ PROBLEM 14-1

Problem 14-2 Identified Perspective

Draw the identified perspective in the following examples. Label where necessary the SP, PP, VP, VPL, HL, and GL. Make sure your drawing remains within the boundaries.

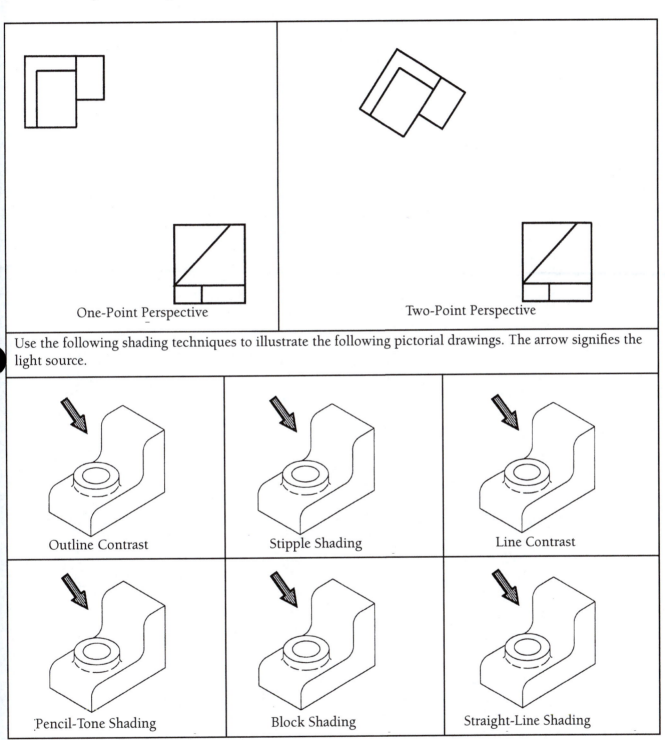

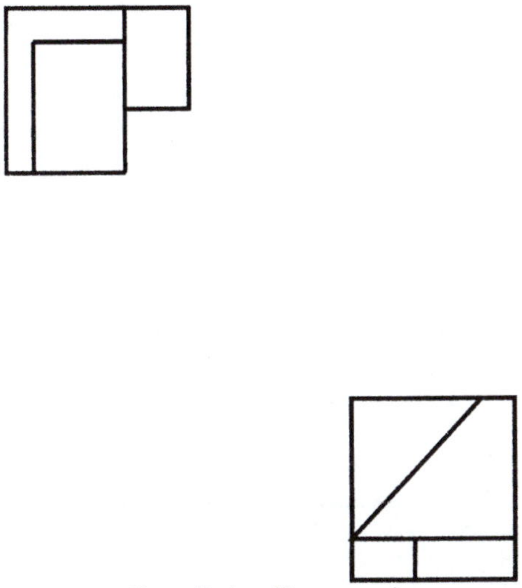
One-Point Perspective

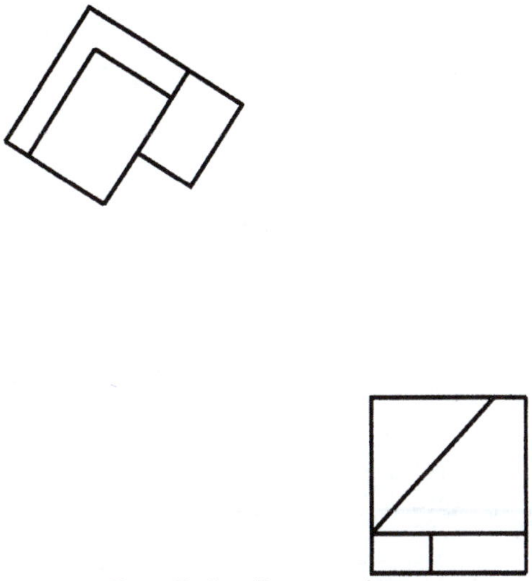
Two-Point Perspective

Use the following shading techniques to illustrate the following pictorial drawings. The arrow signifies the light source.

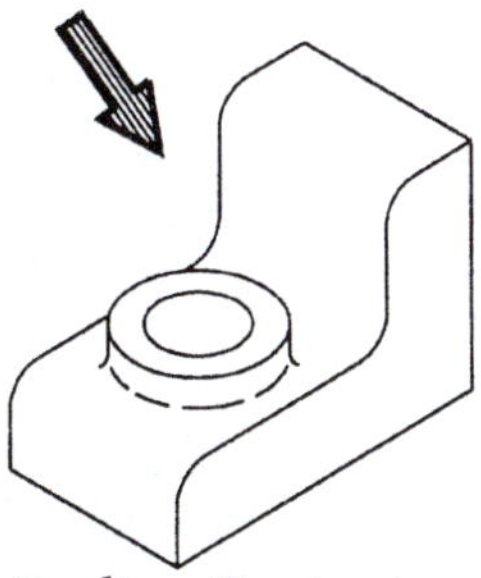
Outline Contrast

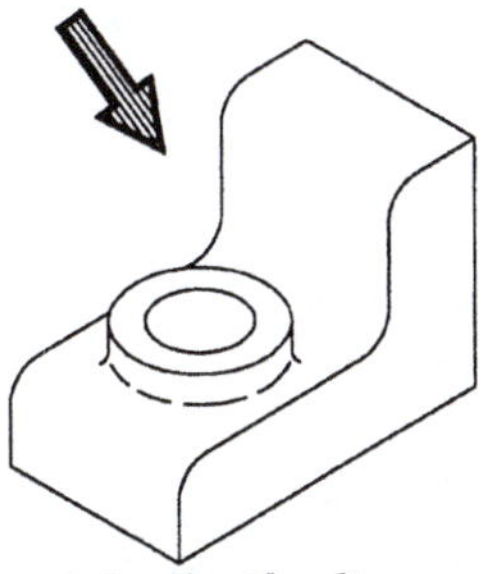
Stipple Shading

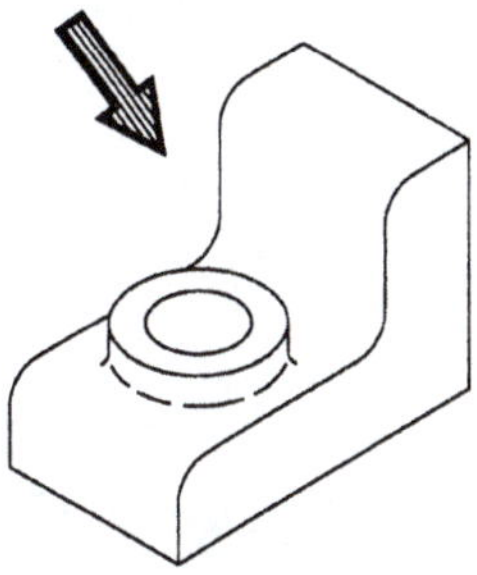
Line Contrast

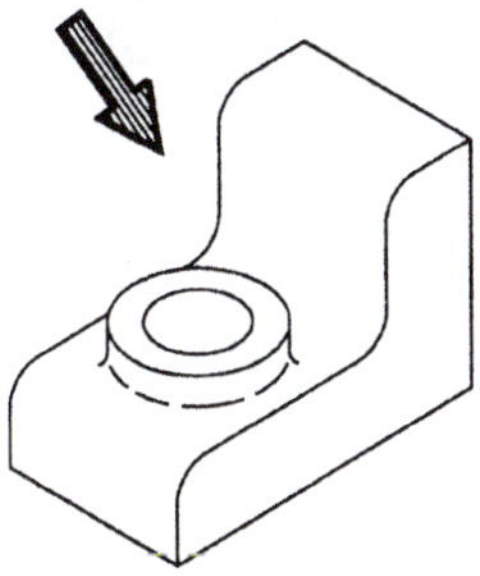
Pencil-Tone Shading

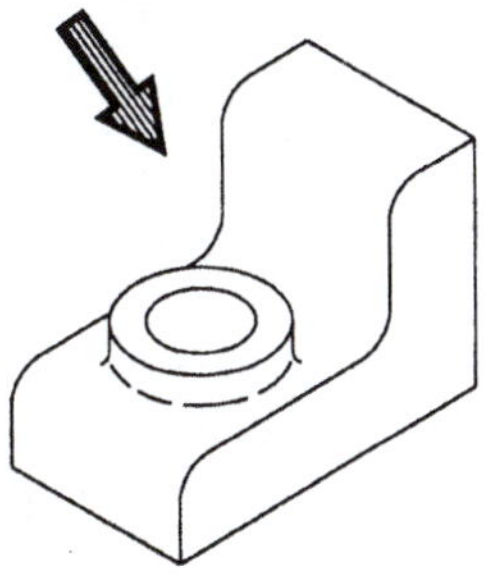
Block Shading

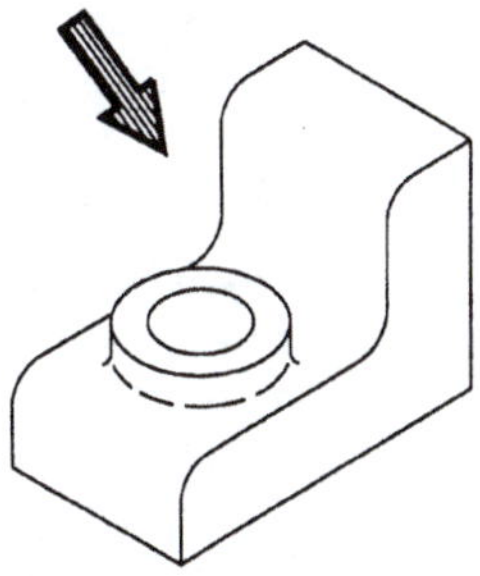
Straight-Line Shading

COURSE________ STUDENT________________________ DATE________ PROBLEM 14-2

SECTION 4

Working Drawings

CHAPTER 15

Working Drawings

Problem 15-1 Working Drawing Definitions

Define the following:

Working drawings ____

____

Detail drawings ____

____

Assemblies ____

____

Part list ____

____

Monodetail drawing ____

____

Multidetail drawing ____

____

COURSE ________	STUDENT ______________________	DATE ________	PROBLEM 15-1

Problem 15-2 Monodetail Drawing Assemblies

The practice of producing a product in-house or by outsourcing has been a controversial topic that people have debated for some time. In the following space, list three cons and pros for outsourcing and in-house production of monodetail drawing assemblies.

COURSE________	STUDENT________________________________	DATE________	PROBLEM 15-2

CHAPTER 16

Mechanisms: Linkages, Cams, Gears, and Bearings

Problem 16-1 Linkages

1. What type of linkage is shown in the problem below?
2. In the linkage on the right, does the rocker arm oscillate or rotate? (Hint: Is the rocker arm shorter or longer than the crank arm?)
3. Determine the rotation or angle of oscillation of the rocker. Be sure to show construction lines.

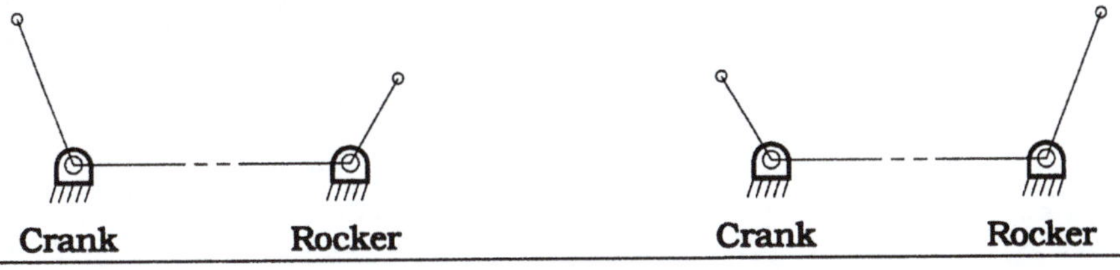

In the slider crank mechanism below, the connecting rod length is 4.1 inches. Show the slider's extreme right and left positions. Determine the stroke of the piston. Show all construction lines.

The example below involves a combination of four-bar linkage and a slider mechanism. Determine the path of point S (on the slider mechanism) using the following information: Point X is the pivot point of the crank arm which measures 0.8 inches. The connecting rod measures 3.5 in. The pivot point of the rocker arm (point Z) is located to the right of point X. The rocker arm measures 1.3 in. Point Y is the pivot point of the slider arm. The distance from the end of the crank arm to point S on the slider arm is 1.8 in. Draw to scale and show major construction lines.

COURSE ________	STUDENT ________________________	DATE ________	PROBLEM 16-1

Problem 16-2 Displacement Diagram

Given: Displacement diagram grid.

Lay out a diagram with the following specifications:

Rise 1.00" 90° harmonic motion, dwell 30°, rise .50", 60° harmonic motion to highest point on cam, dwell 30°, fall to starting level 135° uniform acceleration, dwell 15°, clockwise rotation.

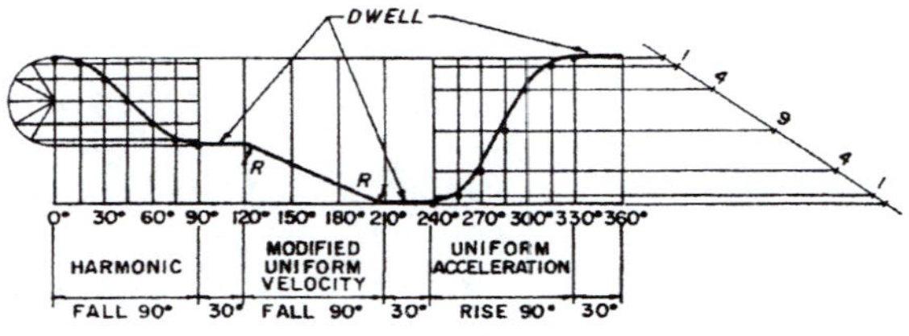

Combination of motions

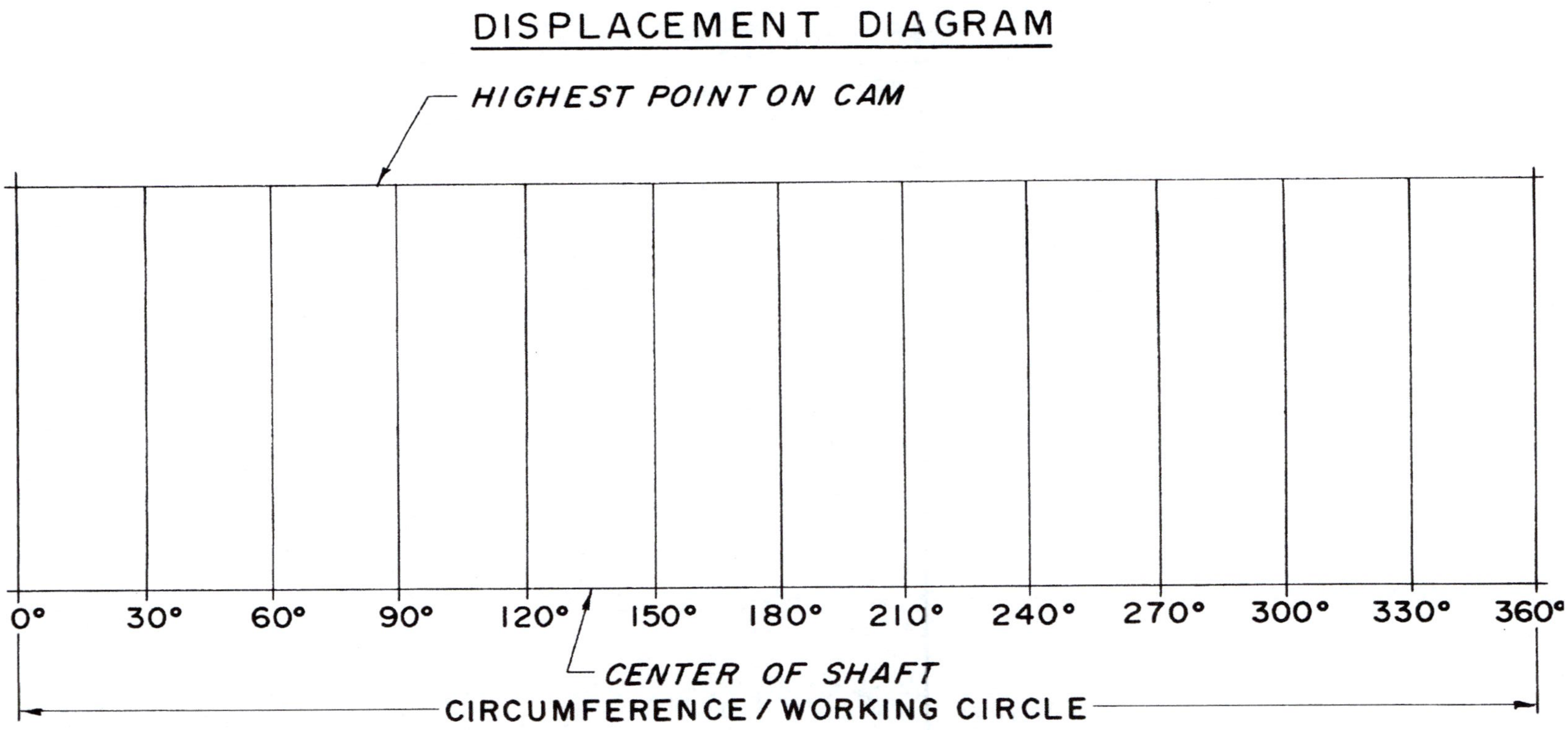

COURSE________ STUDENT____________________ DATE________ PROBLEM 16-2

Problem 16-3 Cam Follower

Construct and fully dimension the cam described in the cam diagram below.

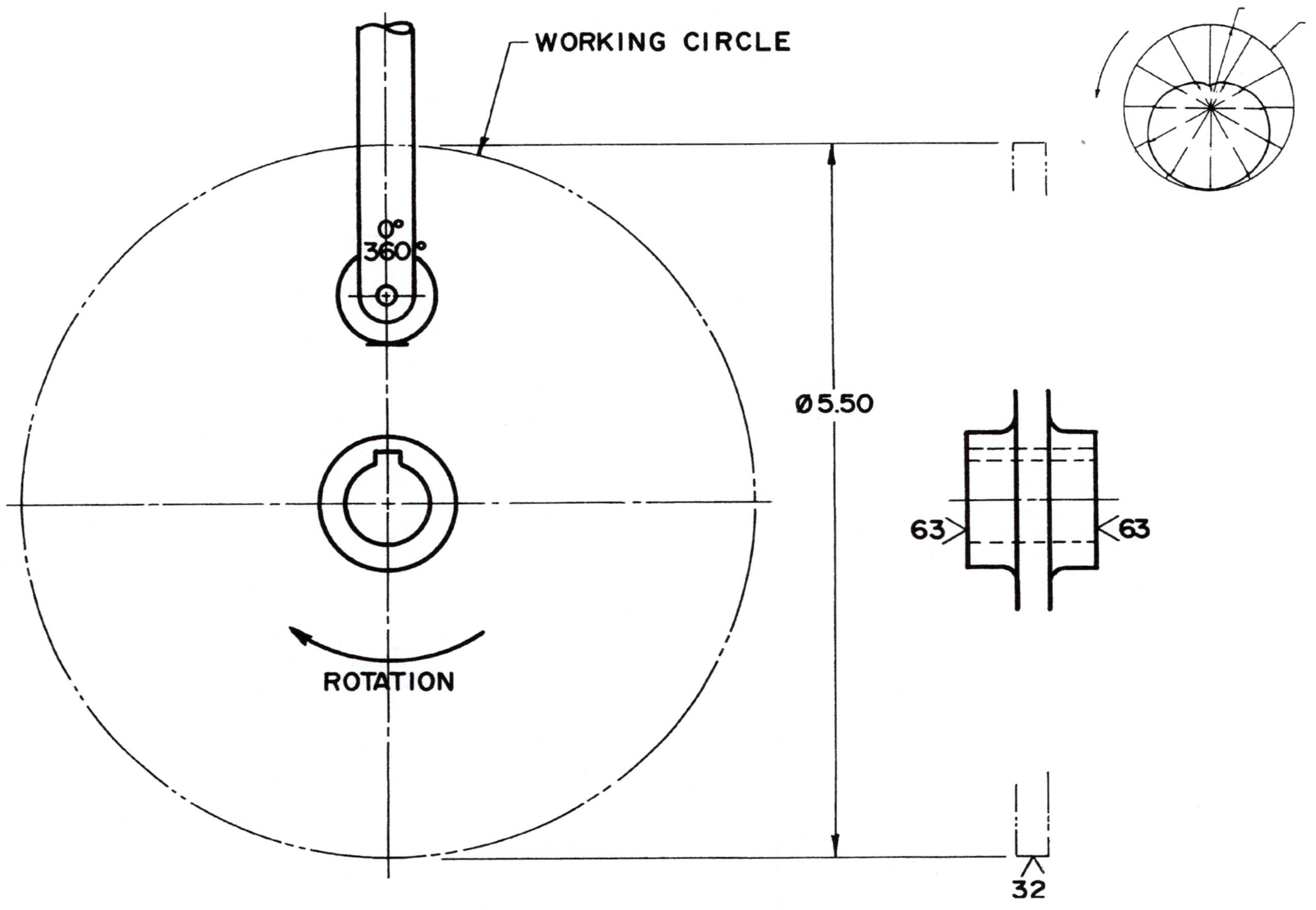

COURSE______ STUDENT______ DATE______ PROBLEM 16-3

CHAPTER 17

Belt and Chain Drives

Problem 17-1 Chain Drives

What specific type of chain is shown? In what kind of situation is it used?

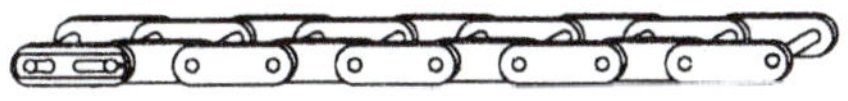

____

____

____

Identify this image. Under what conditions is it used?

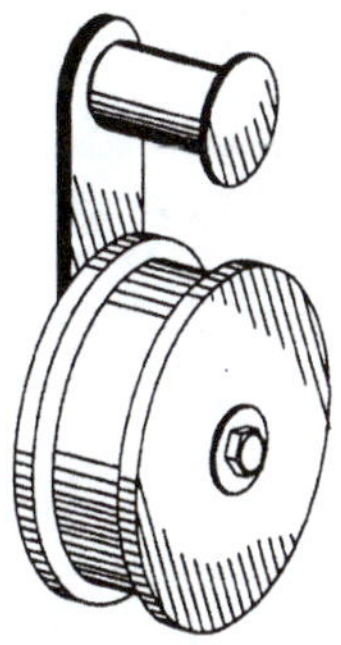

____

____

____

This is a cross section of what specific type of belt?

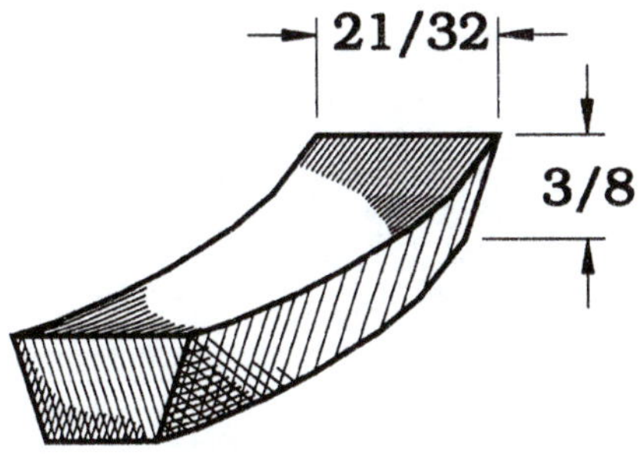

____

____

____

What type of sprocket is shown?

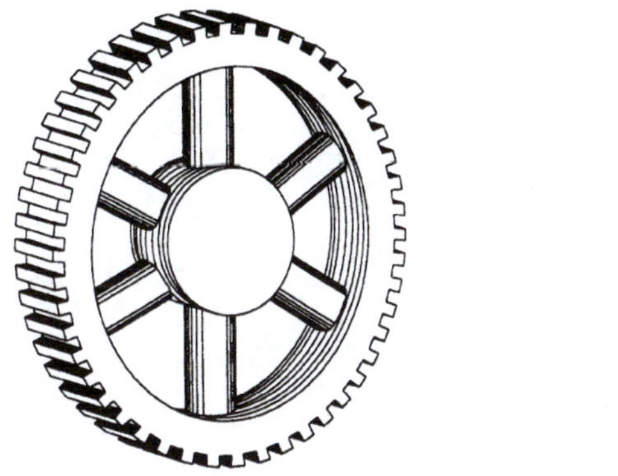

____

____

____

What specific type of chain is shown, and in what category is this chain?

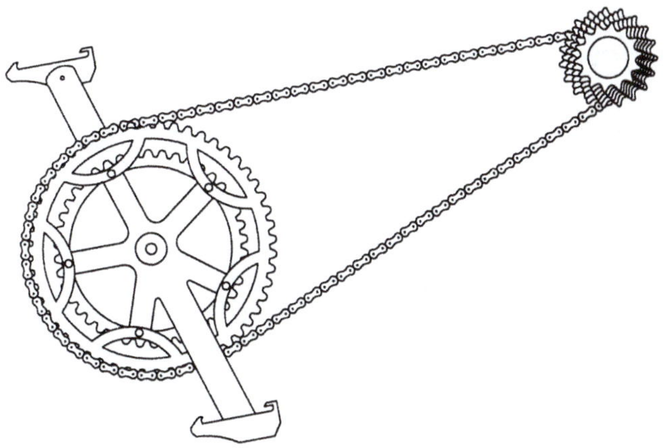

____

____

____

COURSE__________ STUDENT______________________________ DATE__________ PROBLEM 17-1

Problem 17-2 Belt Drives

Answer the following questions.

A. In what situations are flat belts most likely used?

B. What is the pitch line of a belt?

C. Describe the size and position of an inside idler belt.

D. Define *interpolate*.

E. Describe *belt velocity*. When does it become important?

F. What type of shift arrangements should be avoided and why?

G. When designing roller chain drives, what is the preferred center distance between shafts?

H. Describe *service factor*.

COURSE ________ STUDENT ______________________ DATE ________ PROBLEM 17-2

Provide the word or phrase that best fits each of the following statements.

I. These are the most expensive precision chains to manufacture. They are used where high speed and quiet operation are required.

____

____

J. These have a notched underside that contacts a pulley with the same design on the circumference.

____

____

K. These chains are designed for smooth, free-running operation at high speeds and high power.

____

____

L. This is the least expensive of the precision chains, and it is designed to carry heavier loads than nonprecision chains.

____

____

M. This is a combination design between the detachable and the offset sidebar roller chain.

____

____

N. These are commonly used for control mechanisms for light-duty, low-power applications.

____

____

COURSE________ STUDENT______________________ DATE________ PROBLEM 17-2

CHAPTER 18

Welding Processes and Representations

Problem 18-1 Welding Symbols

Fill in all spaces and indicate the meaning of each number.

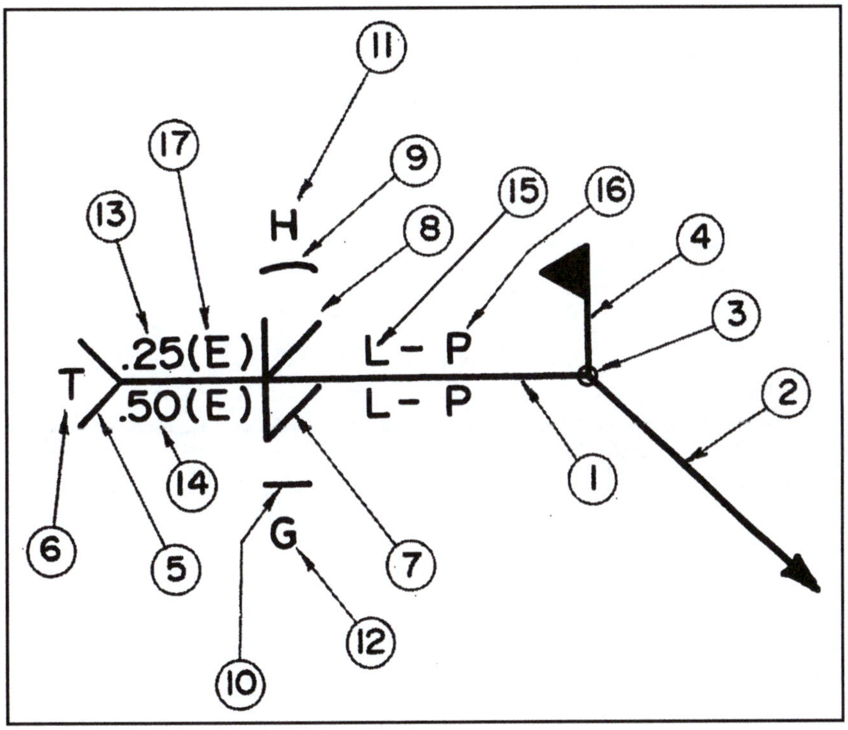

1 ______________________________

2 ______________________________

3 ______________________________

4 ______________________________

5 ______________________________

6 ______________________________

7 ______________________________

8 ______________________________

9 ______________________________

10 ______________________________

11 ______________________________

12 ______________________________

13 ______________________________

14 ______________________________

15 ______________________________

16 ______________________________

17 ______________________________

COURSE ________	STUDENT ________________	DATE ________	PROBLEM 18-1

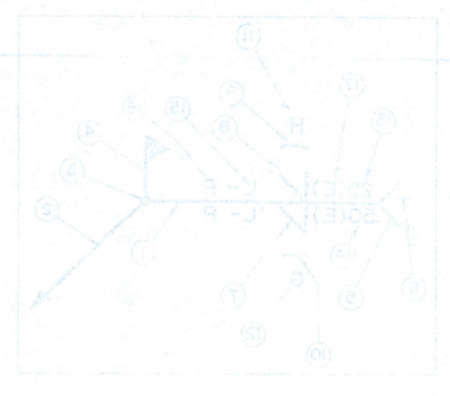

Problem 18-2 Welding Symbols

Complete the welding symbols in order to achieve the illustrated requirement.

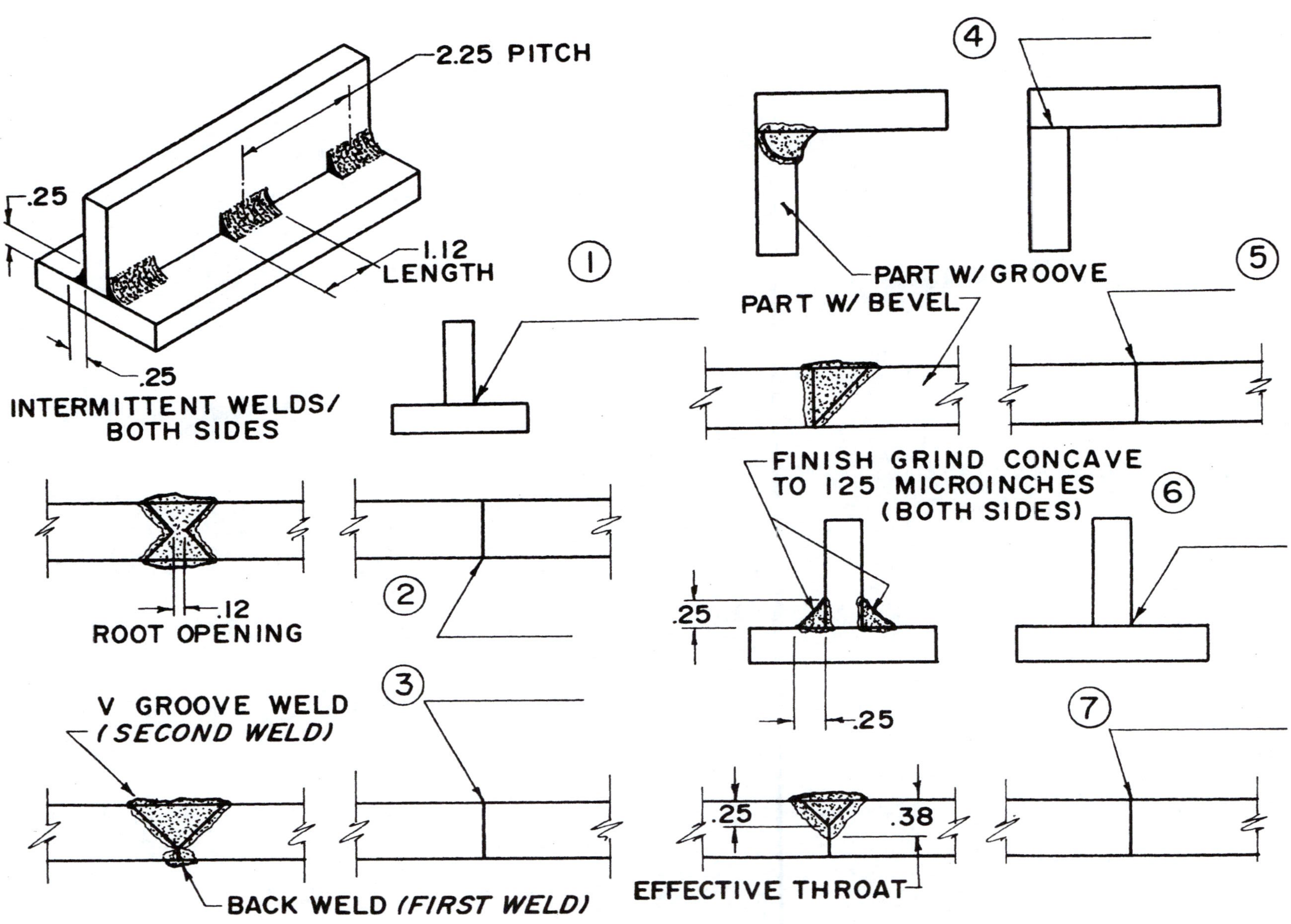

COURSE________ STUDENT____________________ DATE_______ PROBLEM 18-2

Problem 18-3 Changing the Casting Weldment

Refer to the isometric drawing of a cast steel bearing and do the following:

1. Change the casting into a weldment and prepare the front and side views in the space indicated. Half scale. Show weld symbol but do not dimension the object.
2. Change the 38 and 35 dimensions each to 36.5 mm.
3. Add four Φ 15-mm holes in the base at the centers of the R12s.
4. Ignore Section A-A.

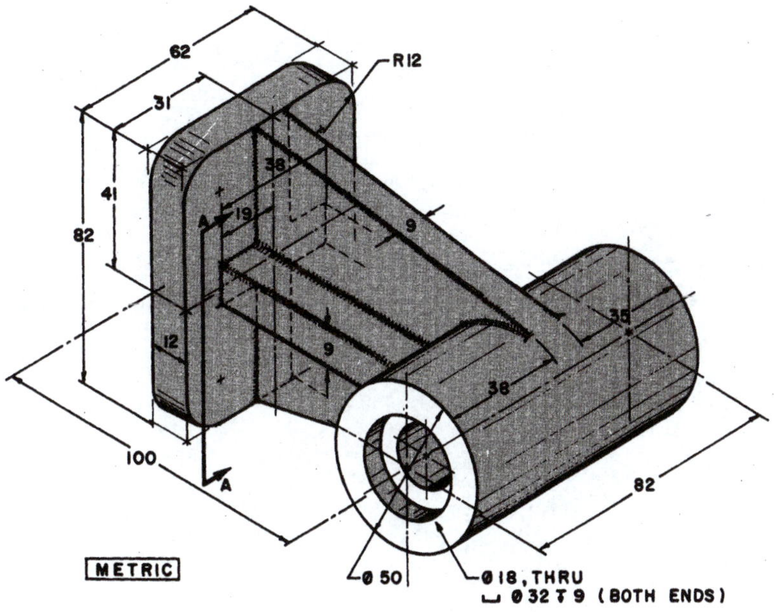

COURSE	STUDENT	DATE	PROBLEM 18-3

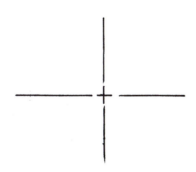

Problem 18-3 Changing the Casting Weldment

Refer to the isometric drawing of a cast steel bearing and do the following:

1. Change the casting into a weldment and prepare the front and side views in the space indicated. Half scale. Show weld symbol but do not dimension the object.
2. Change the 58 and 35 dimensions each to 30 mm.
3. Add four Ø 15-mm holes in the base at the centers of the R12s.
4. Ignore section A-A.

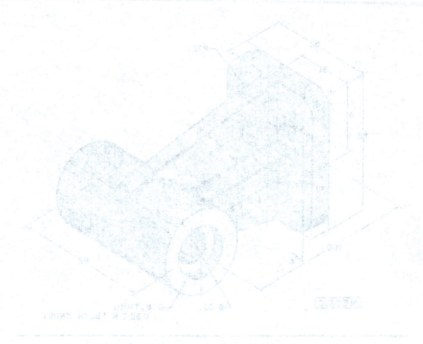

Problem 18-4 Changing the Casting Weldment

Refer to the isometric drawing of a cast steel pipe connector and do the following:

1. Change the casting into a weldment and prepare the front and side views in the space indicated. Half scale.
2. Show weld symbols but do not dimension the object.
3. Ignore Section A-A

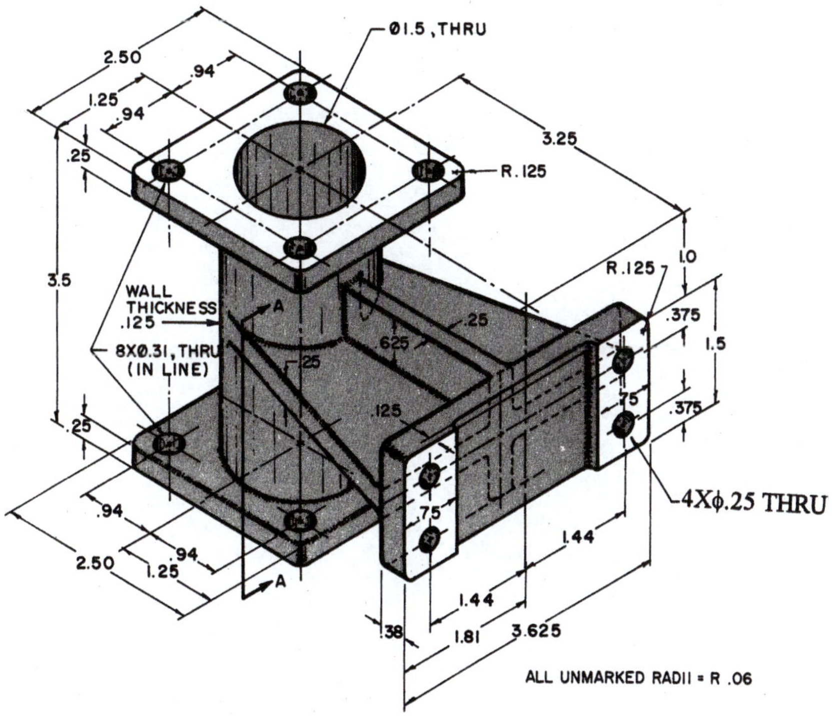

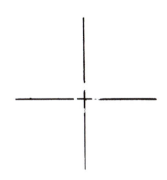

COURSE ________	STUDENT ________________	DATE ________	PROBLEM 18-4

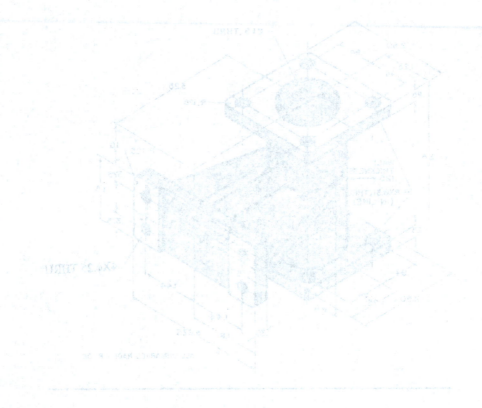

Problem 18-5 Welding Symbols

Draw the symbol required for the following welds.

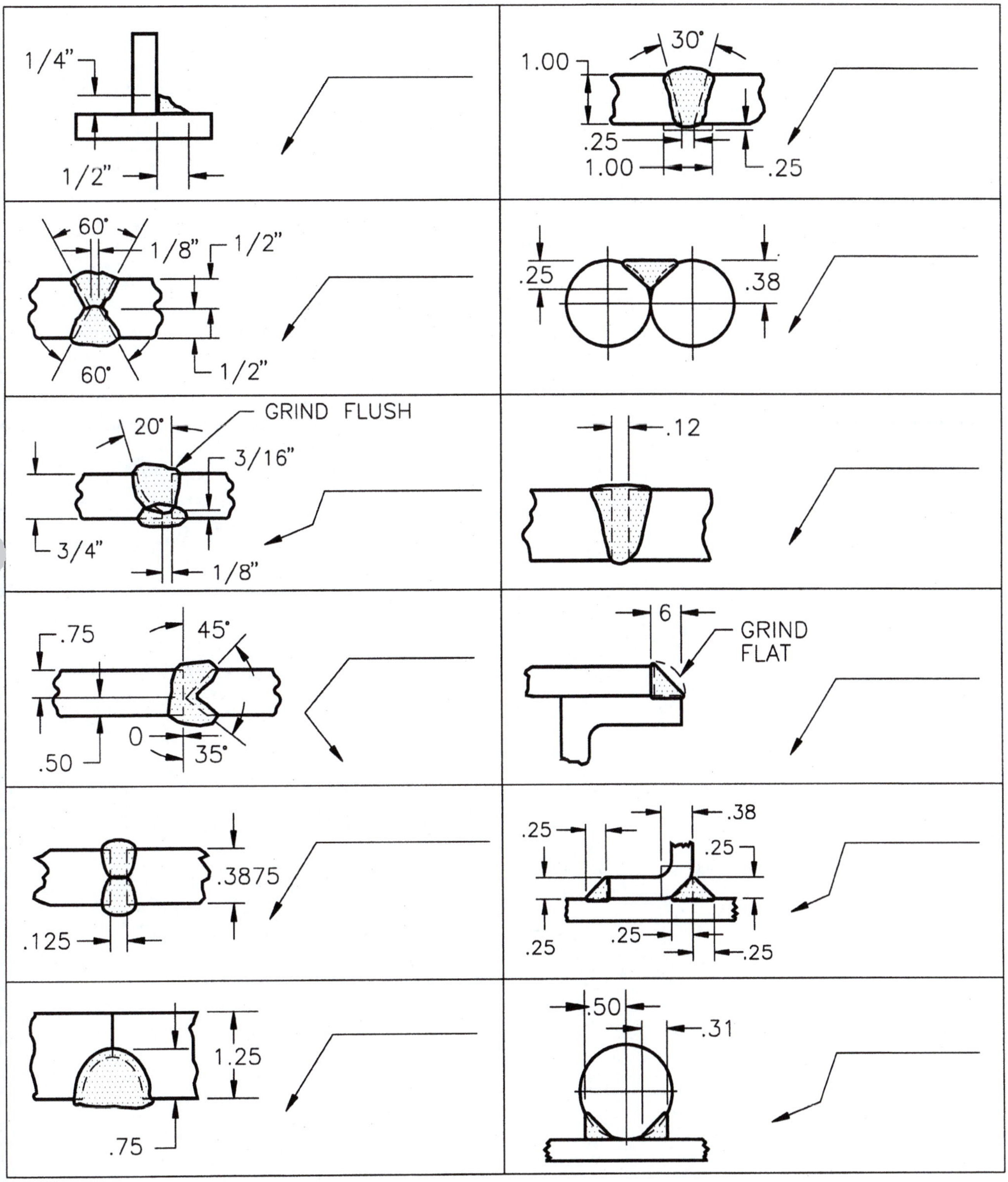

COURSE ________	STUDENT ____________________	DATE ________	PROBLEM 18-5

Problem 18-6 Welding Symbols

Draw the symbol required for the following welds.

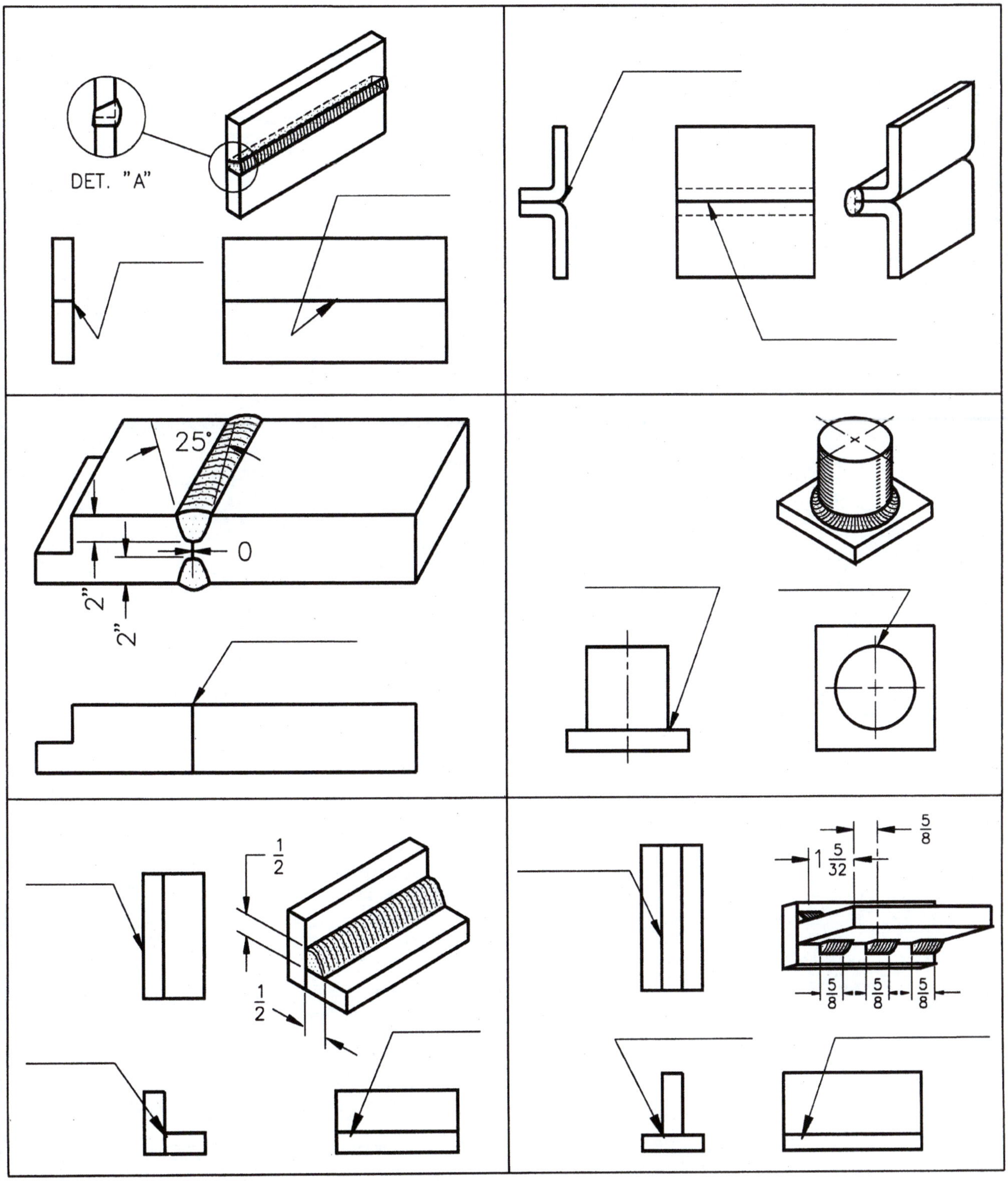

COURSE ________ STUDENT ______________________ DATE ________ PROBLEM 18-6

Problem 18-6 Welding Symbols

Draw the symbols required for the following welds.

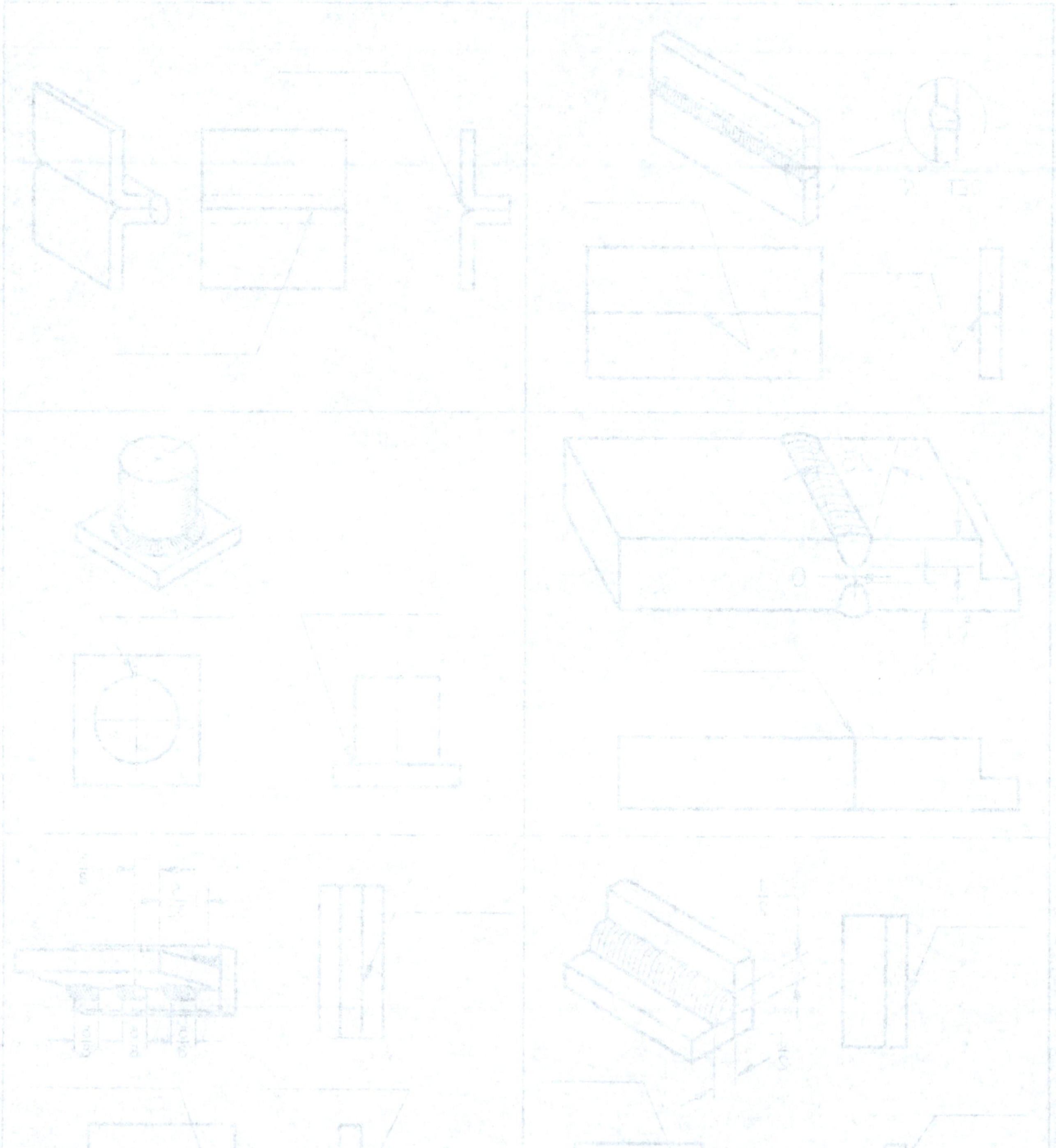

COURSE	STUDENT	DATE	PROBLEM 18-6

Problem 18-7 Fill in the Missing Information

Fill in the missing information and answer the following questions.

Abbreviation	Represents	Type of Welding or Allied Process
RSW	RESISTANCE SPOT WELDING	RESISTANCE WELDING
BB		
		OXYFUEL GAS WELDING
	DIFFUSION WELDING	
		RESISTANCE WELDING
		ARC WELDING
	FORGE WELDING	
TIG		
	ELECTRIC ARC SPRAYING	
	DIP SOLDERING	
	ARC WELDING	
IRS		
CEW		
PGW		
		ARC CUTTING
		SOLDERING
TB		
EBW–NV		

A. What is the difference between metal arc welding and gas metal arc welding?

B. When should a weld length be given?

C. When is a tail added to the welding symbol?

D. What is the concern with oversized welds and undersized welds?

COURSE ________	STUDENT ________	DATE ________	PROBLEM 18-7

Problem 18-8 Picnic Table

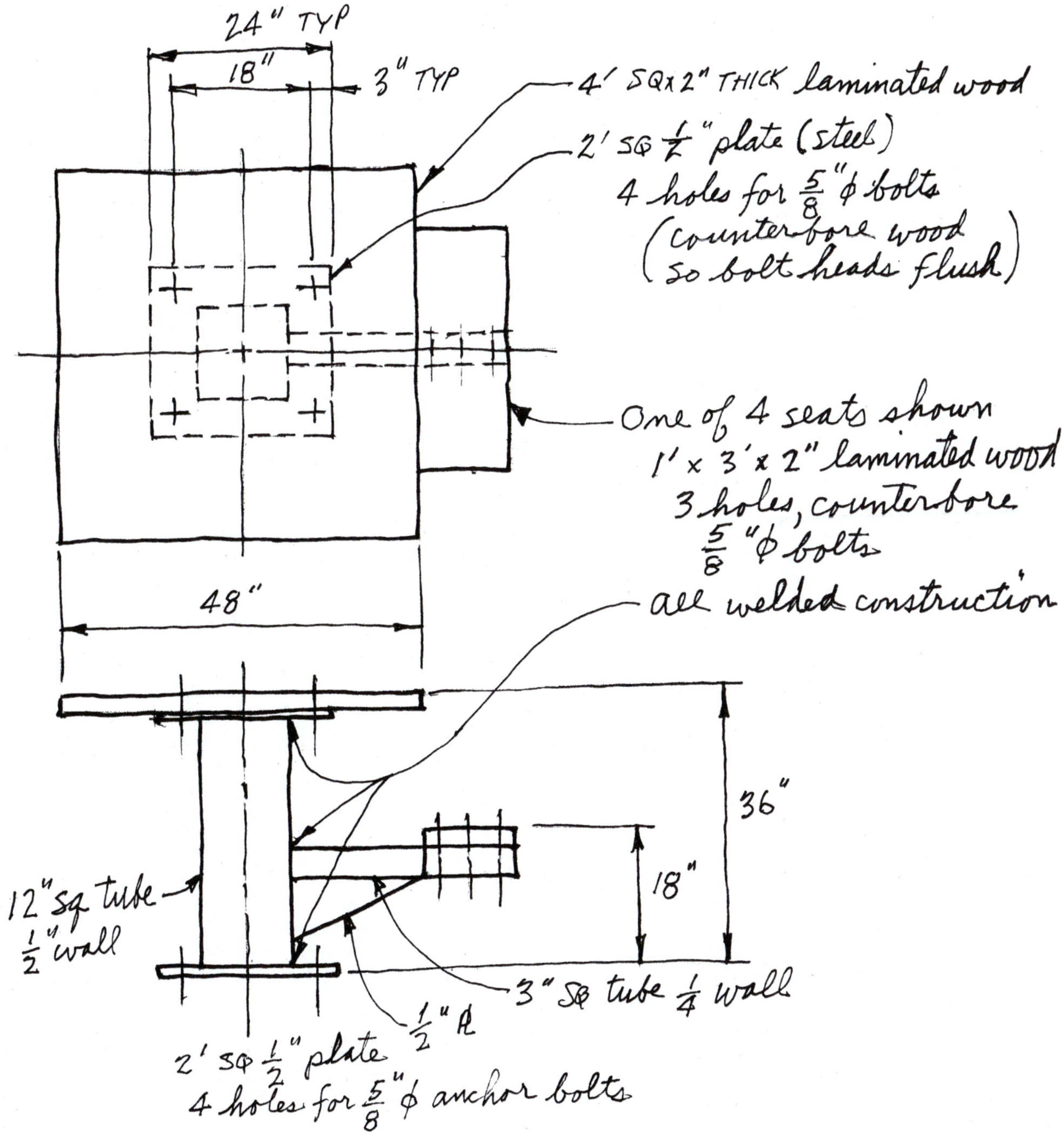

The basic structure of the pictured picnic table is to be a weldment of steel plate and square tubing, and the seats and table top are to be 2" thick laminated wood.

1. Prepare a detail drawing of the steel weldment. Include a parts list.
2. Prepare detail drawings of the table top and the four seats. Include a parts list.
3. Prepare an assembly drawing of the picnic table complete with fasteners and an assembly parts list.

COURSE ________	STUDENT ____________________	DATE ________	PROBLEM 18-8

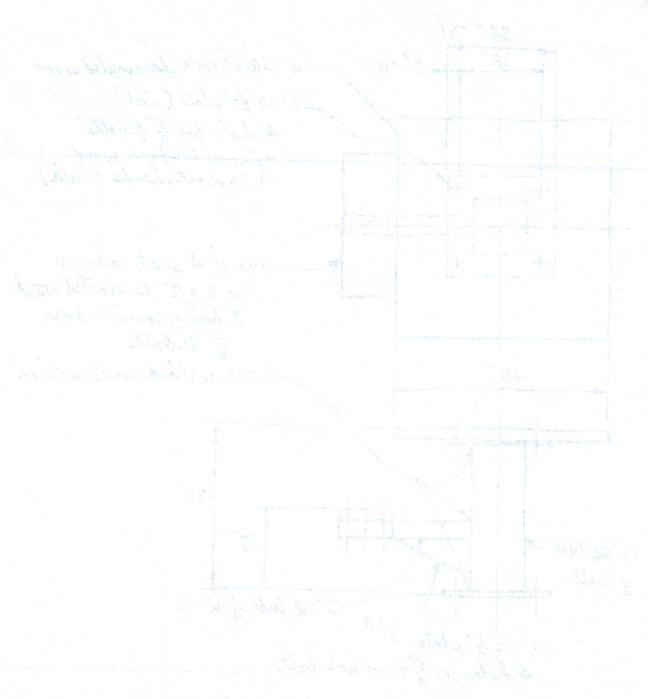

SECTION 5

Specialty Drafting and Design

CHAPTER 19

Precision Steel Metal Drafting

Problem 19-1 Seams and Hems

Follow instructions below.

Identify whether each of the following items is a seam or a hem.

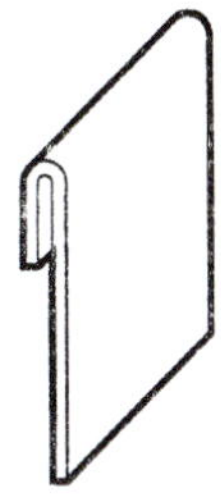
A ______________

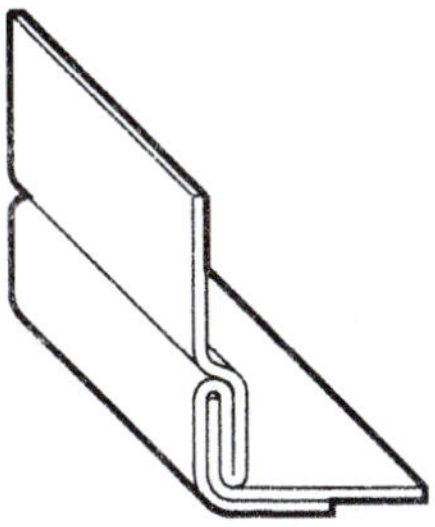
B ______________

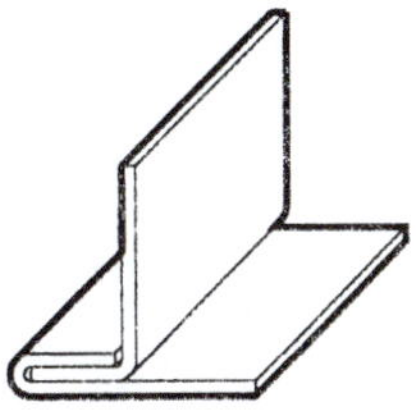
C ______________

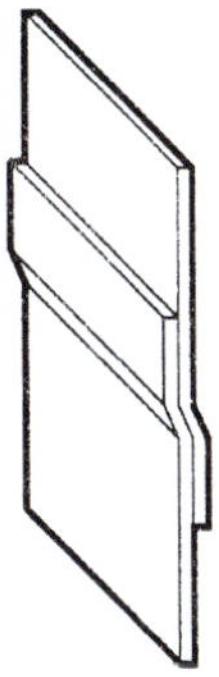
D ______________

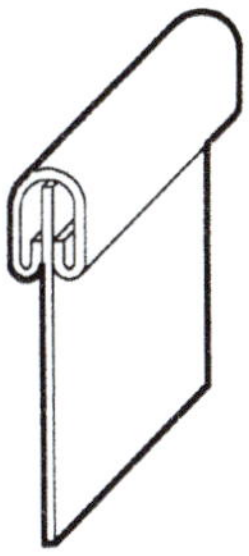
E ______________

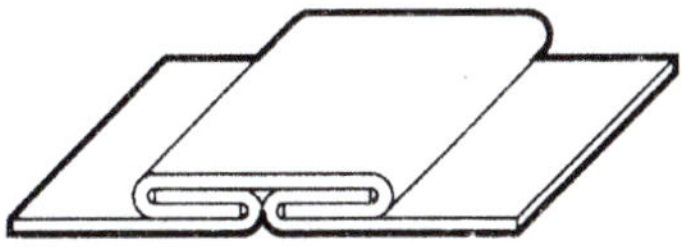
F ______________

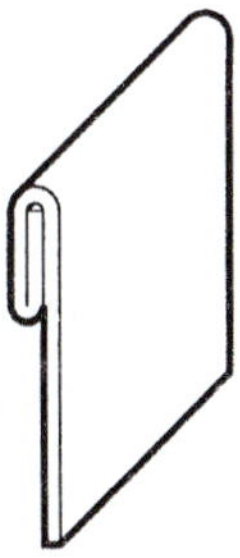
G ______________

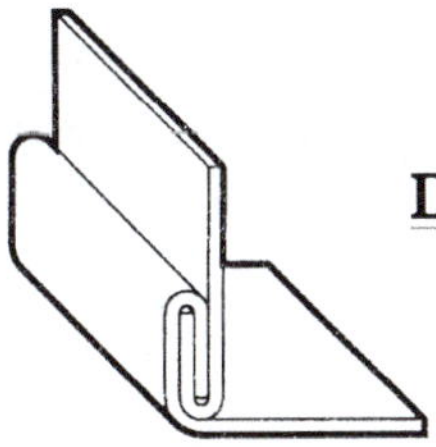
H ______________

I ______________

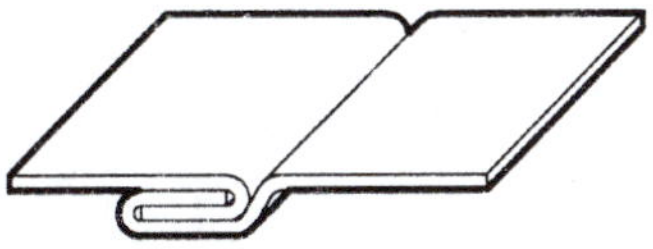
J ______________

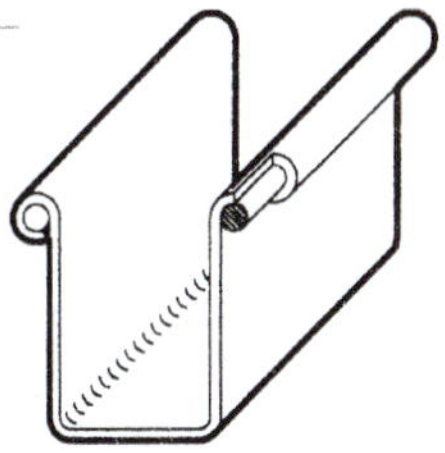
K ______________

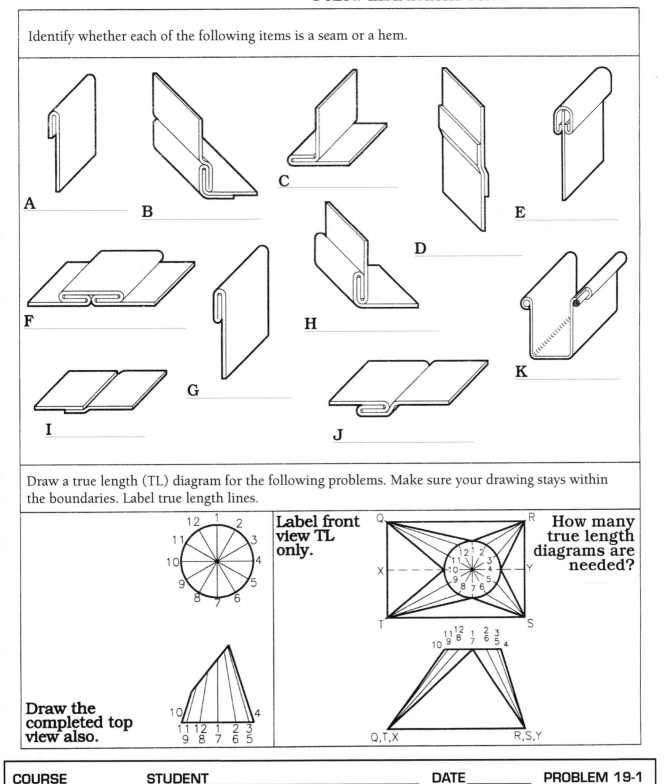

Draw a true length (TL) diagram for the following problems. Make sure your drawing stays within the boundaries. Label true length lines.

Draw the completed top view also.

Label front view TL only.

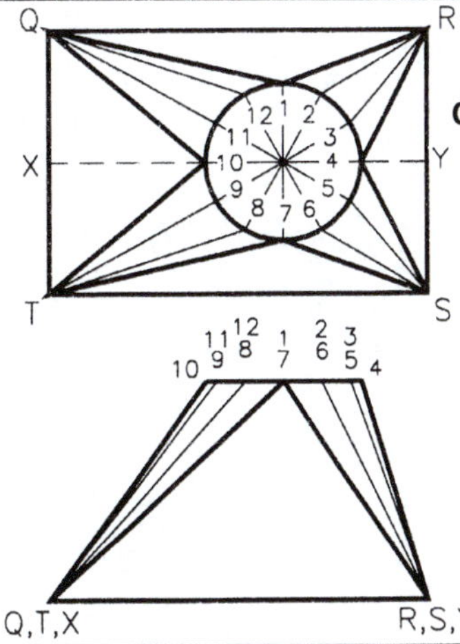

How many true length diagrams are needed?

COURSE ______	STUDENT ______________	DATE ______	PROBLEM 19-1

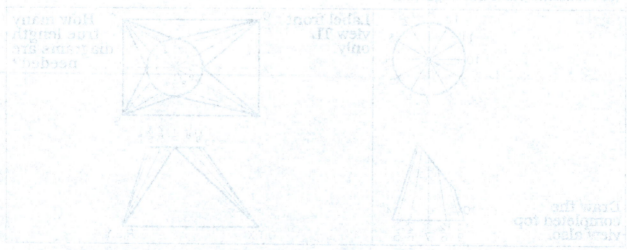

Problem 19-2 Point of Intersection

Determine the points of intersection in the following examples. Show construction lines and final line of intersection. One of these problems has no solution. Briefly explain why.

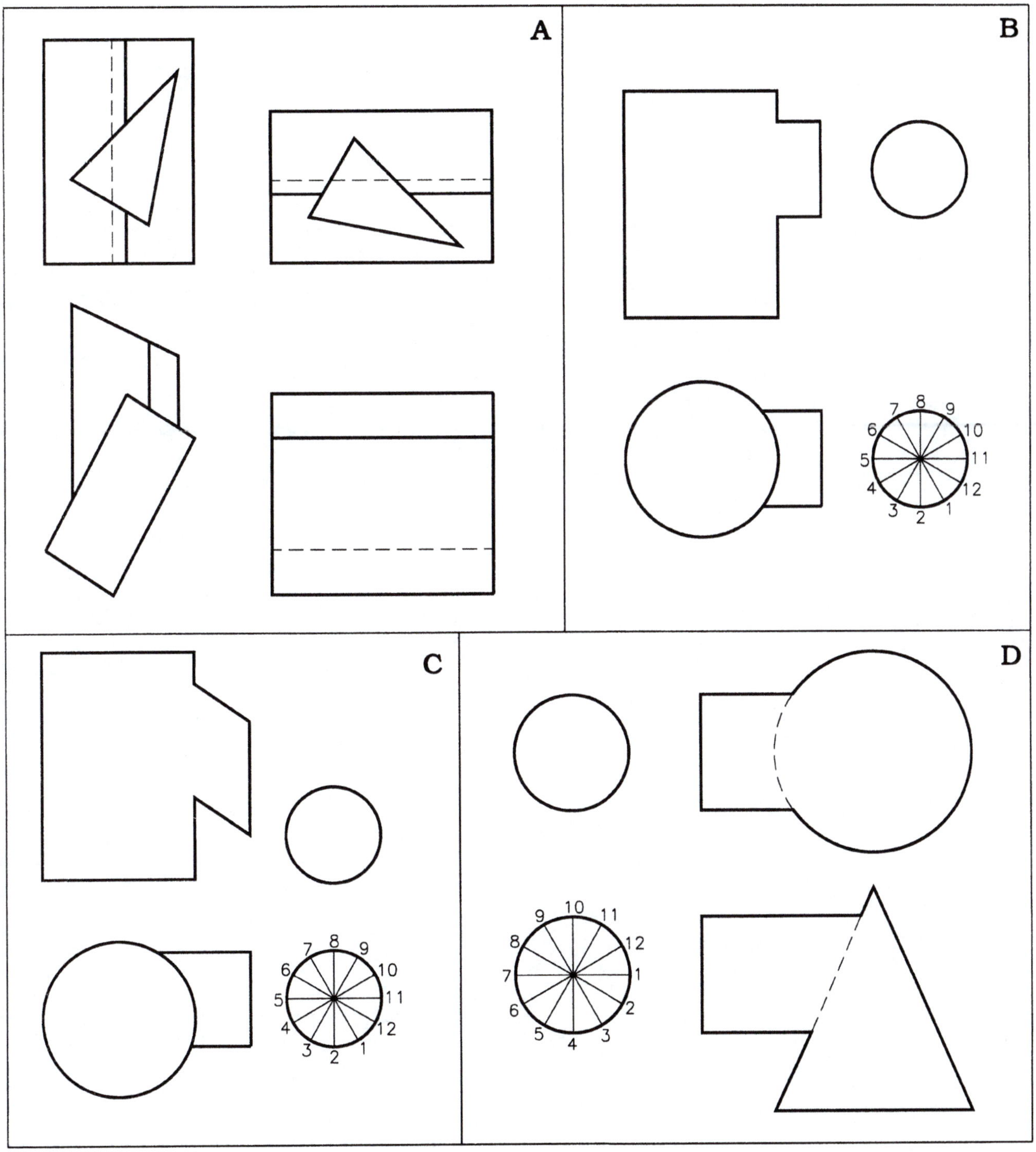

COURSE________ STUDENT____________________ DATE________ PROBLEM 19-2

Problem 19-3 Bend Allowances

A What is a bend allowance and why is it important?

B List the steps involved in constructing the following developments: prism, truncated right cone, transitional fitting.

C Explain the difference between precision and nonprecision sheet metal parts.

D Define the following terms: table, column, row, key, field, and record.

E Name the categories of materials from which a part may be manufactured.

F What are the advantages of linking a database to an AutoCAD entity?

G Given the following information, calculate the bend allowances:

Material = soft brass; radius = 0.125; angle = 45°; thickness = 0.0625

Material = cold-rolled steel; radius = 0.25; angle = 34°16'; thickness = 0.125

Material = soft steel; radius = 0.375; angle = 90°; thickness = 0.0313

H List the steps involved in constructing a development of the following: truncated prism, truncated pyramid, oblique prism.

I What is a radial line development and when is it used?

J What are developments and why are they used?

A. ______________________________

B. ______________________________

C. ______________________________

D. ______________________________

E. ______________________________

F. ______________________________

G. ______________________________

H. ______________________________

I. ______________________________

J. ______________________________

COURSE ________	STUDENT ____________________	DATE ________	PROBLEM 19-3

Problem 19-4 Development

Using AutoCAD, create a development of the following part. This drawing can be found on the CD that accompanies the textbook.

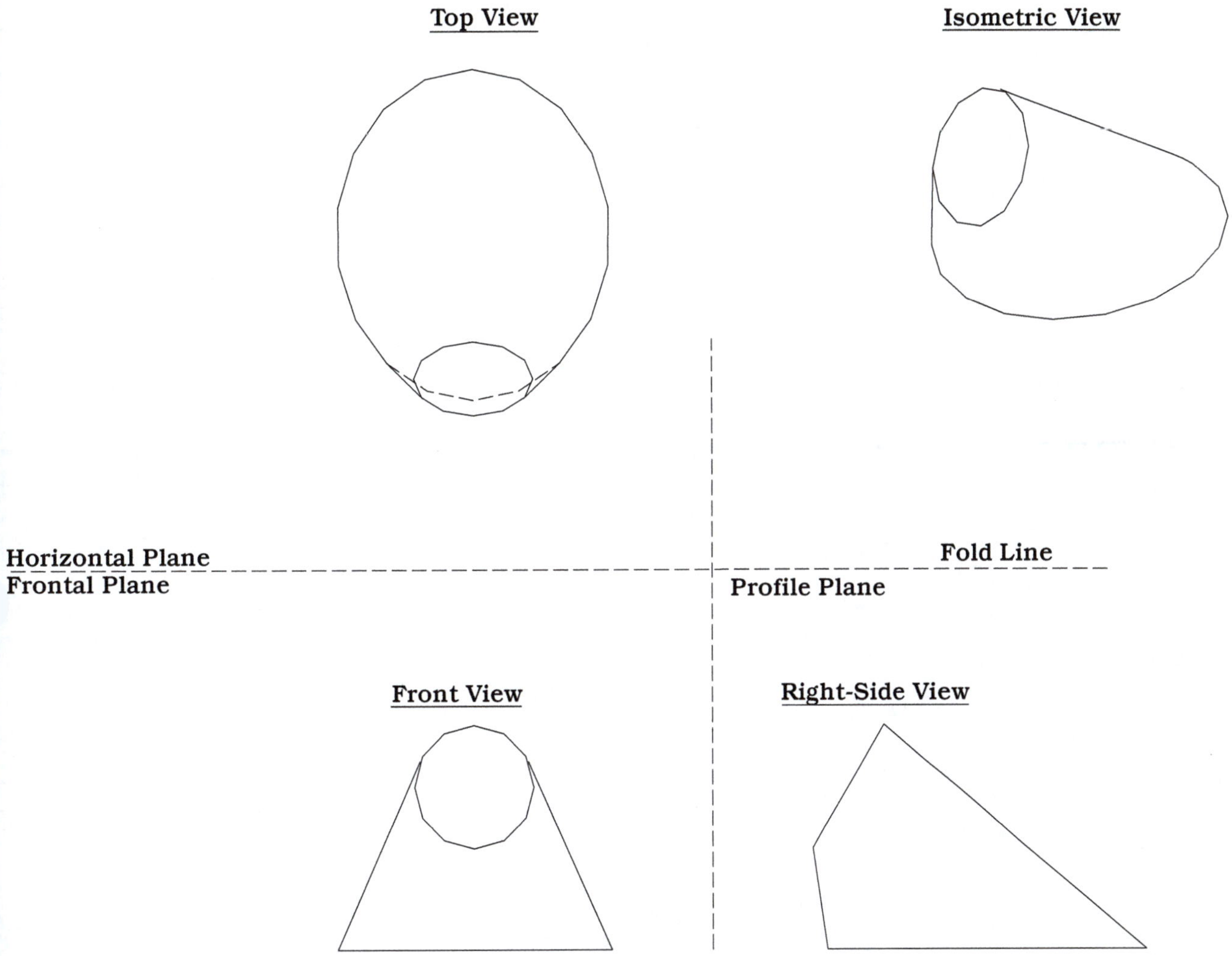

COURSE ________	STUDENT ______________________	DATE ________	PROBLEM 19-4

Problem 19-5 Development

Using AutoCAD, create a development of the following part. This drawing can be found on the CD that accompanies the textbook.

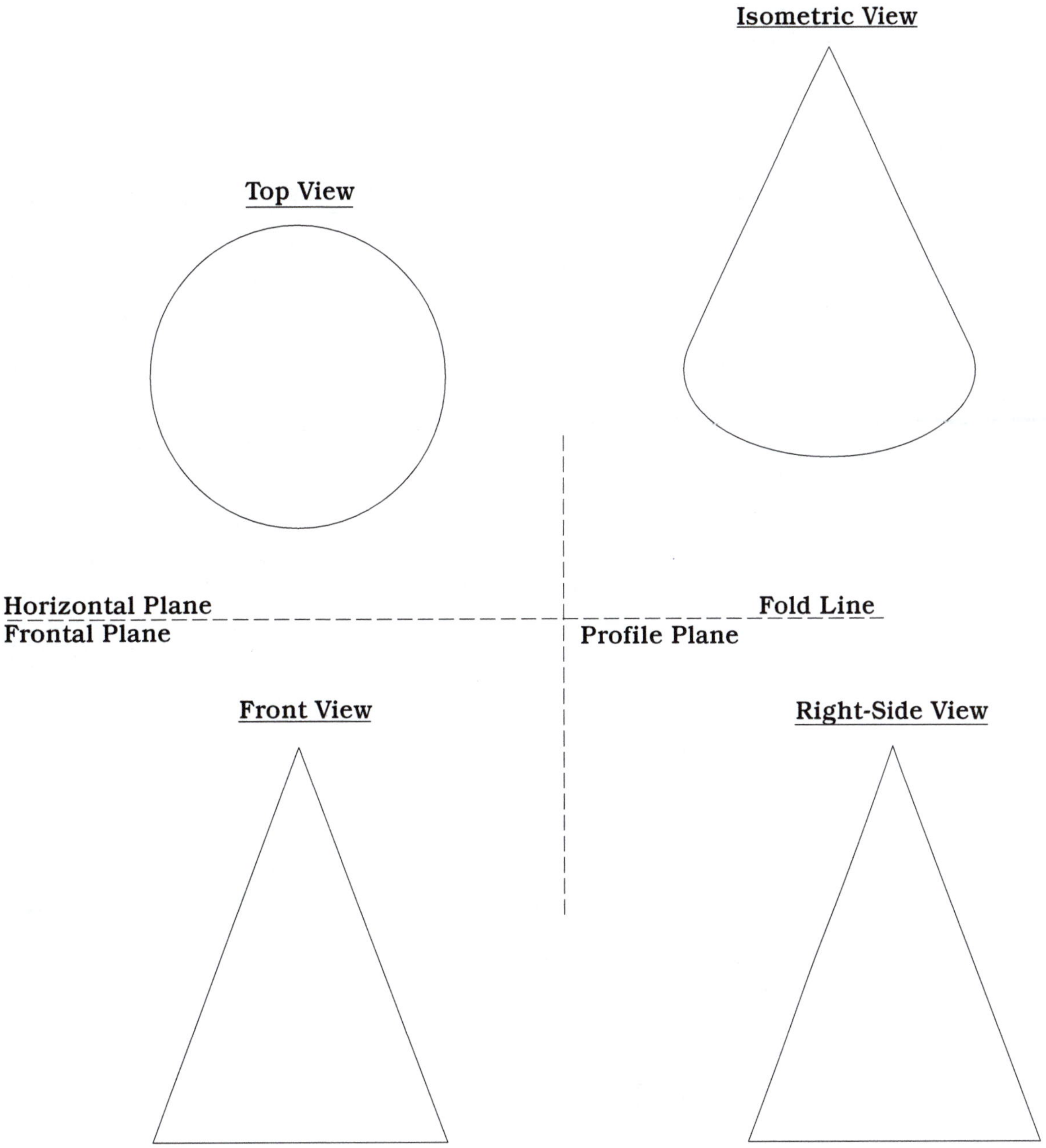

COURSE ________	STUDENT ________________	DATE ________	PROBLEM 19-5

Problem 19-6 Development

Using AutoCAD, create a development of the following sheet metal part.

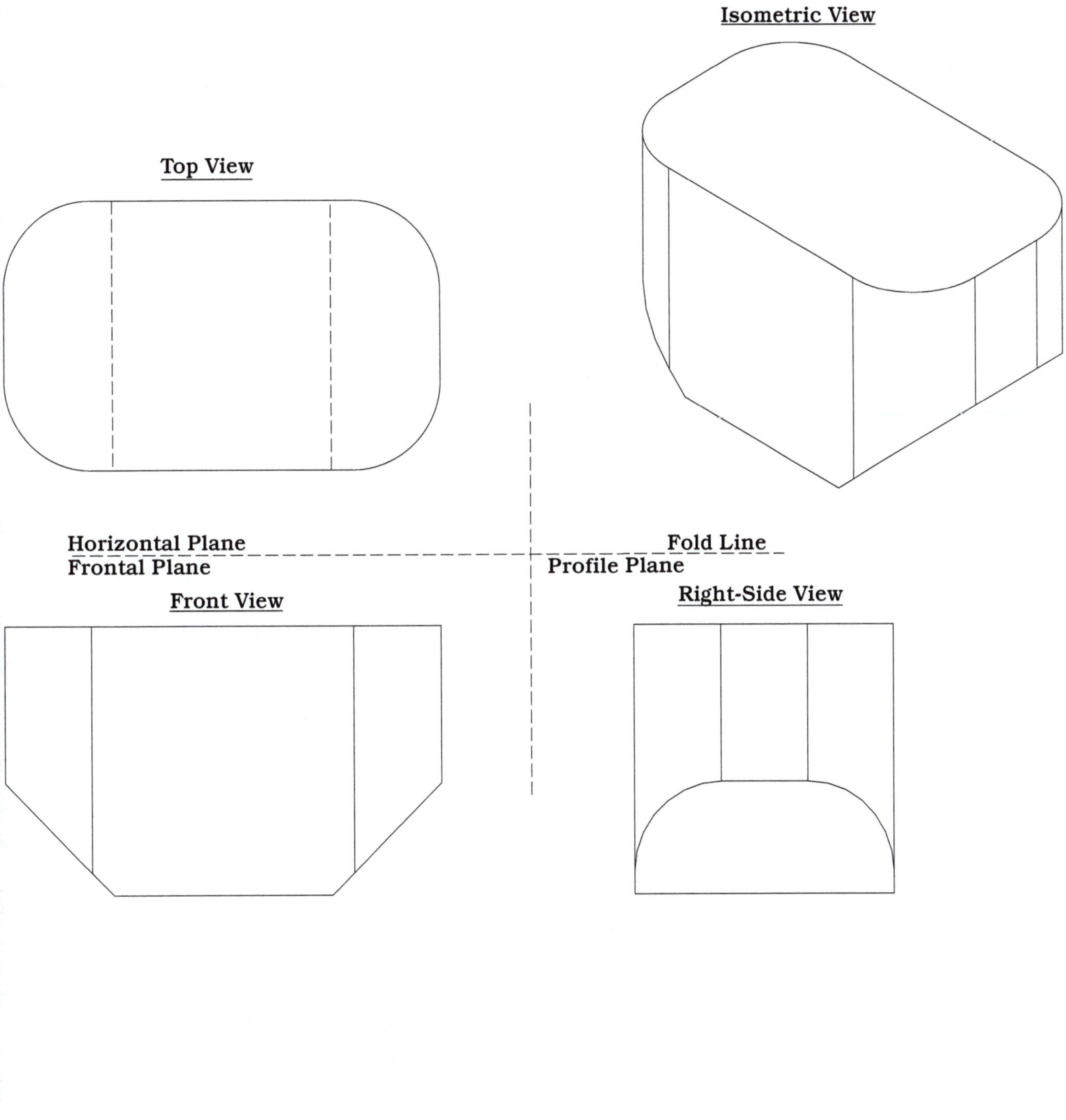

COURSE ________	STUDENT ______________________	DATE ________	PROBLEM 19-6

Problem 19-7 Development

Using AutoCAD, create a development of the following sheet metal part. This drawing can be found on the CD that accompanies the textbook.

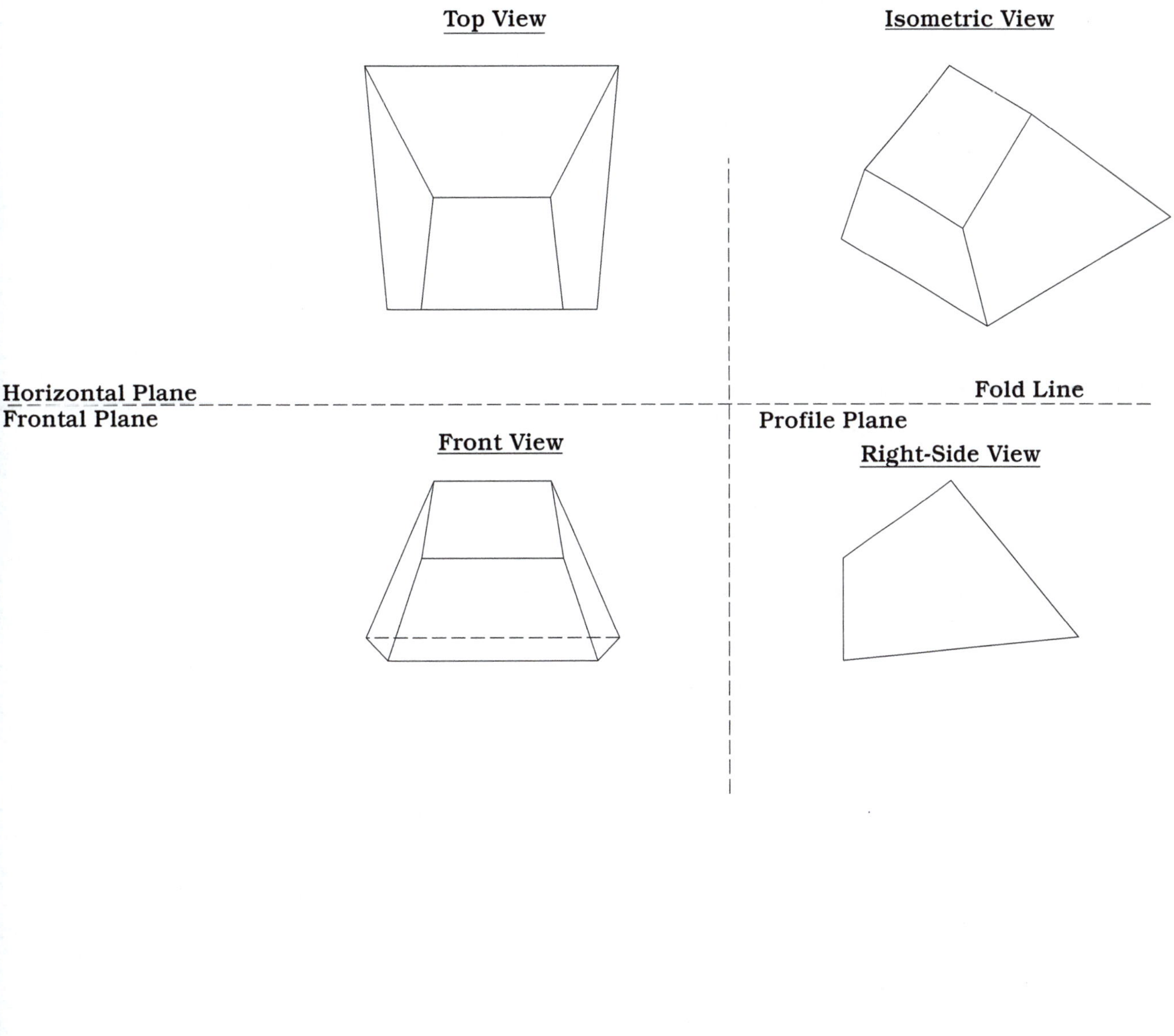

COURSE________ STUDENT________________________ DATE________ PROBLEM 19-7

CHAPTER 20

Electrical and Electronic Drafting

Problem 20-1 Schematic

Using a straightedge, redraw the freehand sketch of the schematic of an overtone oscillator and use the general practices as guidelines.

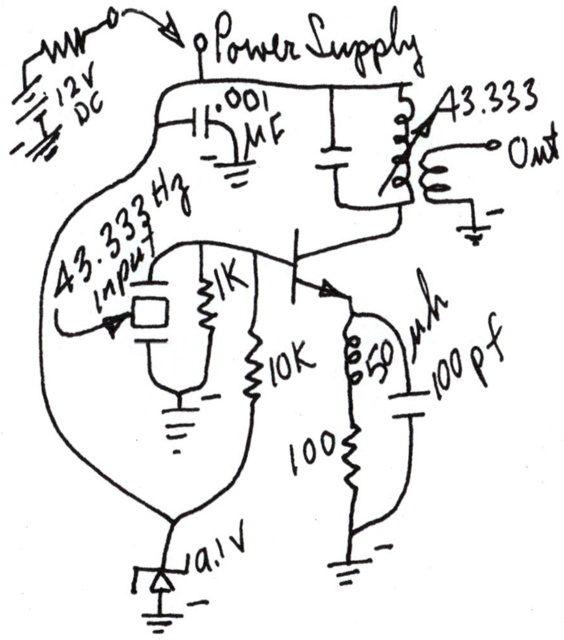

General practices for drawing electronic and electrical schematics:

SIGNALS: Signals should flow from left to right; inputs on left margin, outputs on right margin.

LINES: Spacing between lines should be uniform. Cross over lines as little as possible. If you do cross, break one line—preferably the one traveling in the short dimension of the paper. All lines should be horizontal or vertical.

CONNECTIONS: Positive connections generally go upward, and negative downward.

COURSE ________	STUDENT ______________________	DATE ________	PROBLEM 20-1

Problem 20-2 Start–Stop Control Circuit

Use drawing instruments and redraw the freehand sketch of the schematic of a START–STOP control circuit for an AC motor.

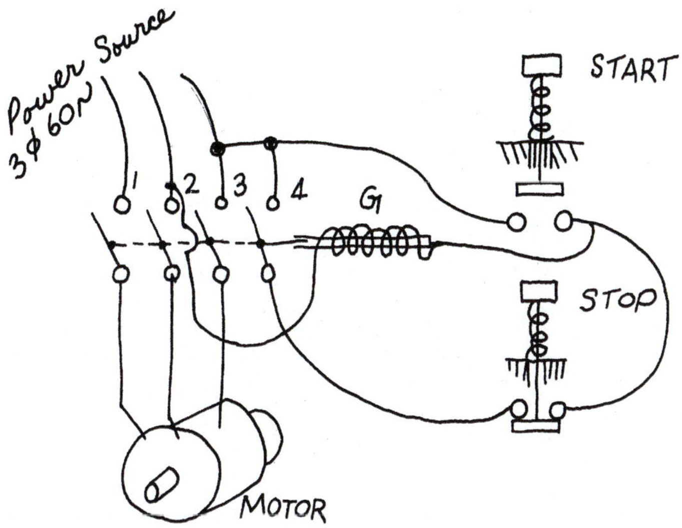

OPERATION: When the spring-loaded normally open START button is pressed, a path for the current is established from line 3 through the solenoid G to line 2. The current energizes the solenoid that causes all four linked switches to close.

Now switch 4 provides a different path for the solenoid current. The current flows through the normally closed STOP switch through the solenoid and finally to line 2.

The current will continue to flow until the STOP button is pressed. The STOP switch opens the current path through the solenoid so all four switches open quickly.

COURSE______	STUDENT______________________	DATE______	PROBLEM 20-2

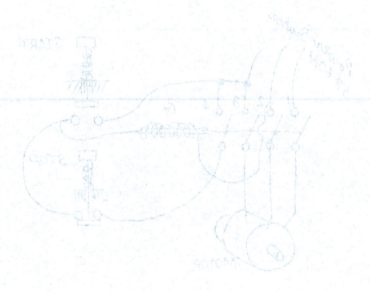

roblem 20-3 Find the Errors

Circle the errors shown in the following drawing and explain why each is an error.

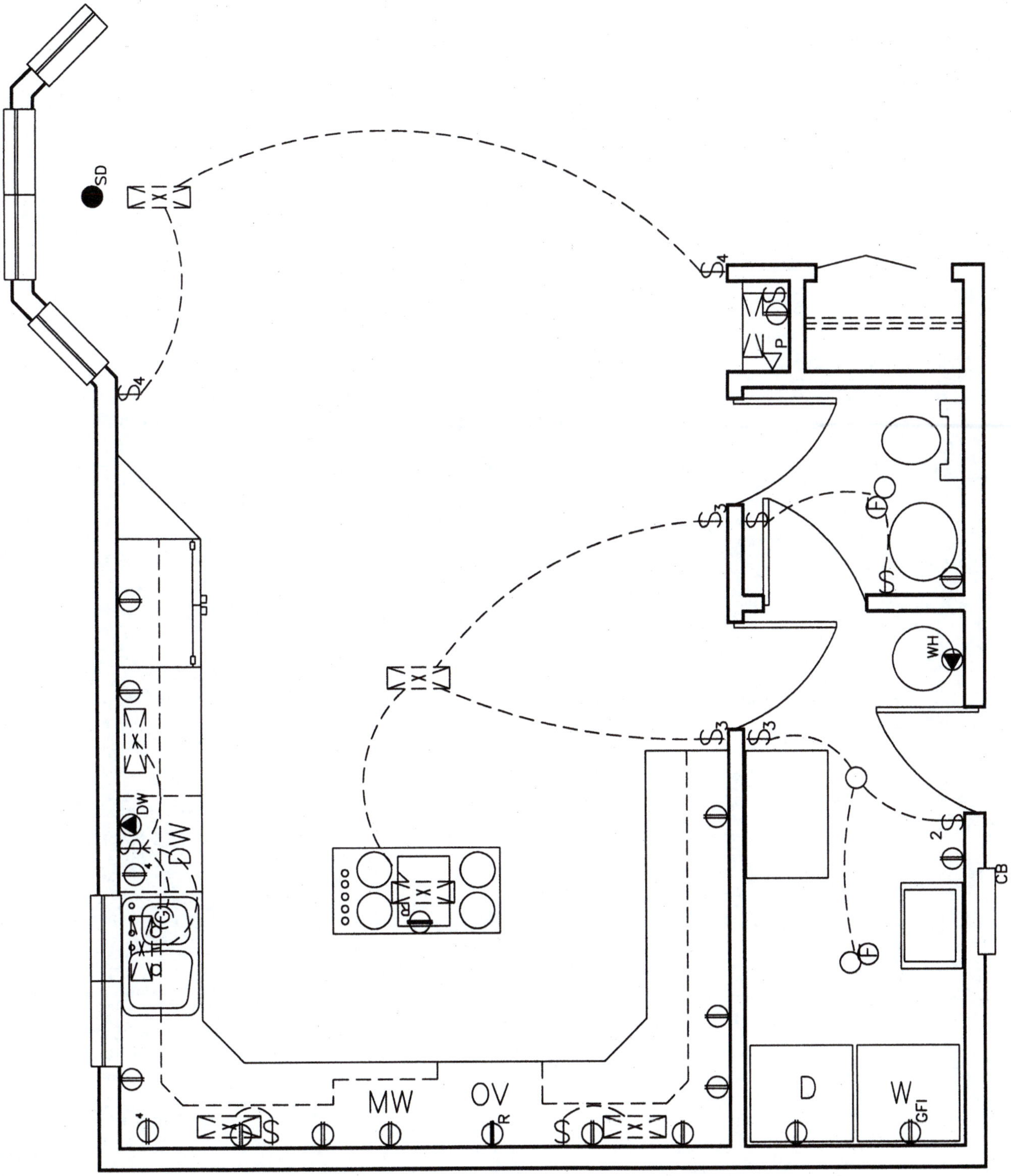

COURSE ________	STUDENT ________________	DATE ________	PROBLEM 20-3

Problem 20-4 Electrical Layout

Given the following information, provide the electrical layout. Be sure to meet the required code.

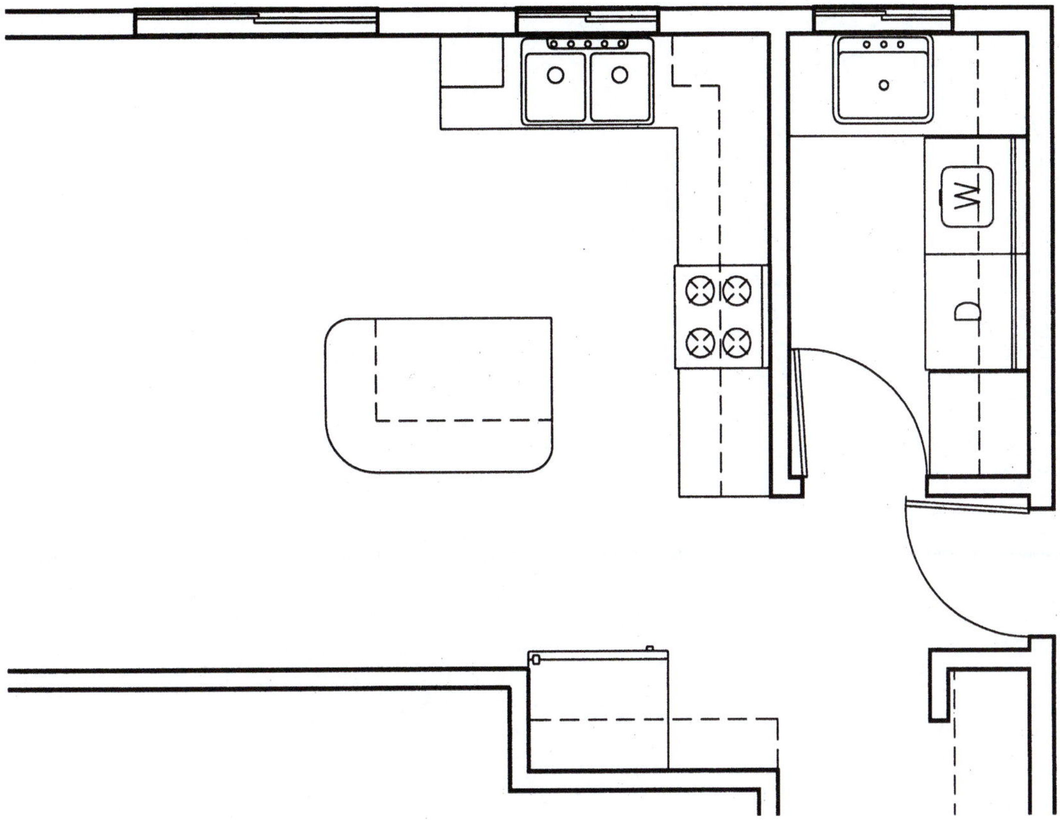

- Provide a 48" × 60" recessed fluorescent fixture centered over the island. Provide three-way switching, with one switch in the dining room, one near the range, and one near the exterior sliding door.
- Provide recessed ceiling-mounted light outside of pantry doors with a single-pole switch by the laundry room door.
- Provide eight recessed lights on the perimeter of the kitchen. A four-way dimmer shall be provided near the east entrance, one near the range, one near the exterior sliding door, and one in the dining room.
- Provide two 110V waterproof outlets on both sides of the kitchen sink. Provide three convenience outlets on the east side of the kitchen wall. Provide recessed fluorescent lights under the cabinets on the island and the east side of the kitchen wall.
- Provide for the garbage disposal unit in the west half of the kitchen sink.
- Provide an exhaust fan over range with a single-pole switch and a note vent to outside air.
- Provide two convenience outlets on the south and west sides of the kitchen island.
- Provide a 48" × 60" fluorescent fixture center in the laundry room with a single-pole switch. Provide an exhaust fan in the laundry room with a single-throw switch and a note vent to outside air. Provide outlets for the washer and electric-operated dryer.
- Provide three waterproof convenience outlets in the laundry room with ground-fault circuit interrupter.

COURSE________	STUDENT________________________	DATE________	PROBLEM 20-4

Problem 20-4 Electrical Layout

Given the following information, provide the electrical layout. Be sure to meet the required codes.

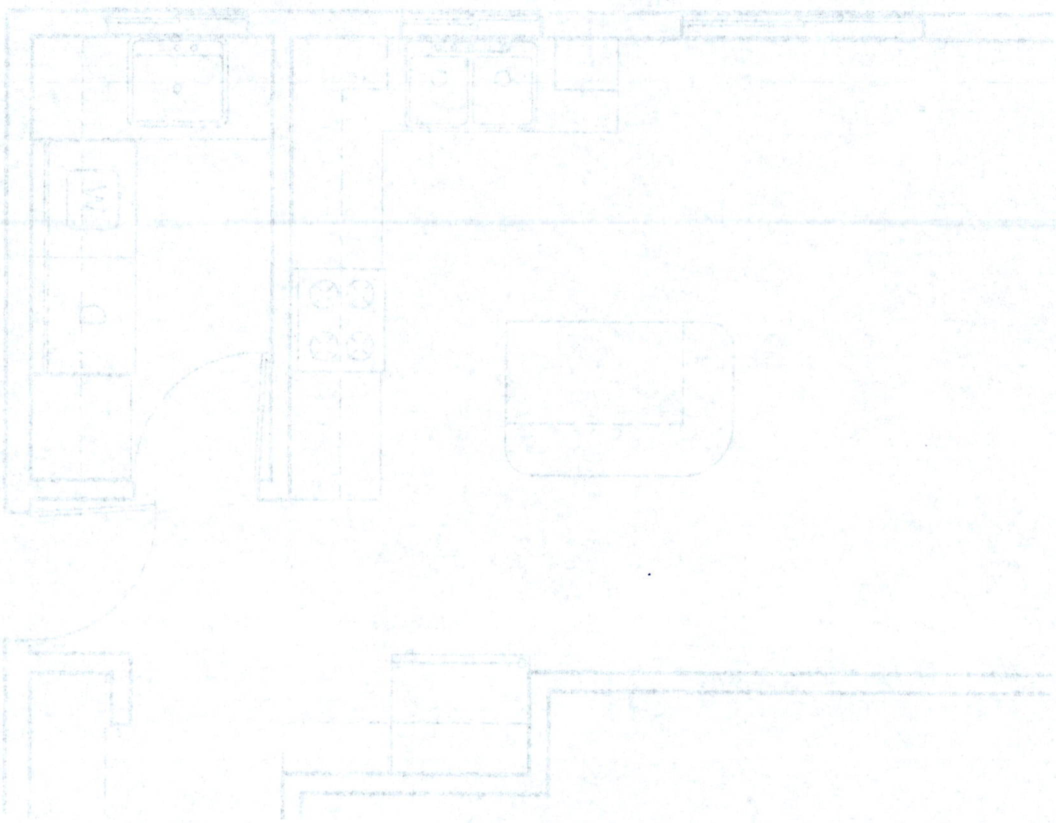

- Provide a 48" × 48" recessed fluorescent fixture centered over the island. Provide three-way switching, one switch in the dining room, one near the range, and one near the exterior sliding door.
- Provide recessed ceiling-mounted light outside of pantry doors with a single-pole switch by the pantry entry door.
- Provide eight recessed lights on the perimeter of the kitchen. A four-way dimmer shall be provided near the east entrance, one near the range, one near the exterior sliding door, and one in the dining room.
- Provide two GFCI waterproof outlets on both sides of the kitchen sink. Provide three convenience outlets on the east side of the kitchen wall. Provide recessed fluorescent lights under the cabinets on the island and the east side of the kitchen wall.
- Provide for the garbage disposal unit in the west half of the kitchen sink.
- Provide an exhaust fan over range with a single-pole switch and a note vent to outside.
- Provide two convenience outlets on the south and west sides of the kitchen island.
- Provide a 48" × 60" fluorescent fixture center in the laundry room with a single-pole switch. Provide an exhaust fan in the laundry room with a single-pole switch and a note vent to outside air. Provide outlets for the washer and electric-operated dryer.
- Provide three waterproof convenience outlets in the laundry room with ground-fault circuit interrupters.

COURSE | STUDENT | DATE | PROBLEM 20-4

Problem 20-5 Electronic Identification

Identify the following components by drawing the symbol for each and by lettering the type of component.

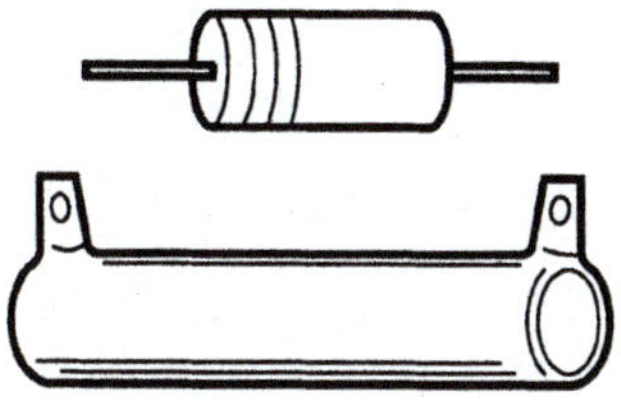

Symbol

Type

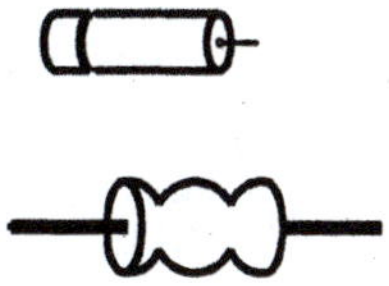

Symbol

Type

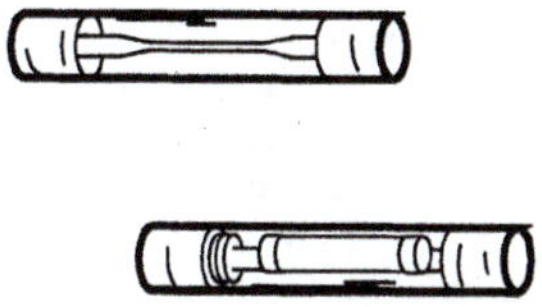

Symbol

Type

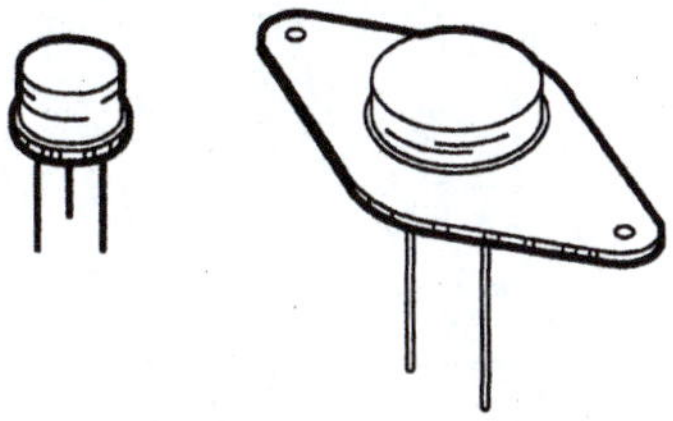

Symbol

Type

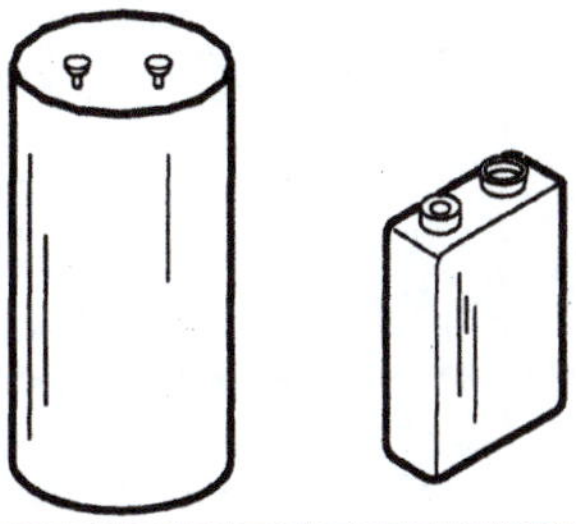

Symbol

Type

Symbol

Type

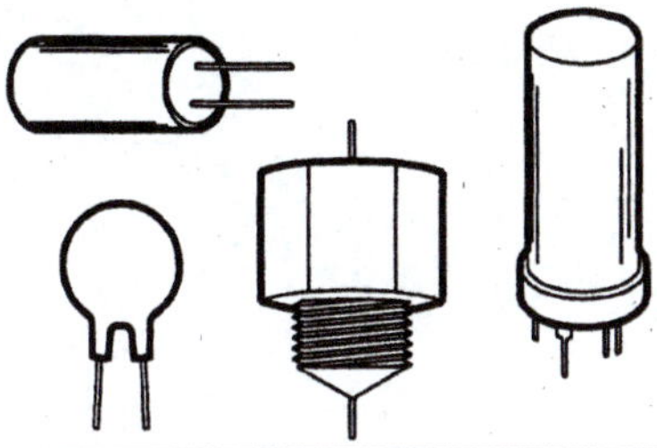

Symbol

Type

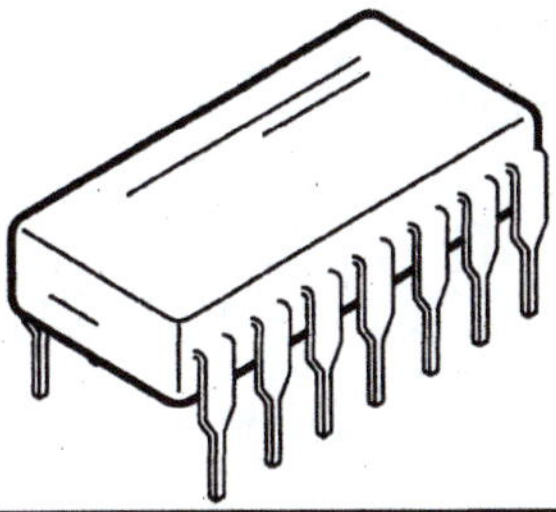

Symbol

Type

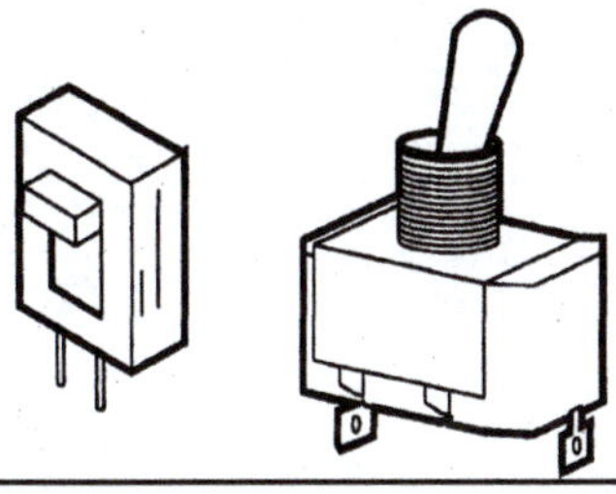

Symbol

Type

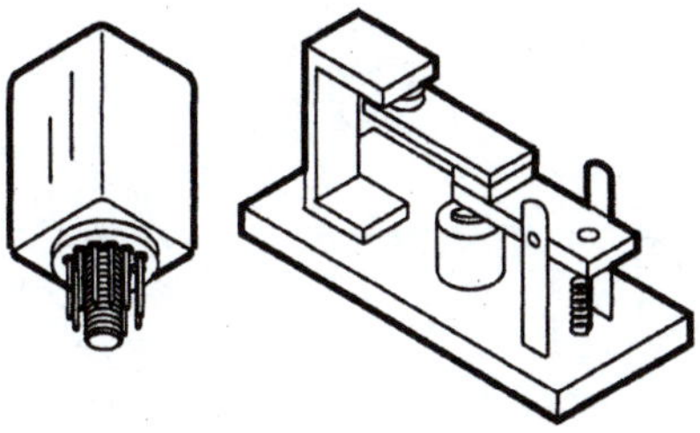

Symbol

Type

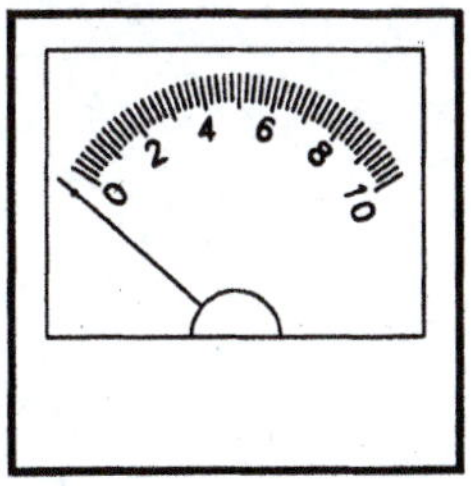

Symbol

Type

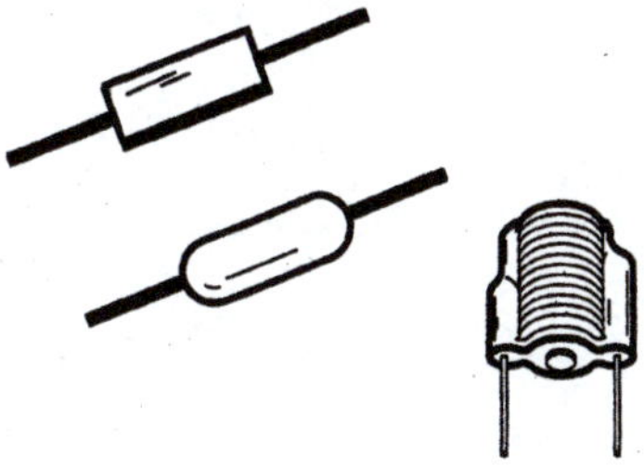

Symbol

Type

COURSE ________ STUDENT ________________________ DATE ________ PROBLEM 20-5

Problem 20-6 Electronic Circuits

Answer the following questions.

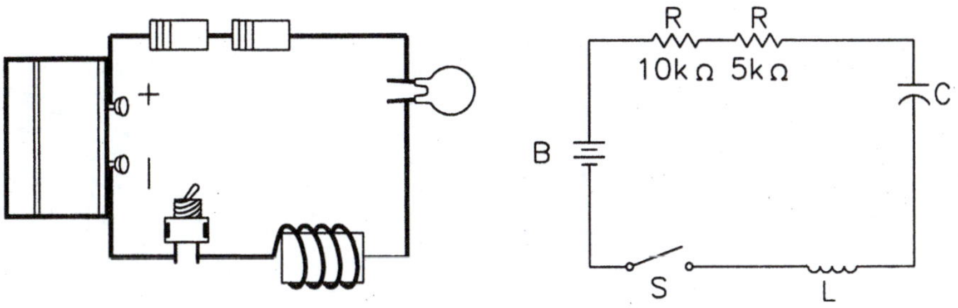

1. How many different kinds of components are there in the circuit? Name them.

2. What are the values of the resistors?

3. How could the schematic be better labeled?

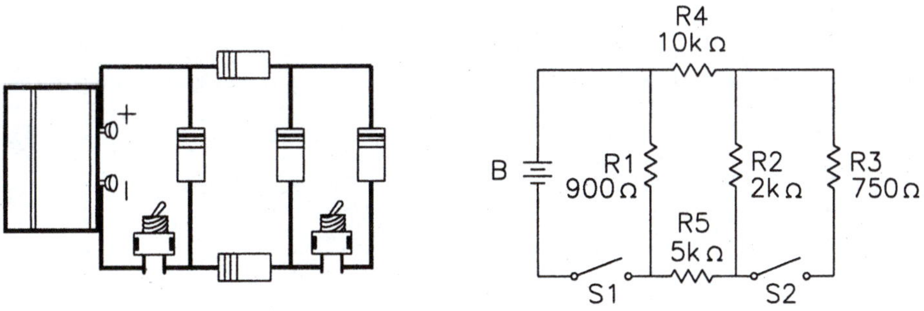

1. How many different kinds of components are there in the circuit? Name them.

2. What are the values of the resistors? Use Ohm's law to determine the unknowns.

Solve for voltage.

$$I = \frac{V}{R}$$

Solve for current (I).

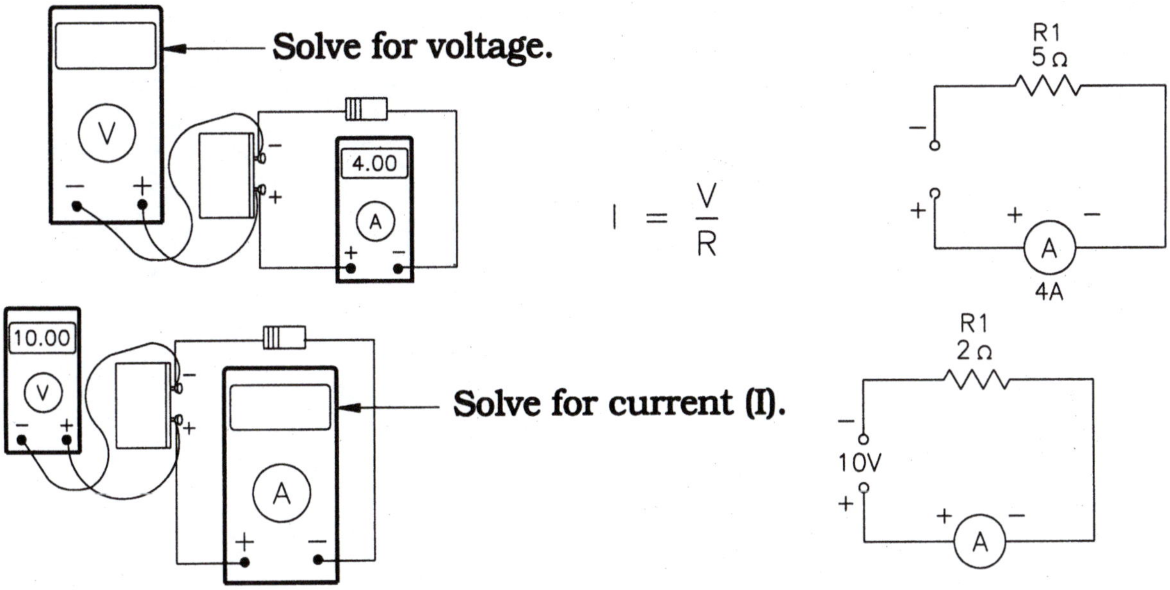

CHAPTER 21

Industrial Process Piping

Problem 21-1 Plumbing Fittings

Review the Supplemental Material for Chapter 21 on the CD that accompanies the textbook. Identify each pipe fitting. Letter each answer in the space provided.

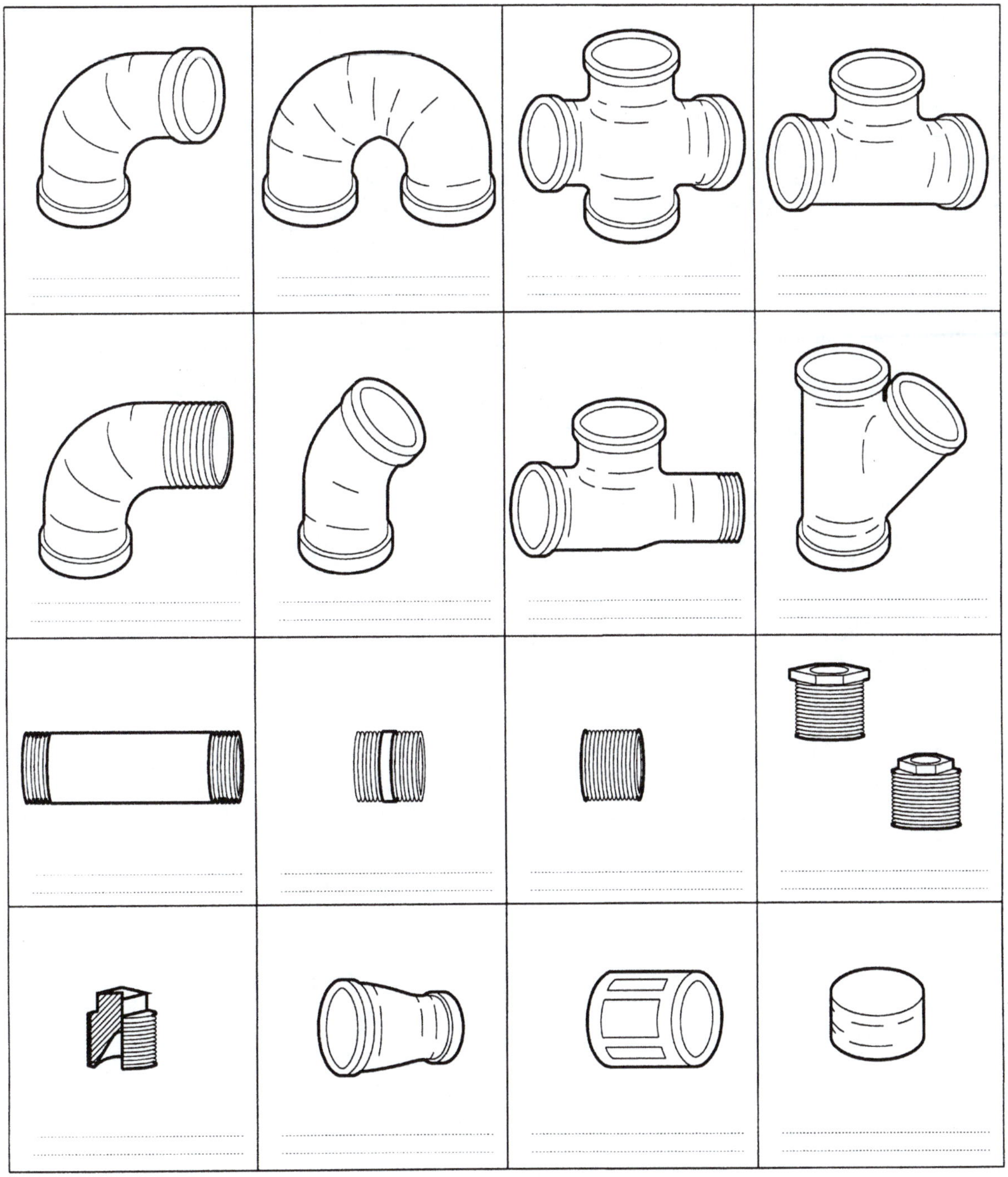

COURSE ________ STUDENT ______________________ DATE ________ PROBLEM 21-1

Problem 21-2 Plumbing Fittings

Review the Supplemental Material for Chapter 21 on the CD that accompanies the textbook. Supply the missing information in the open spaces.

90-degree long radius elbows and concentric .... 6"x4"	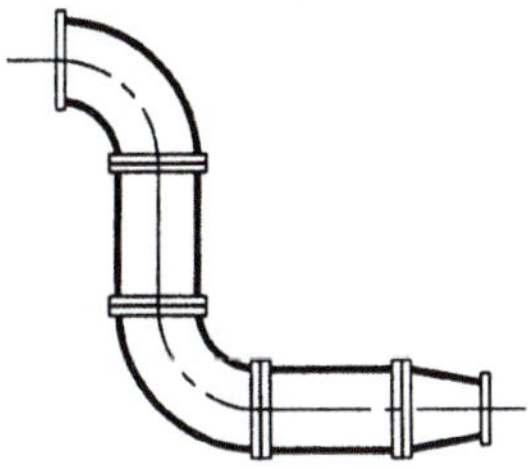	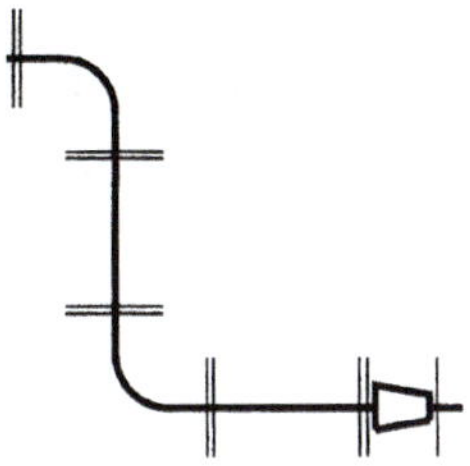	
straight .... 6"x6"x6"	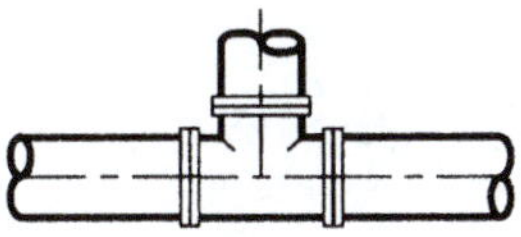		
45 .... lateral 6"x6"x6"	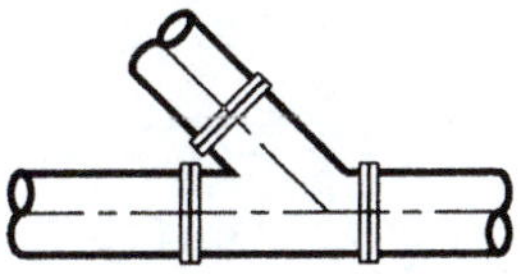		
250# iron body 6" .... valve	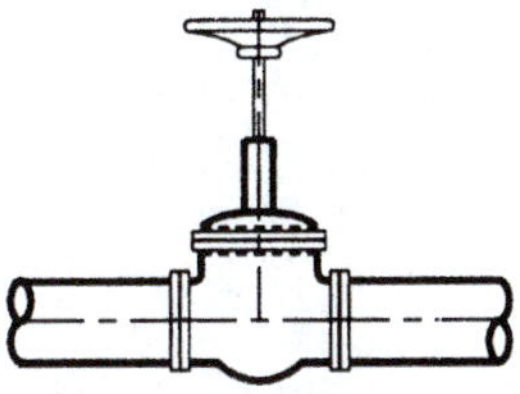	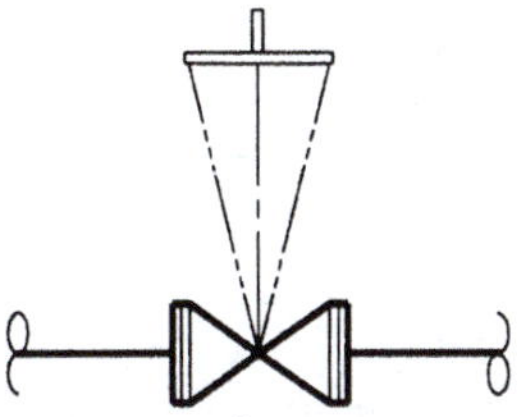	
125# iron body 6" .... valve			
150# cast steel 6" .... valve	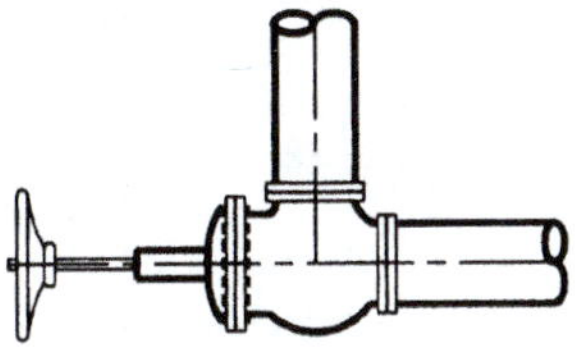	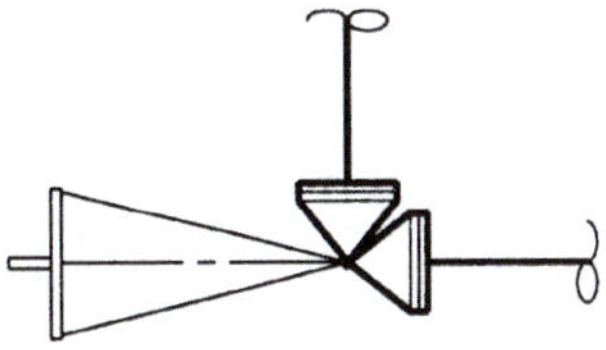	
150# cast iron 6" .... valve	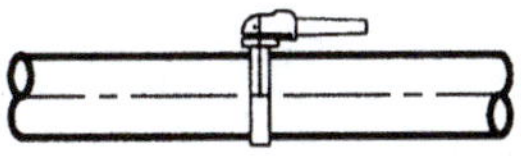		

COURSE________ STUDENT______________________ DATE________ PROBLEM 21-2

Problem 21-3 Hydraulic Circuits

Using a straightedge and a circle template, redraw the freehand sketch of the hydraulic-power schematic shown and label each component. Show the directional valve in the position for pushing the hydraulic piston to the right.

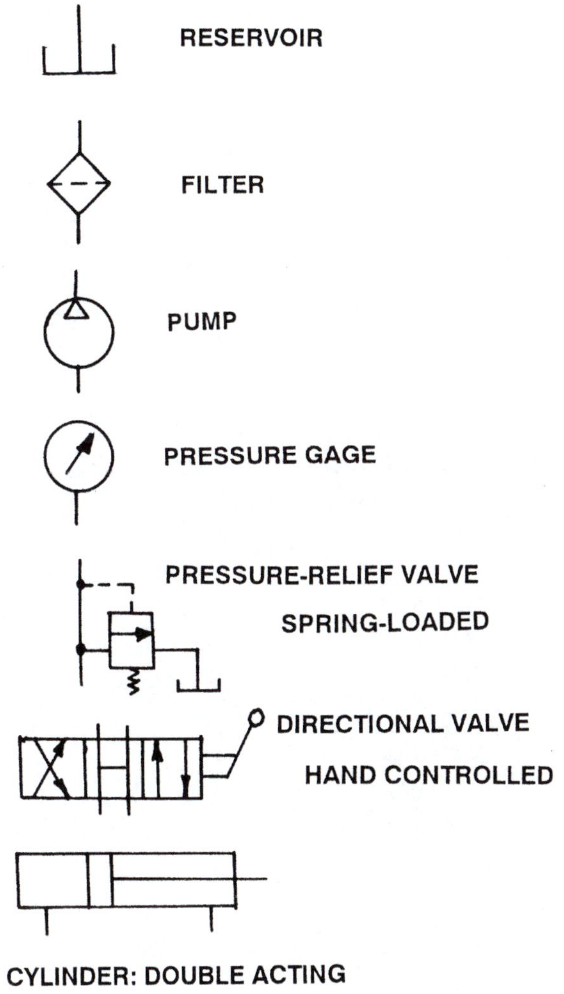

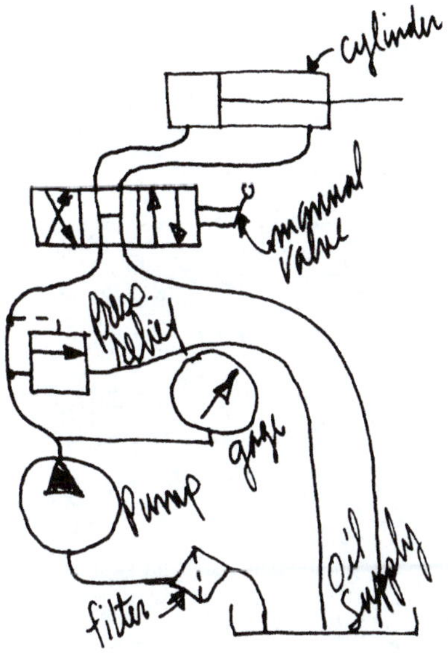

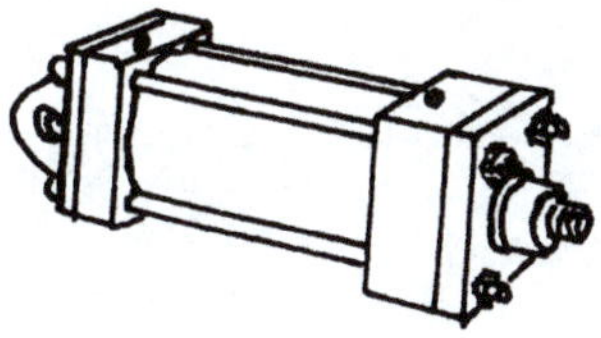

COURSE________ STUDENT________________________________ DATE________ PROBLEM 21-3

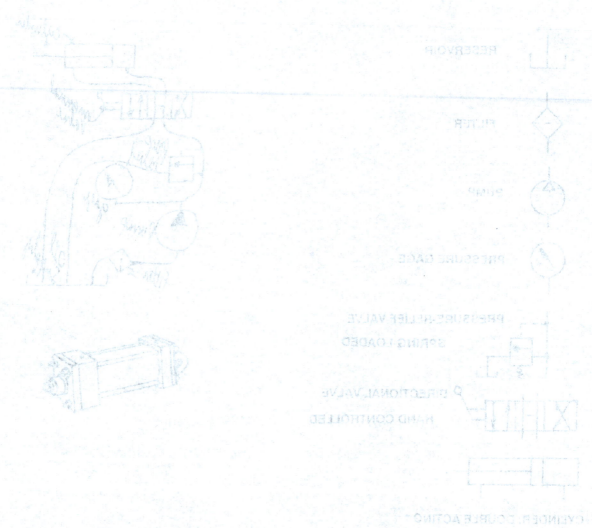

Problem 21-4 Chemical Pot Feeder

Refer to the double-line piping drawing shown and redo the drawing in a single-line orthographic.

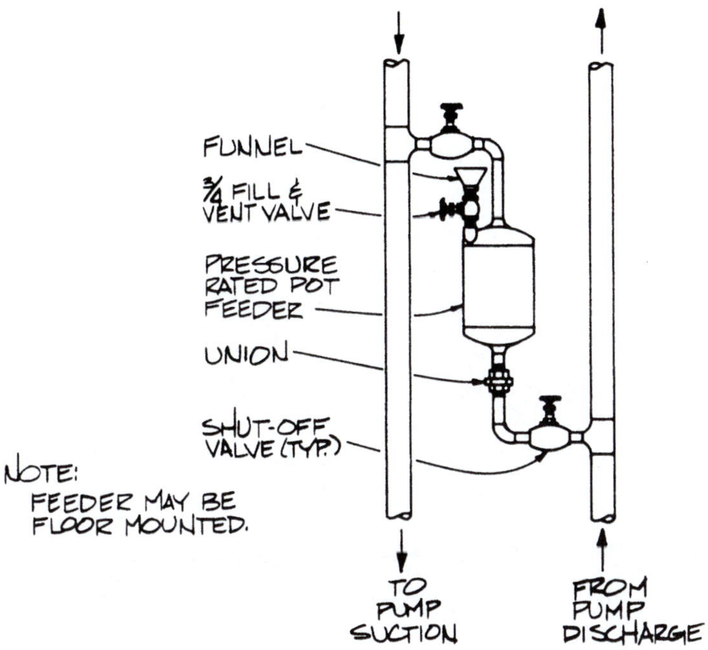

CHEMICAL POT FEEDER DETAIL
N.T.S.

COURSE ________	STUDENT ________________	DATE ________	PROBLEM 21-4

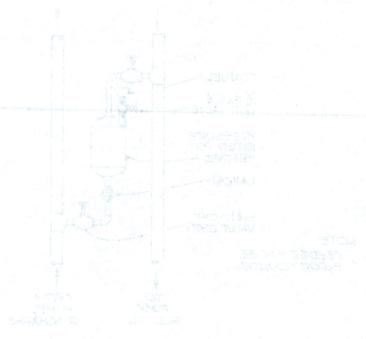

Problem 21-5 Complete the Missing Information

Fill in the missing information. Please note that some of the information cannot be calculated. In this case, give a brief description about why it cannot be calculated. 1/16" gasket used for dimensioning. Review the Supplemental Material for Chapter 21 on the CD that accompanies the textbook.

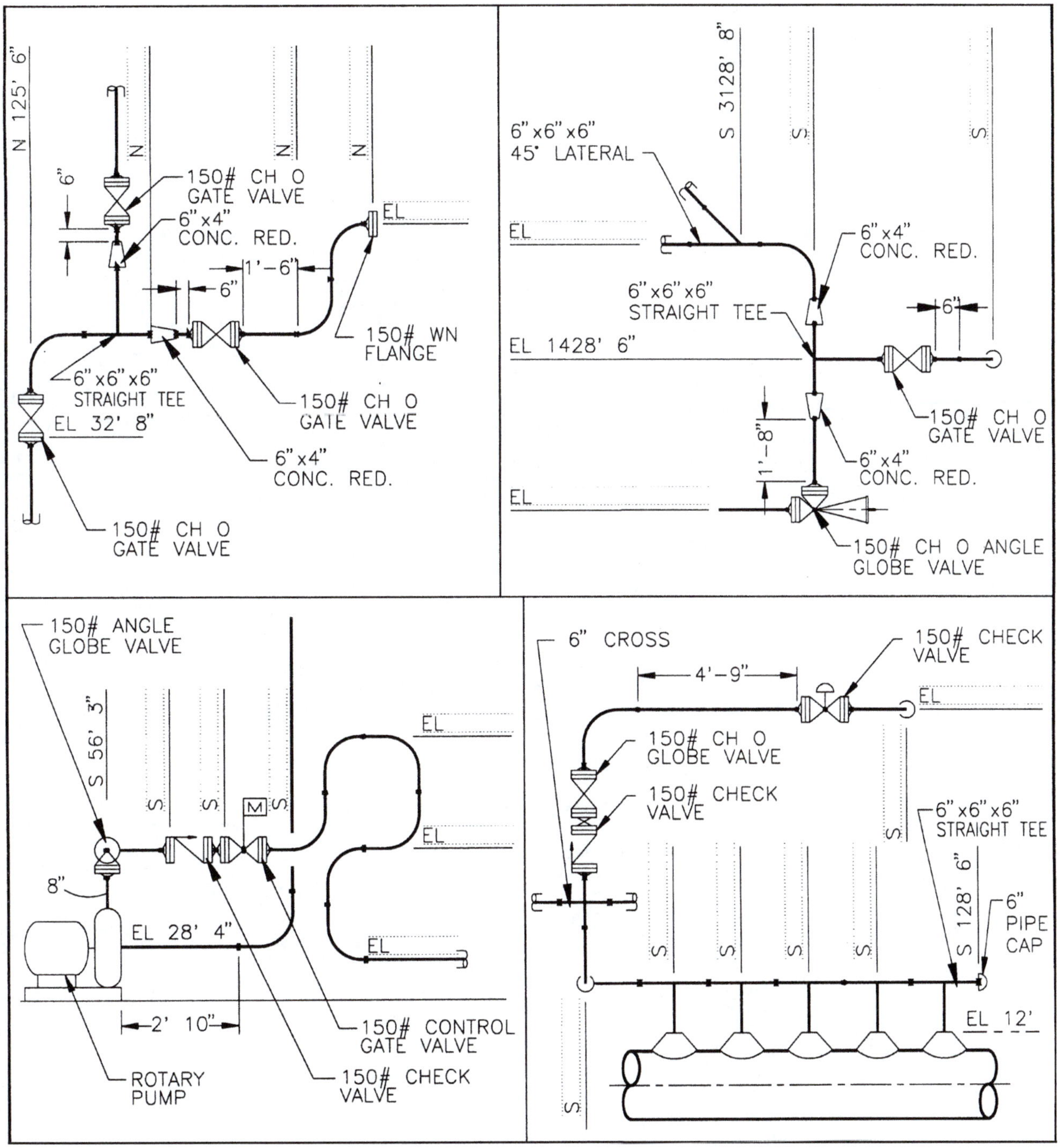

COURSE ________	STUDENT ________	DATE ________	PROBLEM 21-5

CHAPTER 22

Structural Drafting

Problem 22-1 Structural Steel Fabrication Drawing

Use the following structural steel fabrication drawing to answer the questions. Use standard letter techniques.

Notes:
Spec: AISC Latest edition
Matl.: ASTM A36
Open holes 13/16" dia.

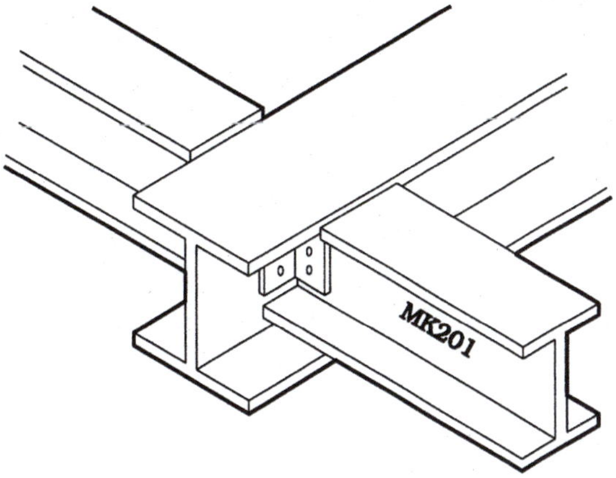

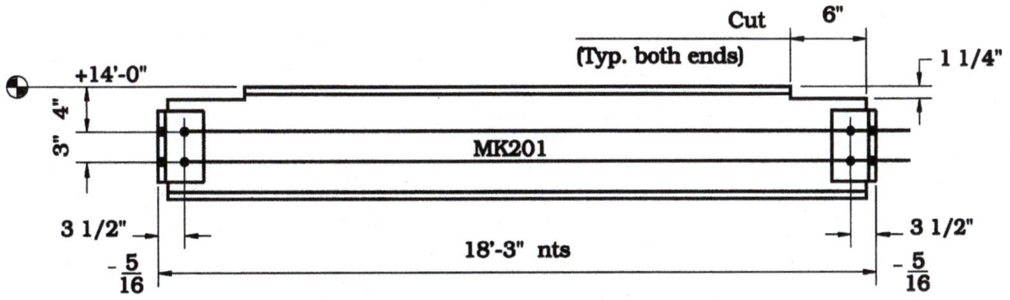

17 BEAMS - W8x35

A. What is the structural shape of the beam? ____________________

B. What is the flange width of the beam? ____________________

C. What is the weight in pounds per linear foot? ____________________

D. What is the elevation of the top of the beam? ____________________

E. What is the overall length of the beam? ____________________

F. What is the total weight of the beam in pounds? ____________________

G. What type of shading is used in the isometric drawing above? ____________________

COURSE ________	STUDENT ____________________	DATE ________	PROBLEM 22-1

Problem 22-2 Site Development

Elements of a plot plan may vary. Locate and identify the following, then check them off.

Typical plot plan items include the following:

____ Plot plan scale
____ Legal description of the property
____ Property line bearings and dimensions
____ North direction
____ Existing and proposed roads
____ Existing and proposed structures
____ Public or private water supply
____ Driveways, patios, walks, and parking areas
____ Topography or land elevations at lot corners
____ Public or private sewage disposal
____ Location of utilities
____ Rain and footing drains and storm sewers
____ Setbacks
____ Specific items on adjacent properties
____ Existing and proposed trees

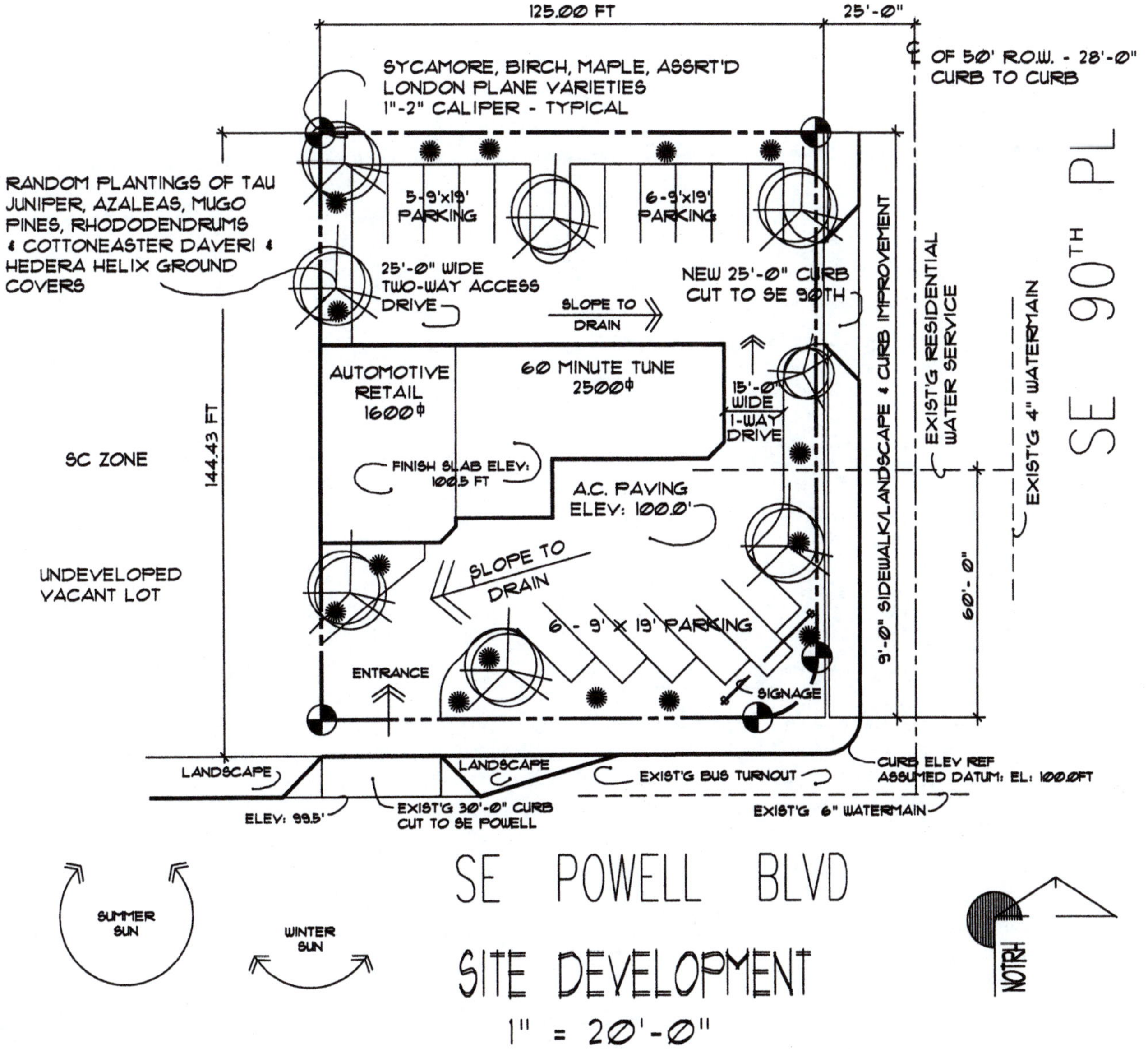

COURSE________ STUDENT______________________ DATE________ PROBLEM 22-2

Problem 22-3 Trusses

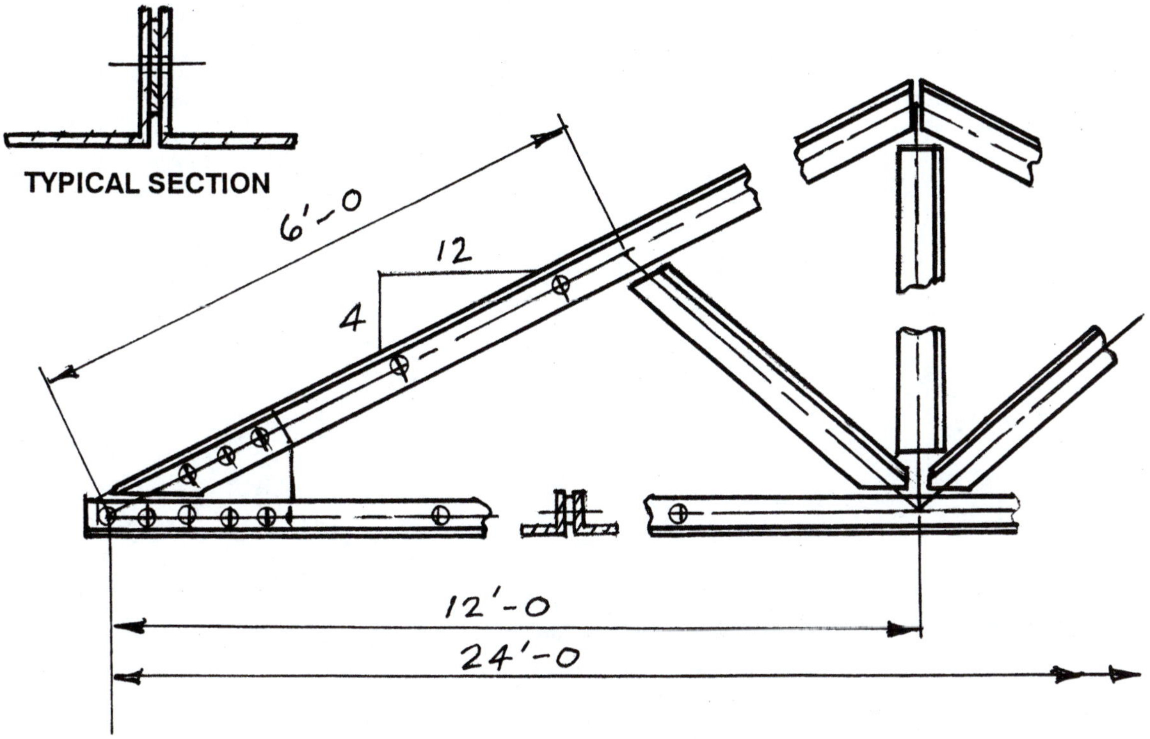

NOT TO SCALE

The truss shown was used in a commercial building and had the following characteristics:

1. Double steel angles ⊥3" X 3" X 1/4"
2. Gusset plates 1/4" steel
3. Hex head bolts ϕ1/2", hex nuts
4. Spacers, ϕ2", located approximately every 2 feet

Lay out the truss with these characteristics (specifications), to a scale of 1/4" = 1'-0, to obtain centerline distances. Lay out joints A, B, C, and D to a scale of 1-1/2" = 1'-0 to obtain lengths of angles, locate bolt holes, and establish size of the gussets. Assume that structural angles may be purchased in lengths of 30'-0.

GENERAL CONSIDERATIONS

A minimum of two bolts per member at a joint is recommended for safety. The locations of the bolts may be adjusted to improve the general appearance of the joint or to facilitate the production process; for example, simplify the shape of the gusset plate by making it rectangular, trapezoidal, or symmetrical.

Prepare a parts list of the angles, gussets, spacers, bolts, and nuts. Use numbers in circles to indicate parts.

COURSE________	STUDENT______________________________	DATE________	PROBLEM 22-3

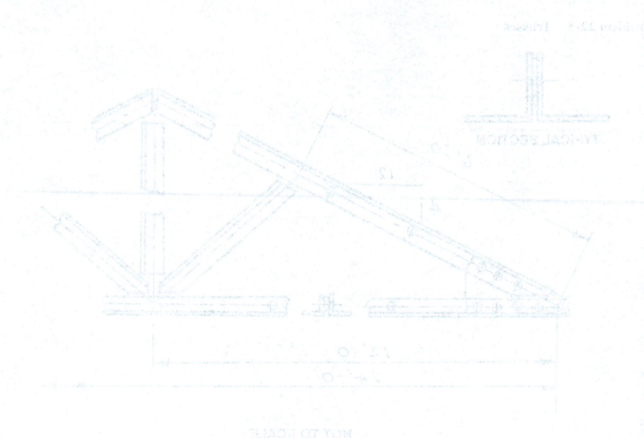

Problem 22-4 Trusses

Joints in trusses may be bolted, riveted, or welded or some combination of the three. An approach for laying out bolted joints for a truss using structural angles is to do the following:

1. Construct the centerlines to scale.
2. Locate the angles using recommended gage distances (GD) and clearances (C); GD = approximately 0.6 X leg of angle; C = 1/4"+.
3. Establish bolt locations using edge distance (ED) = 1.5 X bolt diameter (minimum) and bolt spacing (BS) = 3 X bolt diameter (minimum).
4. Lay out gusset plates using EDs.

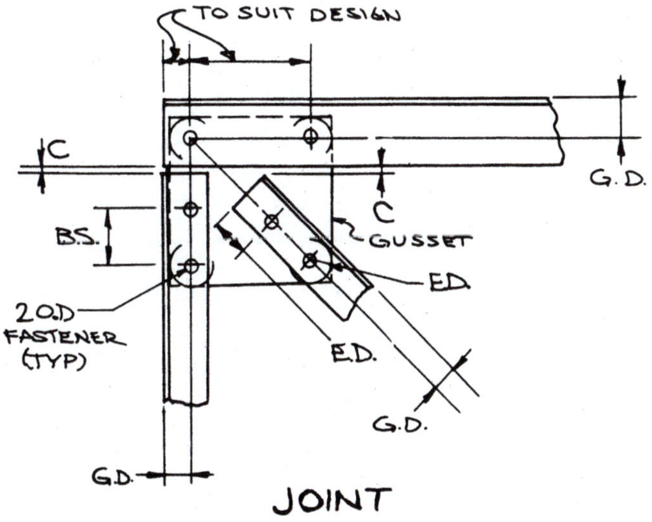

An example joint layout is shown. Note the small radii used to locate edge distances.

COURSE______	STUDENT______	DATE______	PROBLEM 22-4

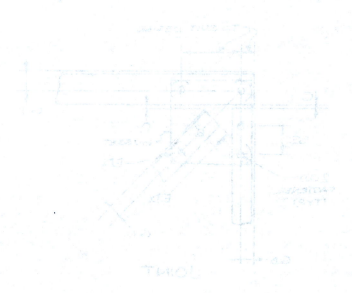

CHAPTER 23

Heating, Ventilating, and Air-Conditioning (HVAC), and Pattern Development

Problem 23-1 Purpose of a Furnace

In the space provided below and in your own words, describe the purpose and function of an HVAC system.

COURSE________	STUDENT________________________________	DATE________	PROBLEM 23-1

Problem 23-2 Steps in Designing an Offset Development

In the following space, list the steps for developing an offset cone development.

COURSE ________	STUDENT ________________________	DATE ________	PROBLEM 23-2

Problem 23-3 Drawing an HVAC System

Using the following sketch, create an HVAC drawing (including plans, schedules, and details).

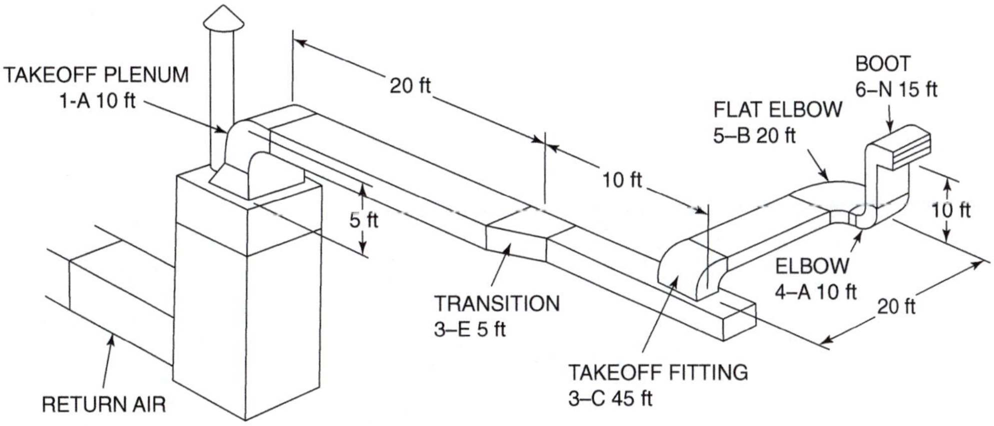

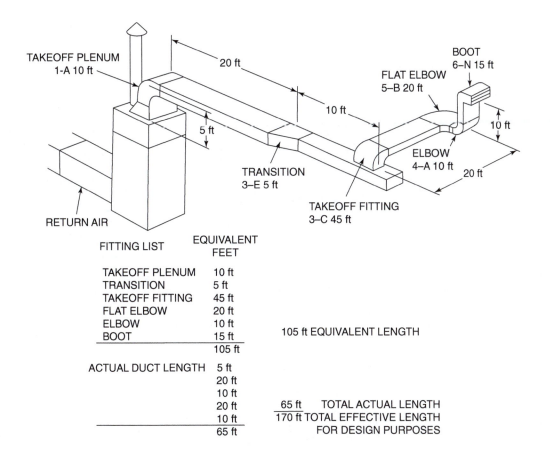

FITTING LIST	EQUIVALENT FEET
TAKEOFF PLENUM	10 ft
TRANSITION	5 ft
TAKEOFF FITTING	45 ft
FLAT ELBOW	20 ft
ELBOW	10 ft
BOOT	15 ft
	105 ft

105 ft EQUIVALENT LENGTH

ACTUAL DUCT LENGTH	
	5 ft
	20 ft
	10 ft
	20 ft
	10 ft
	65 ft

65 ft TOTAL ACTUAL LENGTH
170 ft TOTAL EFFECTIVE LENGTH FOR DESIGN PURPOSES

COURSE________ STUDENT________________________ DATE________ PROBLEM 23-3

CHAPTER 24

Civil Drafting

Problem 24-1 Azimuths, Grade, and Slope

USING AUTOCAD TO CONSTRUCT AZIMUTHS

Using AutoCAD, construct the azimuths listed in the following charts.

USING NORTH AS A REFERENCE

Plotting Azimuths (North)

Line Number	Length	Bearing
1	5.4483	68°9'39"
2	1.5540	40°55'31"
3	3.5729	295°50'30"
4	2.4351	255°58'27"
5	2.3434	219°3'25"
6	3.3965	116°40'15"

Note: Start line 1 at X = 4.6347, Y = 1.9262, Z = 0.0000

Plotting Azimuths (North)

Line Number	Length	Bearing
1	2.2244	148°4'50"
2	5.1765	93°26'6"
3	2.4354	55°50'49"
4	4.3630	15°26'54"
5	3.1514	275°6'20"
6	3.8734	234°7'38"
7	3.1854	257°55'27"
8	2.9952	290°18'26"
9	3.2366	211c42'54"
10	3.5951	83°28'0"

Note: Start line 1 at X = 3.1336, Y = 2.5902, Z = 0.0000

COURSE_________ STUDENT__________________________ DATE_________ PROBLEM 24-1

Plotting Azimuths (North)

Line Number	Length	Bearing
1	3.1085	240°48'33"
2	1.1911	3°58'37"
3	3.5556	308°29'36"
4	3.1471	63°43'10"
5	0.7056	77°34'36"
6	2.5596	120°4'53"
7	2.6917	135°8'0"
8	1.6855	201°33'8"
9	2.0726	309°50'30"

Note: Start line 1 at X = *8.5886,* Y = *4.2541,* Z = *0.0000*

Plotting Azimuths (North)

Line Number	Length	Bearing
1	2.5855	103°45'36"
2	1.7980	153°3'36"
3	2.2126	218°7'26"
4	2.9879	288°34'24"
5	1.9824	319°50'53"
6	2.6173	55°15'27"

Note: Start line 1 at X = *4.6347,* Y = *1.9262,* Z = *0.0000*

Plotting Azimuths (North)

Line Number	Length	Bearing
1	3.4008	281°39'44"
2	2.6269	238°5'56"
3	2.1107	118°59'15"
4	1.6086	118°59'15"
5	3.4992	115°20'35"
6	2.0372	145°34'36"
7	2.3379	64°3'12"
8	2.2824	10°8'54"
9	2.6506	279°31'12"
10	2.4144	299°38'12"

Note: Start line 1 at X = *3.1336,* Y = *2.5902,* Z = *0.0000*

COURSE ________ STUDENT ______________________________ DATE ________ PROBLEM 24-1

Plotting Azimuths (North)

Line Number	Length	Bearing
1	2.6873	79°21'2"
2	2.6344	114°17'16"
3	1.5858	144°6'58"
4	1.4222	231°39'14"
5	3.2639	265°22'31"
6	2.3463	263°33'40"
7	2.4292	318°12'26"
8	2.7133	83°26'53"
9	1.2109	343°17'19"

Note: Start line 1 at X = 8.5886, Y = 4.2541, Z = 0.0000

SLOPE

Determine the slopes of the following lines.

USING SOUTH AS A REFERENCE

Plotting Azimuths (South)

Line Number	Length	Bearing
1	3.1085	240°48'33"
2	1.1911	3°58'37"
3	3.5556	308°29'36"
4	3.1471	63°43'10"
5	0.7056	77°34'36"
6	2.5596	120°4'53"
7	2.6917	135°8'0"
8	1.6855	201°33'8"
9	2.0726	309°50'30"

Note: Start line 1 at X = 8.5886, Y = 4.2541, Z = 0.0000

COURSE ________ STUDENT ____________________ DATE ________ PROBLEM 24-1

Plotting Azimuths (South)

Line Number	Length	Bearing
1	2.5855	103°45'36"
2	1.7980	153°3'36"
3	2.2126	218°7'26"
4	2.9879	288°34'24"
5	1.9824	319°50'53"
6	2.6173	55°15'27"

Note: Start line 1 at X = *4.6347,* Y = *1.9262,* Z = *0.0000*

Plotting Azimuths (North)

Line Number	Length	Bearing
1	2.5855	103°45'36"
2	1.7980	153°3'36"
3	2.2126	218°7'26"
4	2.9879	288°34'24"
5	1.9824	319°50'53"
6	2.6173	55°15'27"

Note: Start line 1 at X = *4.6347,* Y = *1.9262,* Z = *0.0000*

Plotting Azimuths (South)

Line Number	Length	Bearing
1	3.4008	281°39'44"
2	2.6269	238°5'56"
3	2.1107	118°59'15"
4	1.6086	118°59'15"
5	3.4992	115°20'35"
6	2.0372	145°34'36"
7	2.3379	64°3'12"
8	2.2824	10°8'54"
9	2.6506	279°31'12"
10	2.4144	299°38'12"

Note: Start line 1 at X = *3.1336,* Y = *2.5902,* Z = *0.0000*

COURSE_________ STUDENT________________________ DATE_________ PROBLEM 24-1

Plotting Azimuths (South)

Line Number	Length	Bearing
1	2.6873	79°21'2"
2	2.6344	114°17'16"
3	1.5858	144°6'58"
4	1.4222	231°39'14"
5	3.2639	265°22'31"
6	2.3463	263°33'40"
7	2.4292	318°12'26"
8	2.7133	83°26'53"
9	1.2109	343°17'19"

Note: Start line 1 at X = *8.5886,* Y = *4.2541,* Z = *0.0000*

COURSE________ STUDENT________________________ DATE________ PROBLEM 24-1

SECTION 6

Engineering Design

CHAPTER 25

The Engineering Design Process

Problem 25-1 Design Analysis

In the following space, discuss the steps in a design analysis process.

Explain the importance of concurrent engineering and teams in the development of a product.

Explain the design review process.

COURSE ________ STUDENT ________ DATE ________ PROBLEM 25-1